QUALITY MANAGEMENT

(TOTAL QUALITY MANAGEMENT)

(For Engineering and Management Students)

Dr. O.N. Pandey
(Professor, JSSATE, Noida)

Bhupesh Aneja
(Asst. Professor, JSSATE, Noida)

S.K. KATARIA & SONS®
Publisher of Engineering Books
4885/109, Prakash Mahal, Dr. Subhash Bhargav Lane,
Opposite Delhi Medical Association, Daryaganj, New Delhi–110002 (INDIA)
Phone: +91-11-43551243; Mobile: +91-9871775858, 7678425209
e-mail: katariabook@gmail.com, katariabooks@yahoo.com
Website: www.skkatariaandsons.com
(e-books, subscriptions and bulk buying also available on our website)

Published by:

S.K. Kataria & Sons®

4885/109, Prakash Mahal, Dr. Subhash Bhargav Lane,
Opposite Delhi Medical Association, Daryaganj, New Delhi–110002
Phone: +91-11-43551243; Mobile: +91-9871775858, 7678425209
e-mail: katariabook@gmail.com, katariabooks@yahoo.com

First Edition: July, 2011; Second Edition: August, 2012; Reprint: 2025

Price: ₹ 395/-

Laser Typesetting: Plus Computers, Delhi *Printed at :* New A.S. Printing Press, Delhi

Preface to the Second Edition....

The first edition of this book has been in great demand, therefore, it was felt that the book should be further enriched. Accordingly the topics have been revised and additional matters have been added :

1 Dimensions of quality
2. Meaning of quality control
3. Quality concept change
4. Quality functions in services
5. Total productive maintenance
6. Quality function development (QFD)
7. Leadership in quality control
8. Supplier partnership
9. Quality circles
10. Employee involvement
11. Failure mode evaluation and analysis (FMEA)
12. Customer satisfaction
13. Kaizen concept
14. Product liability
15. Zero defect
16. University Question Papers

Further, all chapters have been revised to ensure that the matters are correct and reliable. Any error or ommission noted may be brought to the notice which shall be taken care of in the forthcoming edition of this book.

The authors are thankful to the faculty members and students for making this book so popular.

August, 2012
New Delhi

O.N. Pandey
Bhupesh Aneja

Preface to the First Edition....

There has been lot of progress in industry. However, India has not been able to establish in the Reliability Engineering field due to sellers market conditions. Manufacturers have been able to sell products of whatever quality and reliability they make.

Now, the situation is changing to competitive market as customer expectation is for goods of better quality and reliability. The quality and reliable goods are need of the hour for company survival. Application of reliability engineering concepts leads to better firm survival. The quality and reliability is built in the product by the involvement of all the departments of the company including top management. In view of this, working knowledge of basic concepts and reliability engineering practice is necessary for all executives, production supervisors, sales and personnel.

This book covers all essential quality control techniques needed for designers, engineers, manufacturing plants and service sectors. It gives details of quality concepts, control on purchased product, manufacturing quality, quality management, human factor in quality, control charts, defects diagnosis and prevention, quality standards, total quality management, just-in-time and Taguchi method. Further, it also covers reliability, production planning and control, maintenance functions, general maintenance, predictive maintenance, spare parts management, Pareto principle, SWOT analysis and value engineering.

Appendices cover reliability terms, quality management terms, quality assurance terms, maintenance terms, various standards, normal distribution, cumulative Poisson probabilities and two model question papers.

Authors are very thankful to JSS Academy of Technical Education, Noida for their encouragement leading to such contributions. Authors also appreciate cooperation and helps extended by Mrs. Ranjana Pandey, Mrs. Parul Aneja, relatives and friends.

August, 2011
New Delhi

O.N. Pandey
Bhupesh Aneja

Dedicated to the Parents
of Onkar N. Pandey namely
Smt. Kalavati Devi and
Shri Deo Murti Pandey

and

Dedicated to the Parents
of Bhupesh Aneja namely
Smt. Neelam Aneja and
Shri D.R. Aneja
Who have given so much to the authors

CONTENTS....

1 Chapter One

Quality Concept

CHAPTER OUTLINE

- Introduction
- Dimensions of Quality
- Concept Change
- Evoluation of Quality Control
- Meaning of Quality Control
- Quality Functions in Services

1.1. INTRODUCTION

The American Society for Quality (ASQ) defines the quality as "the totality of features and characteristics of a product and service that bear on its ability to satisfy given needs. " It focused on the features and characteristics of products and services. Quality control and Quality Assurance are still in use, and they are principally used for ensuring the quality of out-going products and services. It aims to fulfil and meet the quality specifications of customers. Various quality management methods which existed in industries can be described as: Inspection in the Pre-World War II era, Quality control in post World War II era, later on Quality Assurance. However, the present day customer demand and the competition require modern quality management systems to consider also the cost, delivery, maintainability, reliability and attractiveness of features of products and services.

Quality become very important tool for a successful business, during revolution of war-shattered industries of Japan in the 1950s. A system of quality management responsive to customer needs, measurable and maintainable become essential for holding on to quality goals, for growth of business. W.E Deming, and Joseph M. Juran led this new quality movement concept in Japan and USA for evolving a modern and comprehensive approach to quality management with focus to meet the needs of customers. New approaches to quality management started to evolve, mainly in Japan, during nineteen fifties and sixties. Japan become the most powerful and reputed industrial powers in post World War II era. Quality become the buzzword in industries globally by taking due from the Japanese. Subsequently the companies all over the world slowly started adopting total quality approach as a means to achieve the customer satisfaction and customer loyalty.

Quality of product signifies the degree of its excellence and fitness for the purpose. Some of the quality characteristics can be specified in quantitative terms but no single characteristics can be used to measure the quality of a product on an absolute scale. The quality of a product means all those activity which are directed to maintaining and improving such as setting of quality targets, appraisal of conformance, taking corrective action where any deviation is noticed and planning for improvements in quality. Quality control practices also evolved to suit the products and methods of production. Important steps in this evolution are craftmanship, supervisors control, inspection, and statistical quality control. There are attributes such as continued fault free service and maintenance costs which are equaly important from the customer point of view. Quality in a measure of the user satisfaction provided by a product, it includes functional efficiency, appearance, ease of installation and operation, safety reliability, maintainability, running and maintenance cost. There are two facts of quality; namely quality of design and quality of conformance. Quality is initially created by the designer in the form of product specifications and manufacturing instructions. The design provides user satisfaction and the product conforms to the design.

Inspection alone cannot build quality into a product unless quality has been designed and manufactured into it. Now, total quality control, quality assurance, and the quality management have been termed as quality control. The quality of the product in a company is determined by the philosophy, commitment and the quality policy of the top management and the extent to which this policy can be put in actual practice.

Quality is perceived differently by different people. But everyone understands meaning of "quality". The customer as a user takes the quality of fit, finish, appearance and performance in a manufactured product. Service quality may be divided on the basis of the degree of satisfaction by the customer who is receiving the service. The dictionary meaning of quality is "the degree of excellence". Such definition is relative in nature. The final test in this evaluation is dependant on the customer. The measurable characteristics in a product or service are basically translation of the customer needs. Once the specifications are developed ways to measure and monitor the characteristics need to be found. This becomes the basis for continuous improvement in the product or service. The ultimate objective is to ensure that the customer is satisfied and pays for the product or service. Consequently, a reasonable profit for the producer or the service provider. The relationship with a customer is an unending process. The product reliability plays an very important role in developing this relationship with the customer.

Some definitions of quality are:

(*i*) Quality is conformance to requirements.

(*ii*) Quality is fitness for use.

(*iii*) Good quality means a predictable degree of uniformity and dependability with a quality standard suited to the customer.

(*iv*) Quality is the degree to which performance meets expectations.

(*v*) Quality denotes an excellence in goods and services, especially to the degree they conform to requirements and satisfy customers.

(*vi*) Reliability is the probability that system can perform its intended function for a specified interval under stated conditions.

Total quality management (TQM) describes the new concept on Quality control. It deals with the product in its totality. Quality is determined by the combined effects of various departments such as Design, Engineering, Purchase, Production and Inspection. Total quality management is an across-the-company, management function which co-ordinates and controls widespread quality activities with a view to achieving the desired quality of the end-product. This is basically the integrated quality system covering all quality control activities during various phases of product development.

1.2. EVOLUATION OF QUALITY CONTROL

The quality movement started from medieval Europe. Craftsmen organized unions called guilds in the late thirteenth century. The product inspection started in Great Britain in the mid 1750s and brought the Industrial Revolution in the early nineteenth century. The concept of producing interchangeable parts to simplify assembly was introduced in 1798.

Measuring methods and ensuring dimensional consistency ended in the mid 1800's with introduction of go and no go gauges. A go gauge for a hole checks for its lower limit. No go gauges check the upper limit for a hole. The principles of scientific management was introduced around 1900. This emphasized division of labour which focused on productivity. The rise in productivity gave a negative effect on quality.

The moving automobiles assembly line was introduced in 1913. The requirement was that the consistently good quality parts are made available to avoid slowdown of production assembly line. The basic idea of the statistical process control chart were introduced in 1924. This started the era of statistical quality control. The statistical quality control methods were widely used by the mid 1930s.

World War II gave birth to quality in manufacturing industries and military applications. A quality revolution in Japan followed World War II. The Japanese became leader in quality by the 1970s. Japanese manufacturers penetrated in American markets which affected the United States. The emphasis became not only on statistics but approaches which involved the entire organization. This movement led to TQM. Several quality standard followed and one of them is ISO 9000 quality system standards published in 1987.

1.3. DIMENSIONS OF QUALITY

Quality relates to the 'features' or 'characters' or some 'attributes' of a product or service. The attributes or features or characters of the product or service should be such that they meet and fulfil the needs of the customers. For example, a wall clock needs to give accurate time, be durable and weatherproof, be convenient to adjust time, date, etc. If needs of customer are not met the customer either may not buy or complain about the product. That is failure of the quality will cause customer dissatisfaction and leads to quality complaint. Therefore, to satisfy the needs of a customer in a business one needs to ensure and maintain the quality of a product or service Quality does not mean something extra in the product or service for customers. In fact, quality is an essential feature of any product or service which attracts customer to buy that product in order to satisfy his needs. Hence, quality must be built and maintained in any product or service for satisfying the given needs of customers. In order to conform to customer specification, the following parameters must be incorporated and maintained in the quality management process.

(*i*) requirement of customer must be clearly understood.

(*ii*) setting quality goals.

(*iii*) planning and designing appropriate quality feature in the product or services.

(*iv*) measuring and checking for conformance to specification.

(*v*) maintaining the quality characteristics by appropriate means and measures.

In simplest term quality control is the control of quality during manufacturing. Raw materials, suppliers, parts or finished products etc. as accepted or rejected by inspection. If control is effective, there is less need for inspection. Quality control gives the causes for variation in the characteristics of products. It also gives solutions by which these variations can be controlled. It achieves economy and objective. It meets procedure and dynamics in its operation.

An excellent product or service that fulfills or exceds our expectations is a quality product or service. These expectations are based on the intended use and the selling price. A customer expects a different performance from a plain steel washer than from a chrome-plated steel washer because they are a different grade. Quality can be quantified as follows:

$$Q = \frac{P}{E}$$

where; Q = Quality, P = Performance, E = Expectation

A more definitive definition of quality is given in ISO 9000 : 2000. It is defined as the degree to which a set of inherent characteristics fulfills requirements. Degree means that quality can be used with adjectives such as poor, good and excellent. Inherent is defined as existing in something, especially as a permanent characteristics. Characteristics can be quantitative or qualitative. Requirement is a need or expectation that is stated; generally implied by the organization, and other interested parties; or obligatory.

Quality control includes statistical sampling, testing before acceptance and the examination of quality characteristics in finished products. Quality symbolize many aspects of customer needs such as:

(*i*) Performance

(*ii*) Features

(*iii*) Durability

(*iv*) Reliability

(*v*) Serviceability

(*vi*) Appearance
(*vii*) Uniformity
(*viii*) Consistency and confirmance
(*ix*) Safety
(*x*) Time
(*xi*) Customer service
(*xii*) Comparability

Some dimensions of quality are given below, which are described in the Table 1.1.

Table 1.1. Dimensions of Quality

S.No.	Dimensions	Description
1.	Performance	How well the product or service performs the intended function such as fuel economy of cars, durability of batteries etc.
2.	Features	Special features which appeal to customer such as design of seats in a car, look and colour of product etc.
3.	Durability	Time duration or amount of use before being replaced or repaired.
4.	Reliability	Failure, repair or expected time of fault-free operation.
5.	Serviceability	Convenient low cost repair and maintenance, and resolving the customer complaints.
6.	Appearance	Small, sound or other effort which is felt by human senses.
7.	Uniformity	Very little variations among different products of same model.
8.	Consistency and conformance	Conforming to standard as per documentation.
9.	Safety	Human safety, health and environment conservation.
10.	Time	Delays in supply/service.
11.	Customer service	After sales service and response to customer complaints.
12.	Compatability	Product compatibility and interferes with pheripherals or other attachments, power source etc.

Thus, the dimension of quality has customer-focus, value-added products and services, and vision of globalization and world-class quality. It has been developed due to hard work of industry leaders and academics for several decades. These pioneers systematically examined what ailed a company's performance, what needed to be done to eradicate the ills, and developed many path-breaking tolls and techniques to give effect to their recommendations for overcoming the deficiencies in the approach to maintain quality in all spheres of company activities. Quality was identified as the key factor for superior business performance and a means to business excellence.

1.4. MEANING OF QUALITY CONTROL

The meaning of quality changed dramatically in the late 1970s. Earlier quality was viewed as something that needed to be inspected and corrected. Moreover in the 1970s and 1980s many U.S industries lost market share to foreign competition. These foreign competitors were providing lower priced products with considerably higher quality. To survive companies had to make major changes in their quality programs. Today successful companies understand that quality provides a competitive advantage. They put the customer first and define quality as meeting or exceeding customer expectations.

Companies in every line of business are focusing on improving quality in order to be more competitive. In many industries quality excellence has become a standard for doing business.

Quality is closely related to cost and customer needs. Quality is the degree to which it fulfils the requirements of the customer. Quality means degree of perfection. Quality can be determined by characteristics such as design; size; material; chemical composition; mechanical functioning; workmanship; finish and other properties.

Quality of a product depends on material, man, machines, and manufacturing conditions. The quality control is the systematic control of these factors. A skilled worker will give better quality. Better machines and better material with good quality product.

Quality description is simple but very complicated when it is to be defined precisely. Quality is right attitude to achieve excellence; producing error-free products and services to the customer within time; and satisfying the requirements and expectations of customers.

It should be well understand that customers are of two types namely: internal customer and external customer. Inside the business system, one employee uses the output of another employee. In assembly, when assembly moves, it changes hands. In this context, the successive person is the internal-customer of the product person. One department is also an internal customer of another. Say, packaging department is internal customer to production department, while marketing department is the internal customer to the packaging department. External customer, is the end-user of the product or services. Some more definition of quality are:

(*i*) Quality is how well a product or service meets the targets and tolerances determined by its designer.

(*ii*) Quality is the fitness for use that evaluates how well the product performs for its intended use.

(*iii*) Quality is defined in terms of product or service usefulness for the price paid.

(*iv*) Quality is defined in terms of the support provided after the product or service is purchased.

Quality is understood as the requirement of a customer for fulfilling a given need, which is expressed in the term of specification. The quality management should ensure the conformance to the specification. The quality can be considered as the conformance to specification, where the specification has been developed from the desired needs of the customer. Quality is the means to meet and satisfy the customer, requirements, therefore, its importance has rapidly increased due to competition in business and industries. Satisfying the needs of customer by consistently maintaining the required quality in products and services has become an essential part of business operations. This is for winning the confidence of customer, and gaining the market share under competitions. In view of this, the companies all over the world are quality conscious to ensure that their products and services meet the customer's needs.

There are challenges of quality in planning, producing and meeting total specifications of goods or service that satisfy the customer's needs with respect to features, characteristics, safety, maintainability, usability and affordability.

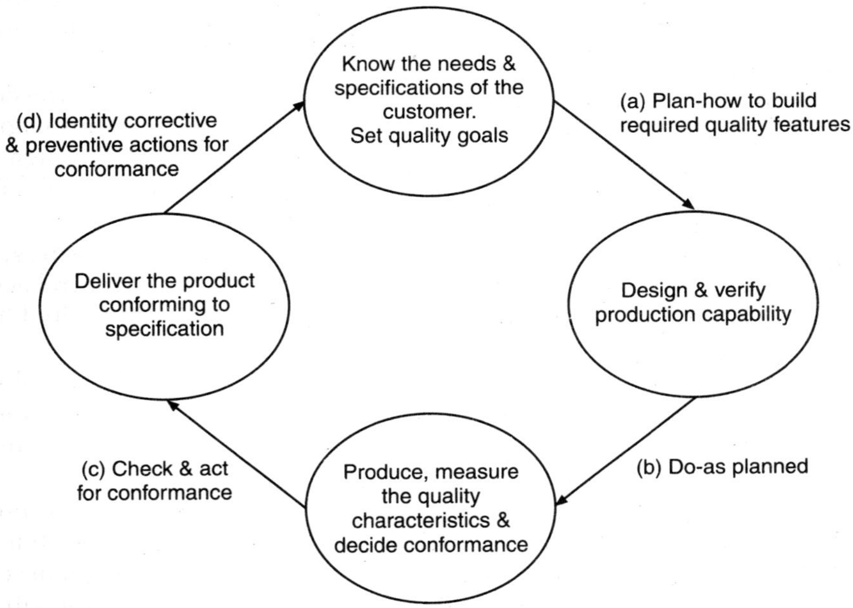

Fig. 1.1. Cycle of Quality Management System

Affordability means value for the money spent to acquire the product or service. This would mean producing at least cost without sacrificing any quality. Meeting this challenge has critical factor for the success of any business or industry. The quality is maintained by setting quality goals after understanding the needs and requirements of users, planning a cost effective process, producing as planned with least variations of parameters, measuring conformance, taking corrective and preventive actions to reduce nonconformance, working for continuous improvement and delivery on time. Thus, severe competitions and changing business environment, setting appropriate quality goals and designing the product and processes to meet the customer's need and expectations have become critical for process.

Quality in manufacturing organizations is different from that of services. Manufactured products can be seen, touched and directly measured such as clothes, computers, food items etc.

"Quality in manufacturing is based on conformance to present standards, performance, features, durability, serviceability, etc."

Service organizations products cannot be seen or touched. Instead, riemed such as delivery of health care, experience of staying at a resort, and studying in university. The intangible nature of the product makes defining quality difficult.

"Quality in services is defined by perceptual factors such as responsiveness to customer needs, courtesy and friendliness of staff, promptness in resolving complaints and atmosphere."

1.5. CONCEPT CHANGE

Quality concept change has taken place to ensure conformance of customer want and consistency in manufacturing for quality. CEO of General Electrics, Jack Welch adopted three initiatives to turn around company. These are quality, services, globalization. Quality initiative was to recreate the processes so that defects and errors did not occur. Normal quality efforts were generally for detecting and correcting quality faults. These were too little and too late for efficient and cost-effective running of the business in the face of global competition. Service initiatives changed GE from highly engineered products to a source of customer-focussed high-value high-quality engineering solutions to problems and requirements of customers. These concepts that turned GE around are not new, but were put in the right perspective and pursued with the right spirit.

There are several individuals who made significant contributions to quality control and improvement. W. Edward Deming is the best-known quality expert in the world. Deming was instrumental in the post-war individual revival of Japan. Japanese industry adopted his methods which resulted in a significant improvement in quality. Deming firmly believed that quality is the responsibility of the management. The Deming philosophy is summarized in the fourteen points. These fourteen points apply to both small and large organizations, to the service industry as well as to manufacturing. They also apply to a division within a company. The fourteen points are as follows:

(*i*) **Create constancy of purpose for improvement of product and service:** This is basically a mission statement which must be understood by all employees, suppliers, and customers.

(*ii*) **Adopt the new philosophy:** Management must understand its responsibilities and take on leadership for change wherein poor workmanship, defective products, or bad service are not acceptable.

(*iii*) **Cease dependence on mass inspection:** Need for inspection must be eliminated by building quality into the product in the first place.

(*iv*) **End the practice of awarding business on the basis of price tags alone:** The objective is the purchase of new tools and other equipment should be to minimize the net cost per hour of operation or per piece produced. A single supplier per item may be the solution for long-term relationship of loyalty and trust.

(*v*) **Improve consistently and forever the system of production and service:** The Deming cycle involves the four-step process of plan, do, check, act. The opportunities for improvement are identified at the plan stage. The theory is tested on a small scale at do stage. The test results are analyzed at the check stage. Implementation of the result is the act stage.

(*vi*) **Institutive training:** All employees must be provided on-the-job training. They must be encouraged to implement the knowledge developed through training.

(*vii*) **Adopt and institute leadership:** Supervisor should help people to do better job using machines, and create an environment where the workers take leadership roles in accomplishing their work.

(*viii*) **Drive out fear:** Workers should be encouraged to ask questions and make suggestions.

(*ix*) **Breakdown barriers between departments:** Research, design material procurement, sales and production personnel must work as a team.

(*x*) **Eliminate slogans, exhortations and targets for the work force:** Asking for zero defects and new levels of productivity create negative effects. Most of the causes of low quality and low productivity are due to system and thus lie beyond the power of the work force.

(*xi*) **Eliminate numerical quotas for the work force and eliminate numerical goals for people in management:** This leads to poor quality. The capabilities of processes and methods to improve them must be learnt.

(*xii*) **Remove barriers that rob people of pride of workmanship:** All employees must be satisfied and motivated. Workers must take pride in their job.

(*xiii*) **Encourage education and self-improvement for everyone:** An organization improves with people who are improving with education.

(*xiv*) **Take action to accomplish the transformation:** Everybody has to involve in the transformation.

The new concept of quality is total quality management or TQM. Total quality management (TQM) means involvement of all aspects of the firm in satisfying the customer. It aims goal that can only be accomplished if the usefulness is recognized by having a partnership environment at all stage of the business, both with the firm. TQM involves:

(*i*) Customer-supplier relationship based on mutual trust and respect. There must be a beneficial strategy for both.

(*ii*) Firm in-house requirements by the customers.

(*iii*) Suppliers are partners in achieving zero-defect situation.

(*iv*) Regular monitoring of supplier processes and products by the customer.

1.6. QUALITY FUNCTIONS IN SERVICES

With the growing influence of service sectors on business and economic development of the country, improvement of quality in service is fast gaining priority. In a developed economy, over 66% of workforce and over 60% of economy is from service sectors. In a survey in USA, it was observed that nearly 90% of quality problems that drive away the customers relate to quality of service, more so in service industries. In other words, major potential of improvement, which is the objective of quality management, lies in service industries and service jobs of manufacturing industries. Many service industries like information technology, banking, communication, healthcare etc. have well developed quality assurance system, and they go for ISO 9000 certification. The important requirements and dimensions of service quality are:

(*i*) Time to serve.
(*ii*) Timeliness of service.
(*iii*) Completeness of the service.
(*iv*) Accuracy of service.
(*v*) Courtesy and behaviour of frontline/delivery personnel.
(*vi*) Consistency of service and service quality.
(*vii*) Responsiveness.
(*viii*) Convenience of getting the service.

There are significant differences between manufacturing industries and service industries, namely:

(*i*) It may be difficult to clearly define in service industry what the customer wants.
(*ii*) The image and brand of the company is a major factor in shaping customer expectations and service requirements.
(*iii*) Establishing and measuring to service standards is difficult in service industries, because the quality standards often depends on the personal behaviour of a service person and expectation of the customer being served, which cannot be reliably measured.
(*iv*) Service is often performed at outside locations where quality cannot be supervised or guided by seniors.
(*v*) The service is typically delivered face to face to a customer, on shot at the customer's locations, on time and with right quality.

Companies that have implemented the principles of quality in their operations, and consistently provided superior services to customers have exceeded in their business throughout their existence. There are many renouned global companies, such as IBM, Toyota Motors, Sony Corporation, Fed Ex, Xerox, etc., who have continued to excel due to their emphasis on quality in services. Of late, many Indian companies are emulating this quality practice in services like TVS Group, Tata Group, Maruti Udyog Ltd, Infosys, TCS etc. Their brand image and business results are showing the merit of the quality approaches.

SUMMARY

1. Quality of product signifies the degree of its excellence and fitness for the purpose.
2. Quality control includes statistical sampling, testing before acceptance and the examination of quality characteristics in finished products. Quality symbolize many aspects of customer needs.
3. There are several individuals who made significant contributions to quality control and improvement. W. Edward Deming is the best-known quality expert in the world. Deming was instrumental in the post-war individual revival of Japan. Japanese industry adopted his methods which resulted in a significant improvement in quality. Deming firmly believed that quality is the responsibility of the management.

4. Major potential of improvement, which is the objective of quality management, lies in service industries and service jobs of manufacturing industries. Many service industries like information technology, banking, communication, healthcare etc. have well developed quality assurance system, and they go for ISO 9000 certification.

EXERCISES

1. Describe evolution of quality control.
2. Explain quality concept change.
3. What do you understand by quality control?
4. What are the dimensions of quality? Explain.
5. What is the meaning of quality control? Explain.
6. Explain quality function in services.

2 CHAPTER TWO

TOTAL QUALITY MANAGEMENT (TQM)

CHAPTER OUTLINE

- Introduction
- Principal Objectives of TQM
- Deming Approach to TQM
- CIT-Quality Excellence Model
- Malcolm Baldrige National Quality Award Model (MBNQA)
- Leadership
- Employee Involvement
- Supplier Partnership
- Total Productive Maintenance (TPM)
- Quality Gurus and TQM Framework
- Basics of TQM
- TQM System and Model
- Customer Satisfaction
- Continuous Process Improvement
- Kaizen

2.1. INTRODUCTION

The challenges of quality for customer satisfaction, organizations needed "total quality" which was in practice in Japan since the mid 1960s. This concept was first developed by Feigenbaum in 1961, based on various studies of "Quality Markets"- like Deming, Juran and Crosby. Feigenbaum stressed the need for a holistic view of the quality management system involving all functions and functionaries of the organization. This new and more comprehensive approach of quality management was known as "Total Quality Control" (TQC). Total quality control of Feigenbaum stressed the need for complete customer satisfaction by offering products and services that are designed and built based on the needs and expectations of customers. It needed integration of systems and efforts of the entire organization through employee involvement, framework and supplier relationship. This initial approach to total quality control was given a more comprehensive form in the mid-1980s by introducing total Quality management models.

An enhancement to the traditional way of doing business is through total quality management (TQM). TQM is a process technique to guarantee survival in world class competition. The actions of management will have to be changed to get the culture and actions of an entire organization be transformed. TQM is generally a common sense. The three words can be understood as:

(*i*) Total for whole

(*ii*) Quality for degree of excellence a product or service.

(*iii*) Management for act, or way of handling, controlling, directing etc.

In fact, TQM is the art of managing the whole to achieve excellence. Thus, TQM is a philosophy and set of guiding principles that represent the foundation of a continuously improving organization. It is the application of methods and human resources to improve all the processes and exceeds customer needs. Fundamental management technique, existing improvement efforts, and technical tools under a disciplined approach are integrated under TQM. TQM has six basic requirements:

(*i*) Effective involvement and utilization of the work force.

(*ii*) Continuous improvement of the processes.

(*iii*) Performance measurements for the processes.

(*iv*) Suppliers as partners.

(*v*) Focus on the customer.

(*vi*) Committed management to provide complete organizational support.

Participation of the management in council must be established the quality programme is a must. A quality council must be established to develop a clear vision, set long-term goals, and direct the program. TQM programs must focus on the customer outside as well as internal customer, customer satisfaction is the most important consideration.

TQM is an organization wide challenge that is everyone's responsibility. Those affected by the plan must be involved in its development and implementation. They understand the powers better than anyone else. Changing behaviour is the goal. People must also think to improve their jobs. People must be empowered at the lowest possible level to perform processes in an optimum manner. A continual effort to improve all processes is a must. A good starting point for quality improvement in projects, on-time delivery, order entry efficiency, billing error rate, customer satisfaction, cycle time, scrap reduction, and suppliers management SPC, benchmarking, quality function usage, ISO 9000 and designed experiments are excellence technical tools in problem solving.

TQM provides a quality product or service to customers to increase productivity and lower cost. Higher quality product and lower price helps to compete in the market place. These events will allow the organization to achieve the objectives of profits and growth. Thus, TQM requires a cultural change wherein old culture change to new culture as follows:

(*i*) Product-oriented to customer oriented.

(*ii*) Priorities from second position of service and cost to first among equals of service and cost.

(*iii*) Decisions from short-term to long term.

(*iv*) Emphasis from detection to prevention.

(*v*) Errors from operations to system.

(*vi*) Responsibility from quality control to everyone.

(vii) Problem solving from manager to teams.

(viii) Procurement from price to life-cycle costs partnership.

(ix) Manager's role from plan, assign, control and enforce to delegate, coach, facilitate and mentor.

TQM provides incentives for employees to identify quality problems. Part of the TQM philosophy is to empower all employees to seek out quality problems and correct them. Workers are empowered to make decisions relative to quality in the production process. They are considered a vital element of the effort to achieve high quality. Their contributions are highly valued, and their suggestions are implemented.

TQM differentiates between external and internal customers. External customers are those that purchase the company's goods and services. Internal customers are employees of the organization who receive goods or services from others in the company.

2.2. QUALITY GURUS AND TQM FRAMEWORK

To fully understand the TQM movement, the philosophies of important personalities who have shaped the evolution of TQM need to be looked into. Their philosophies have contributed to quality of today. Their individual contribution are summarized in Table 2.1.

Table 2.1. Quality Gurus

S.No.	Quality Guru	Main Contribution
1.	Walter A. Shewhart (Grand father of quality control)	Contributed to understanding of process variability. Developed concept of statistical control charts.
2.	W. Edwards Deming (Father of quality control)	Stressed managements responsibility for quality. Developed "14 points" to guide companies in quality improvement.
3.	Joseph M. Juran	Defined quality as "fitness for use". Developed concept of cost of quality.
4.	Armand V. Feigenbaum	Introduced concept of total quality control.
5.	Philip B.Crosby	Coined phrase "quality is free". Introduced concept of zero defects.
6.	Kaoru Ishikawa	Developed cause-and-effect diagrams. Identified concept of "internal customer".
7.	Genichi Taguchi	Focused on product design quality. Developed Taguchi loss function.

The framework for the TQM system is shown in Fig. 2.1 The knowledge provided by Gurus of quality contributed to the development of principles and practices and/or the tools and techniques.

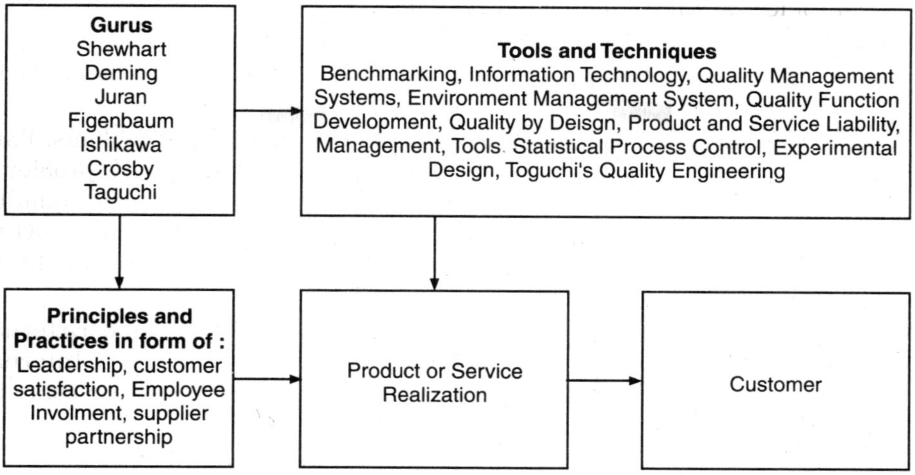

Fig. 2.1. Framework for the TQM system

2.3. PRINCIPAL OBJECTIVES OF TQM AND INTEGRATED MODEL

A total quality oriented organization must have at least following principle objectives. Organization should have system oriented specific objectives as follows:

(*i*) Customer focus, customer delight/satisfaction.

(*ii*) Continuous improvement of a culture of the organization, which must be the way of life.

(*iii*) Focused, continuous and relentness cost reduction.

(*iv*) Focused, continuous and relentless quality improvement.

(*v*) To create an organization whereby everyone is working towards making their organization the best business, and to capitalize on the sense of achievement and working in a world-class organization.

TQM must include a ten dimensional framework as shown in Fig. 2.2.

TQM will have obstacles to its successful implementation. The most common were determined by Robert J Masters after an extensive literature search and the last obstacle added by others. They are as follows:

(*i*) Lack of Management Commitment: There must be a substantial mangement commitment of management time and organizational resource.

(*ii*) Inability to Change Organizational Culture: Changing an organization's culture is difficult as individuals resist change because they become accustomed to doing a particular process. Management must understand that people change when they want to and to meet their own needs. People do not engage in behaviour that serves the organization's values unless adequate reason has been given, people must be moved from a state of fear to trust. Thus, it is difficult for individuals to change their way of doing things. Further, it is much more difficult for an organization to make a cultural change.

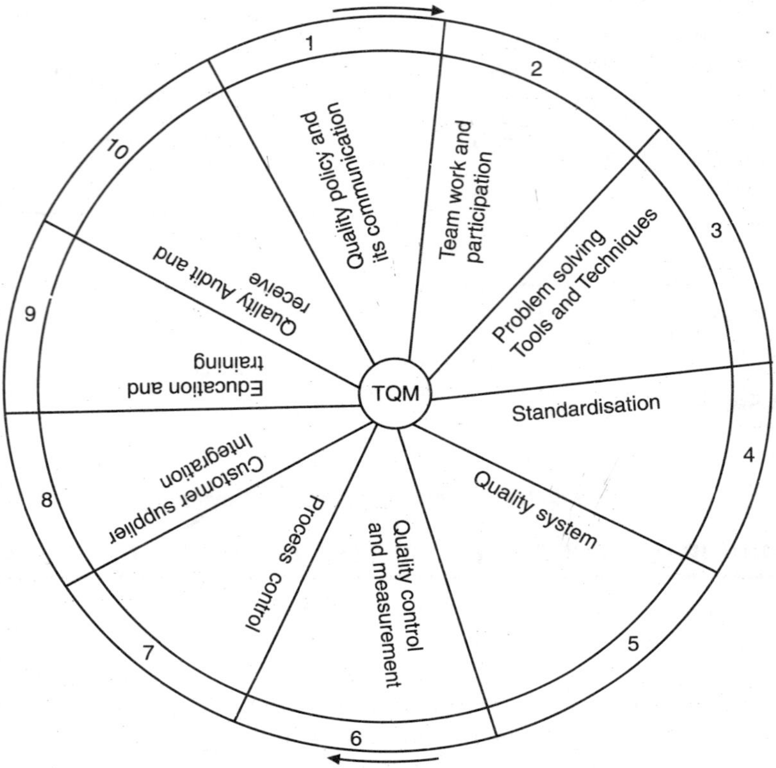

Fig. 2.2. An Integrated TQM Model

(iii) **Improper Planning:** All must be involved in the development of the implementation plan and any modifications that occur as the plan evolves. The two-way communication of ideas by all personnel during the development of the plan and its implementation is of great importance. Customer satisfaction should be the goal rather than financial or sales goals.

(iv) **Lack of Continuous Training and Education:** Needs of training and education must be determined and a plan developed to achieve those needs. Training and education are most effective when senior management conducts the training on the principles of TQM.

(v) **Incompatible Organizational Structure and Isolated Individuals and Departments:** Restructuring to make the organization more responsive to customer needs may be needed. Individuals who do not fall in line with the new philosophy can be asked to leave.

(vi) **Ineffective Measurement Technique and Lack of Access of to Data and Results:** In order to improve a process one needs to measure the effect of improvement ideas. Access to data and quick retrieval is necessary for effective processes.

(vii) **Paying Inadequate Attention to Internal and External Customers:** Understanding the changing needs and expectations of their customers is an effective feedback mechanisms that provide data

for decision making are necessary for this understanding. One way to overcome this obstacle is to give the right people direct access to the customers.

(*viii*) **Inadequate use of Empowerment and Teamwork:** Proper training helps to ensure the team's recommendations to be followed. Individuals should be empowered to make decisions that affect the efficiency of their process or the satisfaction of the their customers.

(*ix*) **Failure to Continually Improve:** A lack of continuous improvement of the processes, product service makes the team inefficient.

The benefits of TQM are improved quality, employees participation, teamwork, working relationships, customer satisfaction, employee satisfaction, productivity, communication, profitability and the market share.

2.4. BASICS OF TQM

TQM foucses on identifying root causes of quality problems and connecting them at the source, as opposed to inspecting the product after it has been made. Not only does TQM covers the entire organization, but it stresses that quality is customer driven. TQM attempts to embed quality in every output of the organization. It is concerned with technical aspects of quality as well as the involvement of people in quality, such as customers, company employes, and suppliers.

Quality is defined as meeting or exceeding customer expectations. The goal is to first identify and then meet customer needs. TQM recognizes that a perfectly produced product has little value if it is not what the customer wants. Therefore, we can say that quality is customer driven.

Companies need to gather information continuously by means of focus groups, market surveys, and customer interviews in order to stay in tune with what customer want. They must always remember that they would not be in business if it were not for their customers.

The Japanese believe that the best and the most lasting changes comes from gradual improvements. To use an analogy, they believe that it is better to take frequent small dose of medicine, then to take one large dose. Continuous improvement, called Kaizen by the Japanese, requires that the company continually strive to be better through learning and problem solving. Because we can never achieve perfection, we must always evaluate our performance and take measures to improve it. There are two approaches that can help companies with continuous improvement namely.

(*i*) the plan-do-study-act (PDSA) cycle

(*ii*) benchmarking

2.4.1. PDSA Cycle

The basic Plan-Do-Study-Act (PDSA) was first developed by Shewhart and then modified by Deming. It explains the activities a company needs to perform in order to incorporate continuous improvement in its operation. The cycle is shown in Fig. 2.3. This cycle shows that continuous improvement is a never-ending process.

(*i*) **Plan:** Evaluate the current process and makes plan based problems. Document all current procedures, collect data, and identify problems. This information should then be studied and used to develop a plan for improvement as well as specific measures to evaluate performance.

(*ii*) **Do:** During the implementation process managers should document all changes made and collect data for evaluation.

(*iii*) **Study:** The data are evaluated to see whether the plan is achieving the goals established in the plan phase.

(*iv*) **Act:** The best way to accomplish this is to communicate the results to other members in the company and then implement the new procedure if it has been successful.

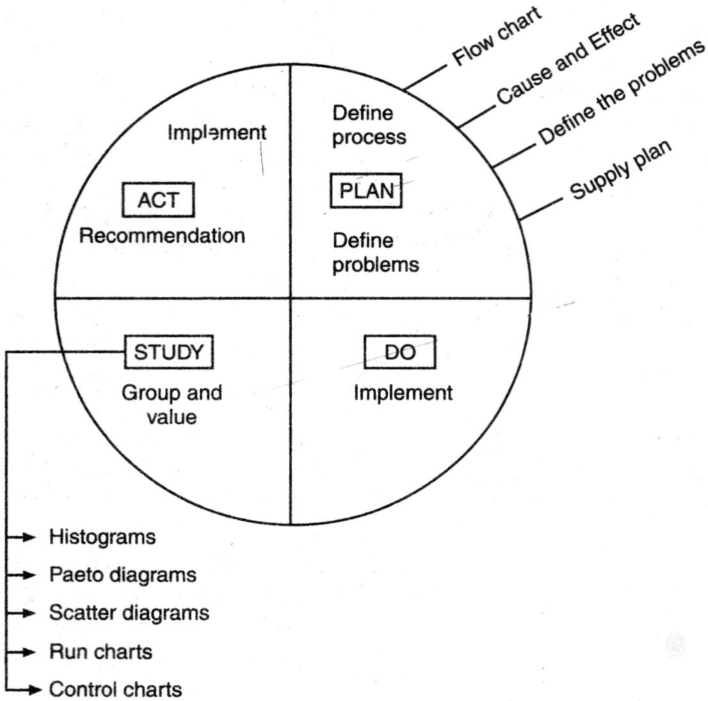

Fig. 2.3. PDSA Cycle

2.4.2. Benchmarking

It is studying the business practices of other companies for purpose of comparison. The ability to learn and study how others do things is an important part of continuous improvement. The benchmark company does not have to be in the same business, as long as it excels at something that the company doing the study wishes to emulate.

Benchmarking is a systematic method which helps organization to measure themselves with respect the best industry practices. The benchmarking is essentially a process of borrowing ideas and adapting them to get the competitive edge. It helps in continuous improvement.

Becnhmarking takes the experience of others and uses it. In the 1800s, Francis Lowell studies British textile mills and got many ideas which he used for improvements. This showed that the benchmarking measures performance against that of best-in-class organizations. It helps to understand how the best in class achieve those performance levels. This information becomes the basis for adaptive creativity and better performance.

Benchmarking a thorough and in depth knowledge of their own processes and as well as the processes of the best-in-class organization. To organize improvement efforts to meet the goal, an understanding of the differences is needed edge. Benchmarking is a tool to achieve business goals and be competitive.

The process involves imitation and adaptation rather than pure invention, therefore, the benchmarking is time and cost efficienct. A working model of an improved process is provided by planning, testing and prototyping effort.

Organizations must continue to innovate as well as imitate for functions that are critical to the business mission. Benchmarking enhances innovation by making organizations to constantly go through the external environment. The information so obtained is used for improvement. Organizations must adapt the process to best fit their own needs and culture. The following six steps are the main techniques of benchmarking:

(*i*) Decision on item of benchmarking.

(*ii*) Current performance understanding.

(*iii*) Planning.

(*iv*) Studying other organizations.

(*v*) Learing from the essential data created.

(*vi*) Application of the outcomes.

If the process personnel are not committed, they may ignore the results. Thus, there is a need of training and justification is needed to get support of the management. To fit the process further, the core activities are tobe consistent. Subsequently, new steps can be to find:

1. The most troublesome processes.
2. Customer satisfaction related processes and non-performing processes.
3. Competitive pressure affecting the organization most.
4. Processes or functions having the highest potential for differentiating the organization from the competition.

It is first necessary to thoroughly understand and document the current process, before comparing practices to outside. It is essential that the organization's performance is well understood by the benchmarker. Benchmarking can be internal, competitive and process. Similar activities are performed in different operating divisions in most of the large firms. An obvious choice to benchmark are the product competitiors. A research has to be due to identify the best firms to benchmark. Competitor, industry or even internal system may have best practices. Large quantities of industry information is compiled by government agencies, government publications and or from government who experts provide the data.

Possible benchmark partners may be found through the planning process. A description of best-in-class process practices and the measurable outcomes of the practices are good inputs. The benchmark findings lead to new goals and objectives. For the development and execution of action plans, the following steps may be needed:

1. Specification of work.
2. Sequence of works.
3. Identification of resource requirements.
4. Work scheduling.
5. Responsibility assignment for each work.
6. Expected results description.
7. Identification of methods of monitoring obtained results.

The execution of the action plan should be consistent with goals and objectives for the end result process superiority. A person executing process in better way than oneself should be found to imitate. Benchmarking leads to persistent criticism that the idea of others ie being copied.

2.5. DEMING APPROACH TO TQM

Deming has pioneered the TQM concept. There are fourteen steps to improve quality as per Deming. These steps are as follows:

1. Create consistency of purpose for improving services and products.
2. Adopt the new philosophy for making the accepted levels of defects, delays etc.
3. Stop reliance on mass inspection as it does not improve and guarantee quality. Teamwork between the firm and its suppliers improves the quality.
4. Stops awarding business with respect to the price.
5. Discover problems and work continuously to improve the system.
6. Take advantage of modern methods of training and also in developing a training programe considering the following:
   (*i*) company objectives
   (*ii*) training goals
   (*iii*) understanding of goals by involving everyone
   (*iv*) orientation of new employees
   (*v*) training of supervisors in statistical thinking
   (*vi*) plan on team-building
   (*vii*) analysis of the teaching need.
7. Institute modern supervision approaches.
8. Eradicate fear so that everyone involved may work to his or her full capacity.
9. Remove department barriers so that everyone can work as a team member.
10. Eliminate items such as goals, posters, and slogans that call for new productivity levels without improvement of methods.

11. Make the organization free of work standards prescribing numeric quotas.
12. Eliminate factors that inhibit employee workman-ship pride.
13. Establish an effective education and training program.
14. Establish a system that will push the above 13 points everyday for continuous improvement.

2.6. TQM SYSTEM AND MODEL

TQM is a system to quality management. It is a complete commitment to quality in all services of the organization. A plan for a quality system is shown in the Fig. 2.4.

Total in TQM stands for an overall integrated approach to all aspects of quality, all domains of system, including, organization, people, resources, hardware/software and even management commitments. TQM sustains on four pillars namely system, management commitment, team work and statistical process control (SQC) tools. Plan for quality system is shown in Fig. 2.4. A total quality management model is shown in Fig. 2.5. The plan for quality system involved following steps:

(*i*) Quality policy by the chief executive.
(*ii*) Detailed TQM objective
(*iii*) Responsibility of the management
(*iv*) Quality system to will the requirement of the customer and of the company. Specify procedures, methods and instructions profitability and growth profile clearly defined.
(*v*) Identify problem areas by regular and its of the operations areas by self assessment and team work.
(*vi*) Improvement programme to achieve astablished objectives.
(*vii*) Program implementation, involvement and the commitment of all employees.
(*viii*) Program monitoring as per agreed timetable.
(*ix*) Audit and receive of overall effectiveness.

TQM is a quality focused culture, journey to achieve excellence in all aspects of organizations activity, with involvement of all members of the organization at all levels of operation. TQM is a management philosophy rather than a quality-technology. TQM began in Japan on the quality philosophy of Dr. W. Edwards Deming who is an American.

2.7. CII-QUALITY EXCELLENCE MODEL

CII-Quality Excellence model award was instituted in the year 1994 by the apex body of the Confederation of Indian Industries (CII) in partnership with the EXIM Bank of India, and called CII-EXIM Business Excellence Award. It recognizes the best company amongst Indian industry which has successfully adopted a total quality approach and achieved performance excellence in its business operations.

The CII-Quality Excellence award is focused to India-specific needs and covers the broad business areas of:

(*a*) Manufacturing/ Engineering sector ; (*b*) Services sector ; (*c*) Education; and (*d*) Governance, Initial winners of this prestigious award in India are: Hewlett-Packard (India) Ltd, Maruti Udyog Ltd., Tata Steel Ltd., and Infosis Technologies Ltd. Briefly the major focus areas of the CII-Quality Excellence model are Customer satisfaction. Product Quality, Plant Reliability and waste elimination. The steps for CII-EXIM model steps are as follows:

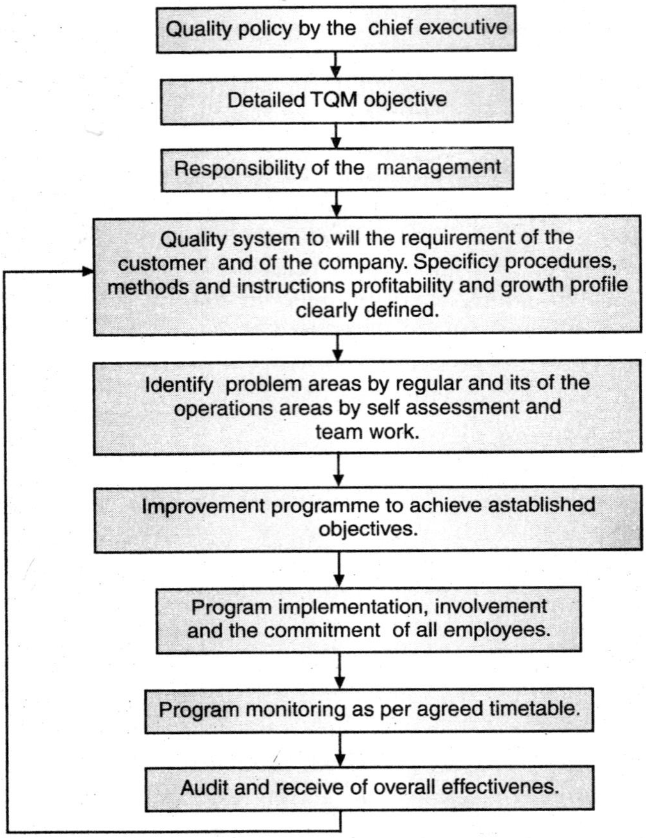

Fig. 2.4. Plan for a Quality System

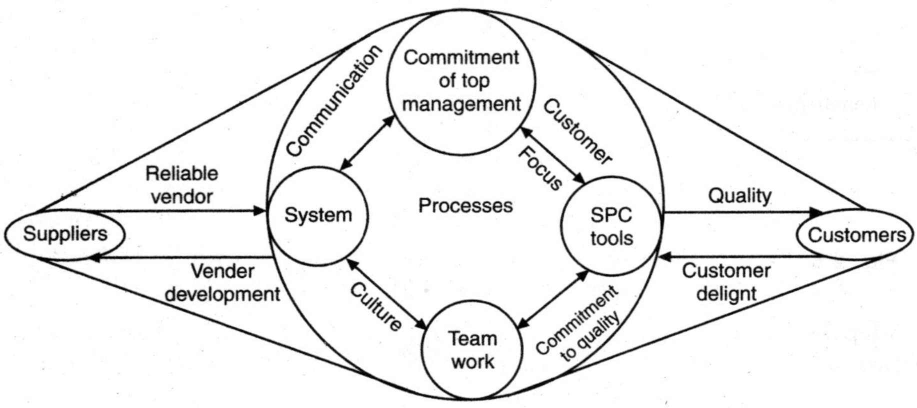

Fig. 2.5. Model on Total Quality Management

(*i*) The application is examined by a team of about six assessors.

(*ii*) Assessors should have undergone training to ensure a high level of consistency in scoring.

(*iii*) Assessors should include academics and quality professionals but the majority should be drawn from the experienced practising managers.

(*iv*) An application is assessed and scored on a scale from 0 to 1000 points.

The results criteria dicides organization achievement or trying to achieve. Enablers criteria are used to understand the methods of achieving.

2.8. MALCOLM BALDRIGE NATIONAL QUALITY AWARD MODEL (MBNQA)

The Deming Application Prize of Japan had innovated the concept of "award" for total quality in 1960s, and this was followed by the Malcolm Baldrige National Quality Award (MBNQA) system for US companies in 1987 for recognizing the efforts towards total quality mangement practices and performances. The Baldrige Award criteria cover a set of seven categories of processes and a number of "items" and "areas" that must be addressed under the TQM system. These seven categories are:

(*i*) Leadership

(*ii*) Strategic planning

(*iii*) Customer and Market focus

(*iv*) Information and Analysis

(*v*) Human Resource Management and Development

(*vi*) Process Management

(*vii*) Business Results.

Figure 2.6, shows these process categories in their respective roles for achieving superior business results.

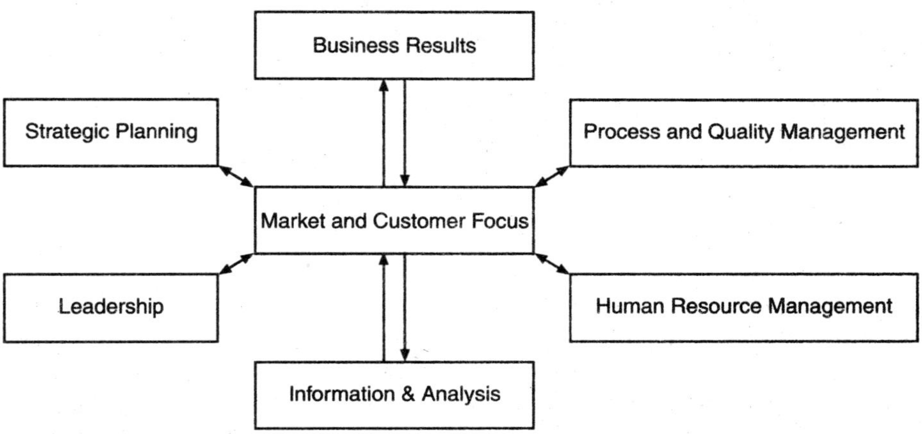

Fig. 2.6. Seven interaction process categories of Malcolm Baldrige National Quality Award (MBNQA) Model

Adoption of these seven process categories of the MBNQA system for TQM practice calls for alignment and reshaping of all internal processes and activities.

This involves changing the organization's thought process, and restructuring of operational processes to ensure that key objectives like customer satisfaction, employee satisfaction and superior business results are accomplished without hindrances.

2.9. LEADERSHIP

The Malcom Baldrige National Quality Award has defined leadership with core values and visionary leadership. There are 12 behaviours or characteristics that successful quality leaders demonstrate:

(*i*) Give priority attention to external and internal customers and their needs.

(*ii*) Empower, rather than control, subordinates.

(*iii*) Emphasize improvement rather than maintenance.

(*iv*) Emphasize prevention.

(*v*) Encourage collaboration rather than competition.

(*vi*) Train and coach, rather than direct and supervise.

(*vii*) Learn from problems. Problem exists, it is treated as an opportunity rather than something to be minimized or covered up.

(*viii*) Continually try to improve communications.

(*ix*) Continually demonstrate their commitment to quality.

(*x*) Choose suppliers on the basis of quality.

(*xi*) Establish organizational systems to support the quality effort.

(*xii*) Encourage and recognize team effort.

Successful leadership requires an intuitive understanding of human nature such as the basic needs, wants, and abilities of people. A leader needs to understand that people:

(*i*) Need security and independence at the same time.

(*ii*) Are sensitivity to external rewards and punishment and are also strongly self-motivated.

(*iii*) Like to hear a kind word of praise.

(*iv*) Can understand only a few facts at a time.

(*v*) Trust their gut reaction more than statistical data.

(*vi*) Distrust if the words are inconsistent with the leader's actions.

Highly Effective People: 7 Habits

(*i*) Are proactive by taking responsibility and showing the ability to choose the response to a situation.

(*ii*) Begin with the end in mind of an image, picture, or paradigm of the end of your life or your frame of reference.

(*iii*) Put first things first by practicing self management day-by-day, moment-by-moment management of time.

(*iv*) Think win-win frame of mind and the heart that constantly seeks mutual benefits in all human interactions.

(*v*) Seek first to understand, then to be understood.

(*vi*) Synergy that the whole is greater than the hearts. That is togetherness can accomplish more than one can accomplish alone.

(*vii*) Sharpen the saw that is preserving and enhancing the greatest asset one has which is self.

Leadership has to be ethical. Ethics is a body of principles or standards of human conduct that governs the behaviour of individuals and organizations. It is knowing what is the right thing to do and is learned when one is growing up, or at a later date during an organization's ethics training program. An ethics management program needs to address pressure, opportunity and attitude. Managing ethical behaviour requires commitment, new policies and procedure a continuous improvement, and investment in appraisal prevention and promotion. Quality in dependent an ethical behaviour. Doing what is right in the first place is a process way to reduce costs, improve competitiveness, and create customer satisfaction. Many companies are hiring ethics consultants to help them achieve their goals.

Senior management must practice the philosophy of Management by Wandering Around (MBWA). Management should get out of the office and visit customer, suppliers, departments, within the organization. Unity of purpose is key to a leadership system. Here values and concepts provide that unity of purpose. An organization's senior leaders need to set direction and create a customer orientation, clean and visible quality values, and high expectations values, direction and expectation need to address all stakeholders. They need to ensure the creation of strategies, systems, and methods for achieving excellence.

2.10. CUSTOMER SATISFACTION

Customers that are satisfied will increase in numbers, buy more, and buy more frequently. Satisfied customers also pay their bills promptly, which greatly improve cash-flow the lifeblood of any organization. Figure 2.7 shows the importance of customer in an organization.

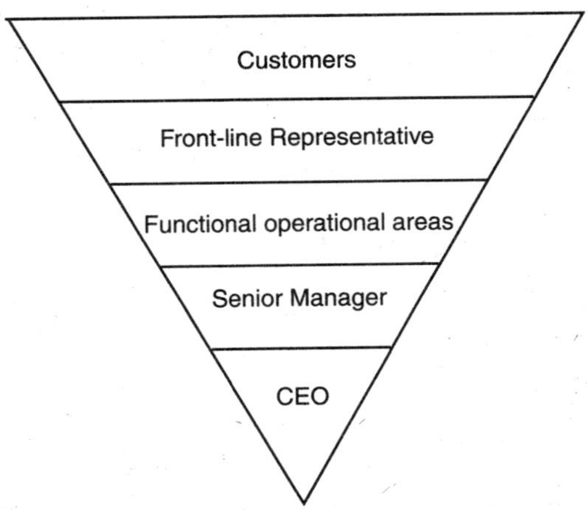

Fig. 2.7. Importance of customer in an organization

There are two distinct types of customer-external and internal. An external customer can be defined in many ways, suchas the one who uses the product or service, the one who purchases the product or service, or the one who influences the sale of the product or service. All processes have outputs, which are used by internal and external customers, and inputs, which are provided by internal or external suppliers. Each supplier perform work that produces some services or products that is used by another customer. Figure 2.8 shows link in the customer/supplier chain, where every chain ends with an external customer and starts with an external suppliers.

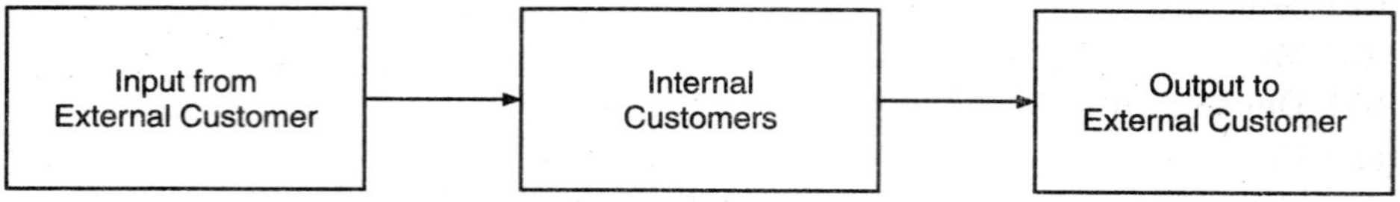

Fig. 2.8. Customer/suppliers chain

There is no acceptable quality level because the customer's needs, values and expectations are constantly changing and becoming more demanding. Before making a major purchase, some people check consumer magazines that rate product quality. Customer is willing to pay a higher price to obtain value. Each customer's concept of value is continually changing. Ongoing efforts must be made by everyone having contact with customers to identify, verify and update each customer's perception of value in relation to each product-and service.

2.11. EMPLOYEE INVOLVEMENT

Employee involvement is one approach to improving quality and productivity. Knowledge of motivation helps us to understand the utilization of employee involvement to achieve process improvement. One of the first and most popular motivational theories was developed by Abraham Maslow as shown in Fig. 2.9.

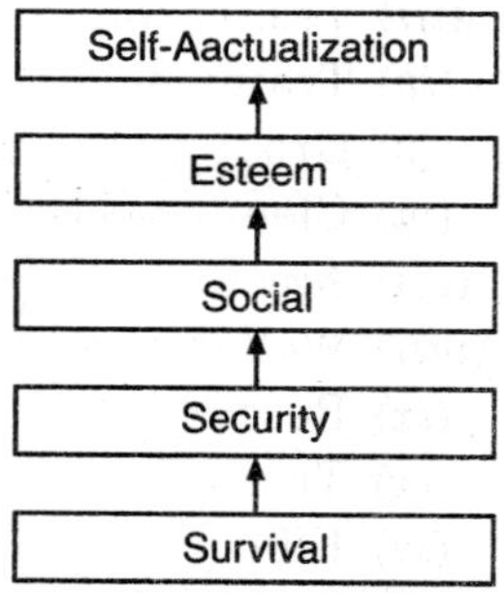

Fig. 2.9. Maslow's Hierarchy of needs

Frederick Herzberg extended the general work of Maslow by his theory on employee motivation. He found that people were motivated by recognition, responsibility, achievement, advancement and the work itself. He showed that bad feelings were associated with low salary, minimal fringe benefits, poor working conditions, ill-defined organizational policies, and mediocre technical supervision. Table 2.2, shows employee wants and manager perception of employee wants. Employee wants tend to follow the theories of Maslow and Herzberg.

Table 2.2. Employees wants (Courtesy Theodore B. Kinni)

S.No.	Factor
1.	Good pay
2.	Promotion/growth
3.	Good working conditions
4.	Job security
5.	Interesting work
6.	Appreciation
7.	Involvement
8.	Loyalty to employees
9.	Tasteful discipline
10.	Help with personal problems

Concept to achieve a motivated work force are:

(*i*) Know thyself
(*ii*) Know your employees
(*iii*) Establish a positive attitude
(*iv*) Share the goals
(*v*) Monitor progress
(*vi*) Develop interesting work
(*vii*) Communicate effectively
(*viii*) Celebrate success

Employee involvement is optimized by the use of teams. Team work because many heads are more knowledgeable than one. Each member of the team has special abilities that can be used to solve problems. Teams may be called by different names and have slightly different characteristics to accommodate a particular organization for example, process improvement team, cross-functional team, natural work teams, self-directed / self-managed work teams. In order for a team to be effective, it should have certain characteristics listed below:

(*i*) Sponsor
(*ii*) Team character
(*iii*) Team composition
(*iv*) Training
(*v*) Ground rules
(*vi*) Clear objectives
(*vii*) Accountability
(*viii*) Well-defined decision production
(*ix*) Resources
(*x*) Trust
(*xi*) Effective problem solving
(*xii*) Open communication
(*xiii*) Appropriate leadership
(*xiv*) Balanced participation
(*xv*) Cohesiveness

Performance appraisal is done to let employees know how they are doing and provide a basis for promotions, salary increases, counselling, and other purposes related to an employee's future. The appraisal should point out strengths and weaknesses as well as how performance can be improved. Performance appraisals should be viewed as a positive way to get employee involved. Some improvement suggestions of performance appraisal are:

(*i*) Use rating scales.
(*ii*) Require work team or group evaluation that are at least equal in emphasis to individual focused evaluations.
(*iii*) Require more frequent performance reviews.
(*iv*) Promotion decisions should be made by an independent administrative process.
(*v*) Use peer feedback as an index of internal customer satisfaction.
(*vi*) Include evaluation for process improvement in addition to results.

2.12. CONTINUOUS PROCESS IMPROVEMENT

Organizations should strive to achieve perfection by continuously improving the business and production processes. Continuous process improvement is designed to utilize the resources of the organization to achieve a quality-driven culture. Business processes such as purchasing, engineering, accounting and marketing are areas where nonconformance can represent an opportunity for substantial improvement. Figure 2.10, shows a process models

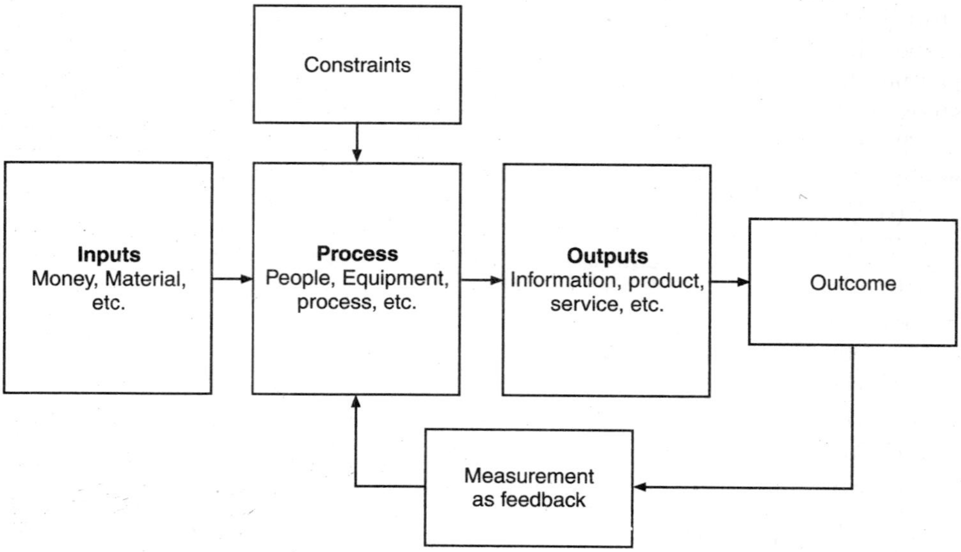

Fig. 2.10. Process model

Process improvement achieves the greatest results when it operator within the framework of the problem-solving method. In the long form, a systematic approach will yield the greatest benefits.

An organization requires performance measurements to achieve one or more of the following seven objectives:

(*i*) Baseline trends.
(*ii*) Which processes to be improved.
(*iii*) Process gains and losses.
(*iv*) Actual performance.
(*v*) Information for evaluation.
(*vi*) Information to make decisions.
(*vii*) Overall performance.

Some items that can be measured are human resources, customers, production, research and development, suppliers, marketing/sales, and administration.

2.13. SUPPLIER PARTNERSHIP

An organization spends about 60% of the cost of goods on the purchase of raw materials, components and services. Therefore, supplier quality can substantially affect the overall cost of the product or service. Thus, for high-quality products and services customer needs to work with suppliers in a partnering atmosphere to achieve the same quality level as attained within the organization. Deming, advocated single suppliers for each item to help develop a long-term relationship of loyalty and trust. Another force changing supplier relations was the introduction of the just-in-time (JIT) concept. It calls for raw materials and components to reach the production operation in small quantities when they are needed and not before. Many original equipment manufacturers (OEMs) are developing strategic partnerships with their suppliers. Suppliers are now taking on increased product-development responsibilities. Customer supplier relationships gives mutually beneficial partnerships and joint efforts improve quality, reduce costs and increase market share for both parties.

There are the conditions for solution and evaluation of suppliers. It is essential that supplier:

(*i*) Understands and appreciate the management philosophy.

(*ii*) Has a stable management system.

(*iii*) Maintains high technical standards.

(*iv*) Provide those raw materials and parts which meet the quality specifications.

(*v*) Has the capability to produce the amount of production needed.

(*vi*) Has no danger of breaching corporate secrets.

(*vii*) Meets the delivery dates.

(*viii*) Sincere in implementing the contract provisions.

(*ix*) Has an effective quality system and improvement program.

(*x*) Has a track record of customer satisfaction credibility.

After supplier selection and approval the next step is the certification process. The customer and supplier shall have agreed upon specifications that are mutually developed, justifiable and not ambiguous.

The supplier shall have no non-product related rejection for a stated period of time. The supplier shall have no negative non-product-related incidents for a stated period. The supplier shall have successfully passed on-site system evaluation. The supplier must conduct inspections and tests. The supplier shall have the ability to provide timely inspection and test data.

2.14. KAIZEN

2.14.1. Kaizen Concept

Kaizen is a Japanese word for "Improvement" or "Change for the better". It is a philosophy or practice that focus upon continuous improvement of processes in manufacturing, engineering, supporting business processes, and management. Kaizen are activities that continually improve all functions, and involves all employees from the CEO to the assembly line workers. Kaizen aims to eliminate waste by improving standardized activities and process. Kaizen is now being implemented in many other venues besides just business and productivity.

Kaizen is a daily process. If is also a process that humanizes the workplace, eliminates overly hard work. It teaches people how to perform experiments on their work using the scientific method and how to learn a spot and eliminate waste in business processes. In all, kaizen process is a humanised approach to workers and to increasing productivity. It nurtures the company's human resources as such that there is appreciation and encouragement for kaizen activities. The participation of workers is required in the improvement down to janitorial staff, as well as external stockholders when applicable participate in kaizen.

The Kaizen improvement focus on the use of:

(*i*) Value-added and non-value-added work activities.

(*ii*) Muda, that is identification of the over-production, delay, transportation, processing, inventory wasted motion and defective parts.

(*iii*) Motion study.

(*iv*) Material handling.

(*v*) Documentation of standard operating procedures.

(*vi*) The five S's for workplace means proper arrangement, orderliness, personal cleanliness, cleanup and discipline.

(*vii*) Visual displays that everyone in the plant can use for better communications.

(*viii*) Just-in-time to produce right quantities at the right time and with right resources.

(*ix*) To present or detect errors.

(*x*) Problem solving, communication skills, and conflict resolution.

Kaizen encourages suggestions by operators who continually try to incrementally improve their job or process.

Kaizen yields large results in the form of compound productivity improvement. This differs from the "command and control". This methodology includes changes and results monitoring, then adjusting large scale pre-planning and extensive project scheduling are done by smaller experiments.

The basic theme of Kaizen is that a large number of small improvements over a period will result in substantial improvement in organizational performance. The cumulative effect of these small incremental improvements in turn, leads to quantum improvement in the areas of operations and performance. As a result, Kaizen culture changes the vision of the company and morale of the employees, which are essential for superior performances. The change which always accompanies a successful Kaizen process, has to be made after detailed study and through analysis, including views and suggestions of the people working in the respective areas of business.

It is based on five primary elements:

(*i*) Teamwork

(*ii*) Personal discipline

(*iii*) Improved morale

(*iv*) Suggestion for improvement

(*v*) Quality circle.

The main task of the Kaizen process are to identify the areas of improvement, form the Kaizen team set improvement goals or objectives, set-up review committee, define do's and don't analyze and evaluate results and establish the new standards and procedures for improved results.

2.14.2. Implementation

Kaizen activity cycle can be as follows:

(*i*) Operation is standardised.

(*ii*) Standardised operation cycle time is measured. In-process inventory is determined.

(*iii*) Gauge measurements are done as per requirements.

(*iv*) Innovations are done to meet requirements and increase productivity.

(*v*) Improved operations are standardized.

(*vi*) This cycle is continued.

(*vii*) This cycle is continued.

It is also known as PDCA cycle.

In addition to business applications of the method, it has been popularized into personal development principles. The five main elements of kaizen are:

(*i*) Teamwork

(*ii*) Personal discipline

(*iii*) Improved morale

(*iv*) Quality circles

(*v*) Suggestions for improvements.

2.14.3. Criticisms

Sometimes it is considered that the cost cutting measures come at the expense of fair labor practices and quality products. Important examples are:

(*i*) Accusation of death by overwork at Toyota due to unpaid voluntary quality control meetings held after regular work hours.

(*ii*) Suppliers refusing to accept orders from Toyota or its affiliates due to successive price cuts.

(*iii*) Safety was being put at risk by aggressive cost cutting.

2.15. TOTAL PRODUCTIVE MAINTENANCE (TPM)

Total Productive Maintenance (TPM) is keeping the current plant and equipment at its highest productive level through cooperation of all areas of the organization. The objective is to create a system in which all maintenance activities can be planned and not interfere with the production process. Surprise equipment breakdowns should not occur. In TPM, Total means maintenance and production individuals working together. Productive means goods and services that meet or exceed customers expectations. Maintenance keeps equipment and plant in almost original condition at all times. The TPM goals are:

(*i*) Equipment capacity maintaining and improving.

(*ii*) Maintaining equipment in original condition.

(*iii*) Inputs from all employees.

(*iv*) Continuous improvement teams.

The total quality management uses, failure mode analysis, employee involvement, continuous improvement, just-in-time manufacturing, statistical process control, and experimental design. Total productive maintenance (TPM) is an extension of the Total Quality Management (TQM) philosophy to the maintenance function.

TPM has seven basic steps:

(*i*) Learning the new philosophy by the management.

(*ii*) Promoting the new philosophy by the management.

(*iii*) Training is for everyone.

(*iv*) Improvement areas are identified.

(*v*) Formulation of performance goals.

(*vi*) Implementation plan development.

(*vii*) Establishment of autonomous work groups.

Initially this change will require work by management. Eventually, it will mean less work as all individuals start solving their own problems. Management must first build credibility and the best-way to accomplish that task is to change first and lead the way.

Employees need to learn about the various tools used in performing their task as part of an autonomous work group, Autonomous work groups are established based on the natural flow of activity. The overall goal of the autonomous work group is to reduce the occasions for maintenance activity.

SUMMARY

1. Total quality management (TQM) means involvement of all aspects of the firm in satisfying the customer. It aims goal that can only be accomplished if the usefulness is recognized by having a partnership environment at all stage of the business, both with the firm.
2. Quality is defined as meeting or exceeding customer expectations. The goal is to first identify and then meet customer needs. TQM recognizes that a perfectly produced product has little value if it is not what the customer wants.
3. There is no acceptable quality level because the customer's needs, values and expectations are constantly changing and becoming more demanding. Before making a major purchase, some people check consumer magazines that rate product quality. Customer is willing to pay a higher price to obtain value. Each customer's concept of value is continually changing. Ongoing efforts must be made by everyone having contact with customers to identify, verify and update each customer's perception of value in relation to each product-and service.
4. Kaizen is a Japanese word for "Improvement" or "Change for the better". It is a philosophy or practice that focus upon continuous improvement of processes in manufacturing, engineering, supporting business processes, and management. Kaizen are activities that continually improve all functions, and involves all employees from the CEO to the assembly line workers.
5. Employees need to learn about the various tools used in performing their task as part of an autonomous work group, Autonomous work groups are established based on the natural flow of activity. The overall goal of the autonomous work group is to reduce the occasions for maintenance activity.

EXERCISES

1. What is TQM? Explain in detail.
2. Discuss on some principles of TQM gurus.
3. Explain objectives of TQM.
4. Explain Deming approach to quality.
5. What is PDSA cycle? Explain.
6. Explain benchmarking.
7. Discuss customer satisfaction.
8. What is continuous improvement? Explain.
9. Explain employee involvement.
10. Explain supplier partnership in TQM.
11. Explain Kaizen concept.
12. Describe implementation techniques of Kaizen.
13. Explain CII-Quality Excellence model.
14. What is MBNQA? Explain.
15. Explain TQM system with a model.
16. What is role of leadership in TQM? Explain.
17. Describe total productive maintenance (TPM).

3 CHAPTER Three

QUALITY BY DESIGN

CHAPTER OUTLINE

- Introduction
- Quality Function Deployment
- Product Development Cycle
- Failure Mode Evaluation and Analysis (FMEA)
- Assessment of Customer's Needs
- Formulation of Design Specification
- Preliminary Design
- General Consideration for a Good Design
- Evaluation of the Prototype

3.1. INTRODUCTION

Product quality is created by design. There should be consideration of quality in design of a product which meets customers need at a price acceptable to him. Newer and better products are introduced as a result of technological process. Therefore, the customer has lot of choices. The expectations of customers are to fairly high levels. In such a competitive market, the designers have to keep on improving for higher and higher quality levels.

Quality by design is the practice of using multidisciplinary team to conduct conceptual thinking product design, and production planning all at one time. Design team is composed of specialists from business, production engineering and the customer base. Suppliers of process equipment, purchased parts, and services are also included. Quality of design is a team of specialists who simultaneously design and develop a product to ensure ease of producibility and customers satisfaction. When each of the specialists has early input to the product definition and specifications, cost is minimized and performance is maximized. Thus, better quality products are manufactured for less cost with shorter time to market.

Traditional engineering requires repeat steps of redesign and reverification in order to compile all previous design stages. But in the quality by design engineering model, the initial design phase encompasses all of the aforementioned attributes, which eliminates the need of redesign. There is a longer lead time involved in the traditional method, because each step is performed independently and sequentially, one after another. The quality by design engineering method combines all these steps into one. The product is designed to be successful at each stage of its life cycle.

Quality by design provides manufacturers with the tools and communication and management techniques. Benefits of quality by design are better quality, less work in progress, increased productivity, etc. Quality by design utilizes teams to take advantage of prior experience. These teams include product develops from marketing, design, production, test and logistics.

3.2. QUALITY FUNCTION DEPLOYMENT

It is not enough to design a product well, but the process of manufacturing also has to be designed to produce the product or service that meets customers need and, at the same time, meet the quality goals under the operating conditions. This calls for strategic planning for quality designing, production and conformance in the organization. If the customer requirements are not incorporated in the design stage of a product; costly errors and product defect may creep into the process and cause customer complaints and loss of goodwill in the market place.

Quality Function Deployment (QFD) is a methodology to ensure that customer needs and expectations are attended to and fulfilled throughout the process of design, operations and delivery of the product or services. Steps for QFD process can be split into following sequences.

(*i*) Identify and establish customer attributes.

(*ii*) Working out and listing all technical features that are necessary for meeting customer satisfaction.

(*iii*) Establish a relationship matrix between the customer attributes and technical features.

(*iv*) Rate customer importance about the areas of greatest interest and highest expectations.

(*v*) Evaluate technical features of competitive products, review of own technical features and development of target quality features.

(*vi*) Identify the features and product characteristics that have a strong relationship with customer's needs and have strong selling points.

3.3. FAILURE MODE EVALUATION AND ANALYSIS (FMEA)

The other popular technique for incorporating the customer's voice into the design and manufacturing for conformance to quality is Failure Mode Evaluation and Analysis (FMEA). If the customers are to be satisfied with the product, then manufacturers have the responsibility to ensure that the product does not fail in service and cause inconvenience and irritation. In the quality planning of a product or for new product development, it is necessary to check out thoroughly that the product will not be liable to failure, and cause inconvenience, loss or risk to personal safety.

If the product has a large number of parts or sub-assemblies, FMEA analysis of all units may become very expensive and time consuming. Under such circumstances items that are critical for operation can be identified and FMEA analysis can be done for those units only.

3.4. PRODUCT DEVELOPMENT CYCLE

First, the assessment of the customer's needs are done. Then, the product development and design is done. The main steps in the development cycle are:

(*i*) Customer's needs assessment through market survey and analysis.
(*ii*) Design specification formulation.
(*iii*) Preliminary design of the product.
(*iv*) Review of design.
(*v*) Manufacture of the prototype and its evaluation.
(*vi*) Manufacturing drawings and product specifications preparation.
(*vii*) Pilot production run and changes in the design.
(*viii*) Design release for mass production.

3.5. ASSESSMENT OF CUSTOMER'S NEEDS

The prospective customers are explained the variety of features of the product. Information about their preferences are recorded. If a prior market survey can leak the information about the new product to competitors, then the designers, the marketing organizations and the quality engineers decide the desirable features in the proposed product. This exercise is considered essential to ensure a customer acceptance of the product.

3.6. FORMULATION OF DESIGN SPECIFICATION

The requirements and expectations of the prospective customers translated into a design specification. A design specification is always advisable to formulate in consultation with the quality engineers. Quality engineers knew the technical failing of the previous product. Thus, the drawbacks are avoided in the new design. A design specification should be converted in a comprehensive document. It should fully reflects the demand and expectations of the users. The proposed product quality characteristics should be clearly defined.

3.7. PRELIMINARY DESIGN

With the design specifications the designer is given a free hand in developing the preliminary design of the product. If a particular feature or characteristic cannot be achieved within existing framework of the design specification or other constraints, then *the fact made known to the management.* The following decision can be taken to overcome the problem.

(*i*) Reduction of the required feature.
(*ii*) Finding alternative design.
(*iii*) Addition of compensatory features.

Such decisions involve consultation with quality engineers and the marketing organizations.

3.8. GENERAL CONSIDERATION FOR A GOOD DESIGN

There are some general requirements which are universal in nature.

3.8.1. Appearance

The designer should keep in mind the normal taste of an average customer when designing the external shape and finish. The first impression of the product with the prospective customer is through its shape, finish and packing. The design and finish of the product should be compatible with its function. Visual appeal is important for consumer goods such as household appliances and automobiles. Industrial goods require functional design and essential protective finish. Sharp lines, corners, and abrupt changes of contour should be avoided in the external shape for a good visual effect.

3.8.2. Functional Efficiency

Functional performance is essential characteristic of a product at the time of purchase. The designer has to ensure that the product performs its intended function.

3.8.3. Safety

Safety is essential to achieve to avoid an accident causing damage or injury. The products can be made more safe if a conscious effort is made during design. The designer should consider that the user of the product is careless and design, the product to be accident-proof. Sockets for domestic electrical connections which automatically remain covered when not in use should be incorporated, machine guards may switch off the machine when removed for safety reasons.

3.8.4. Reliability

Product should continue to function satisfactorily during its intended life. This is called reliability. Like other quality parameters, reliability also is essentially built-in during design. The reliability of an equipment depends upon the number of components. The lesser the number of components the lesser will be the chance of any of them failing and move reliable will be the equipment.

3.8.5. Maintainability

Some failures can take place during the usage. The equipment design should incorporate ease in repair. If the life of certain components is much less than the intended life of the equipment, then such component should have easy accessible in the product for early replacement. In complex equipment, an indication may be automatically given when a system or subsystem fails. The operator may be able to take quick remedial measures to avoid possible further damage.

3.8.6. Ease of Production

Designer should consider the process by which the products components will be manufactured during mass production. Facilities should be provided for holding the components during polishing as well as for suspension and good electrical contact during plating. Less number of parts with compatible tolerancess ensure easy assembly. The mating parts may include locating holes and lays for self-alignment. The knowledge of production processes is essential for the designers. A production engineer may be included in the design team so that expert advice on production matters is easily available.

3.8.7. Standardization

Designer should use components from the existing design of the previous model or other similar standard products used for other applications. Such standard parts are normally available in the market at low cost. This saves time in design and avoids testing to assess their reliability. This is good for maintainability also.

3.8.8. Review of Design

The preliminary design is subjected to a critical review by quality engineers, production engineers and sales personnel etc. The design engineer explains the salient features of the design with the help of a model. Each functional representative is allowed to comment on the design. Say, the quality engineer highlights the quality problems which were faced in the similar previous product during manufacture. The production engineer may indicate non-availability of a machine or processing equipment with the standard of accuracy desired in the design. He may suggest some modifications in the design to manufacture the product by an available alternative process to avoid capital investment in new equipment. The sales manager gives the customer reaction on the features provided in the design. The user may give a high importance to a feature which may not been provided in the design.

Some changes in the design are desirable and these changes or modifications are incorporated, when the design is on the drawing-board. The preliminary design may also be subjected to "value analysis" for reducing the cost of the product.

3.9. EVALUATION OF THE PROTOTYPE

The evaluation of the prototype is carried out according to a well planned evaluation programme. The plans clearly lay down aspects to be studied and specific points on which information is required. The evaluation programme is coordinated by an agency different from the designer to maintain the objectivity of evaluation. Normally, the Quality Manager is assigned the job of prototype evaluation. Production and marketing is also associated with evaluation programme. Certain changes or modifications may be necessary in the design as a result of evaluation. Finally, all the features are subjected to confirmatory evaluation to ensure that the shortcomings of the design observed earlier are eliminated. This process of design and evaluation is continued until a fully satisfactory design is evolved.

3.9.1. Manufacturing Drawings

The drawing reflect the final design which emerged after evaluation studies. Even if the drawings are a precise reflection of the approved prototype, the matter does not end here. The tolerances of various components are selected for sound technological reasons. *The use of tolerance data or similar previous products and the statistical analysis of tolerances help to provide a tolerancing practice.*

3.9.2. Product Specification

The product specification precisely lays the quality levels which are to be achieved in the mass production. The specifications forms the basis for all trade enquires for the manufacture and supply of the product. It also serves as the standard of reference for quality management during manufacture, inspection and shipment. The specification clearly define all the important characteristics to be achieved and also lay down general guidelines for measurements. Reference is made to the organizational or rational standards wherever possible. If a particular measurement is peculiar to the product and is not covered in widely used standards then the details of such tests are given for the guidance of the production and inspection.

The design department prepares draft-specifications on the standard format used by the company. This draft is discussed with Quality Engineer and manufacturing departments to make further changes. Final specifications is issued by the Quality Department.

3.9.3. Design Changes During Production

Even when the design has been release for production, it cannot be considered to be absolutely final. Some problems faced during production may call for changes in design. The implications and repercussions on quality parameters are considered by the design and quality engineering department. Subsequently, necessary amendments should be carried out in all the affected drawings and specifications. Normally the copies of drawings and specifications are made. A suitable procedure is followed and communicated to all concerned. The amended copies of the drawings are issued to the concerned sections and are asked to return the old copies to the design office.

3.10. QUALITY IN ENGINEERING AND DESIGN

There is a risk of design error or manufacturing problems at the later stage, if quality dimensions are not introduced early in the design stage of product development as per customer expectation. The problem design should be of exact quality that customer expects. Quality design function (QFD) provides step-by-step approach to build quality in the product for doing engineering of quality characteristics in measurable terms. Taguchi "Quality Loss Function" analysis fine tune this approach to correct designing and setting measurable engineering parameters. This is achieved by formalizing the design process into three distinct stages which are system design, parameter design and tolerance design. Steps of Taguchi for quality designing and engineering are:

(i) **System Design:** Step to determine the desirable functions, prototype designing and parameter identification for manufacture prototype.

(ii) **Parameter Designing:** Find out the factors affecting desirable product functioning. Also, distinguished controllable and uncontrollable factors. The design are made robust by the combination of factors least sensitive to changes.

(iii) **Tolerance design:** Decide engineering tolerances using parameter designing for minimum variation achievable from the manufacturing line. The design calls for a change to achieve the quality objectives, if parameter design obtained is not acceptable. A high precision CNC machine, or superior material inputs, tighter tolerances, etc. may be changes necessary to achieve.

To ensure that the product gives customer satisfaction improvement, actions can be taken during the design stage itself. Simultaneously the company should not do not improvements in the parts beyond a level such that there is no return on investment.

Costly errors and products defects may creep in, if this type of quality approach is not initiated at the design stage of a product. This may lead to high cost of quality and undesirable product features. It has been observed that the design costs only 5% of the product cost whereas the manufacturing process costs 70% of the manufacturing costs. It is clear that the quality in designing and engineering has very large saving in the cost of a product.

The design has to be accurate for the intended functions and made manufacturable with least variations, so that the quality product is manufactured. The quality synergy between functions and processes interlinking the manufacturing activities from start to finish, must be supported by quality functions and manufacturing.

Conformance to specification as per design, and cost are the primary quality concerns of manufacturing. Specifications are established by the design quality specifications and not the design quality. Any deviations from these specifications implies reduction in quality. The process capability must be supported by the manufacturing system and control of the variations at each step of manufacturing.

Quality functions in manufacturing and service should be handled by the different types of quality issues. The customer driven approach of modern quality management systems should be used to work out the quality. The details of quality management procedures may differ between manufacturing and service industries. Total quality management has been successfully applied globally, whereever there is a customer to receive the benefits of results or services. In manufacturing sectors, quality issues may differ in some raspect from service sectors. Customer satisfaction in manufacturing industries relates to performance, reliability and consistency of the product, conformance to specification, durability, serviceability, aesthetics, and some perceived quality like features, brand, etc. Quality concern is conformance to specifications. This should not be taken quality expectations much beyond conformance. A product should have superior design features. Superior product needs to be free from defects and manufactured quality must be cost-effective and attractive to customer. That is the quality is achieved through improved product design and product manufacturing. The automobile manufacturing, precision industries, watch making, etc. are some of the examples.

There should be the concept of "internal customer" within the chain of activities for production, and JIT manufacturing which forces the unit to "plan and do" as exactly required and demanded for competitive manufacturing. They are the most significant contributions of total quality movement towards maintaining the consistent delivery of new quality dimensions to customers. The objective of producing the right part in the right place at the right time and just in time (JIT) is a cost effective manufacturing process because its operational philosophy helps to continually build quality into the processes and the products. The whole system falls through, if it is not continually improved quality through a great self correcting system of manufacturing.

SUMMARY

1. Product quality is created by design. There should be consideration of quality in design of a product which meets customers need at a price acceptable to him. Newer and better products are introduced as a result of technological process. Therefore, the customer has lot of choices.
2. First, the assessment of the customer's needs are done. Then, the product development and design is done.
3. The requirements and expectations of the prospective customers translated into a design specification.
4. The evaluation of the prototype is carried out according to a well planned evaluation programme. The plans clearly lay down aspects to be studied and specific points on which information is required.
5. Even when the design has been released for production, it cannot be considered to be absolutely final. Some problems faced during production may call for changes in design.

EXERCISES

1. Discuss quality concept in design.
2. Explain review of design.
3. How do you assess customer needs?
4. Discuss preliminary design.
5. Explain product development cycle.
6. How is formulation of design specification done?
7. Explain general considerations for a good design.
8. How does quality engineering help? Explain.
9. What do you understand by internal customers? Explain.

4 CHAPTER Four

CONTROL ON PURCHASED PRODUCT

CHAPTER OUTLINE

- Introduction
- Strategic Sourcing
- Evaluation of Suppliers
- Development of Sources
- Procurement Process
- Empowering Teams
- Capacity Verification

4.1. INTRODUCTION

The organization needs to purchase a variety of raw materials and products for further processing or consumption. The quality of the product is important and is given due consideration during purchase. If the purchased material is required for further processing, then its quality affects the quality of end product. The physical properties of metal determine the durability of the component. Thus, control of the purchased product is essential.

All manufacturing companies have some form of quality control set up for assuring the quality of their own products. If the user is a sales agency, a government department or a service organization then it has to have a special organization for quality assurance of purchased products.

Procurement of new product should be done through the quality control organisation which can make comparative evaluation of the products which can meet the general requirements of the company by analysis of the technical specifications of competing products.

Additional features to meet the special requirement of the company may be specified. Many times, it is more economical to get some of the components as per their own design manufactured by ancillary companies or subcontractors. The specifications of such components have to be prepared in consultation with company's design engineers. A particular design manufactured from trade if detailed specification covering all important quality characteristics are prepared.

While deciding on the actual quantity to be purchased at a time, consideration is also given to the economy which can be achieved by bulk purchase. Consideration should also be given to the possible deterioration of the product quality during storage. Delivery of the product may have to be phased in accordance with the actual consumption rate keeping a safe margin for unforseen interruptions in the supply. Selection of viable sourses of supply requires evaluation of suppliers with respect to their ability to maintain the quality of the product at the desired level, as well as their production capacity. Evalution of the firm being considered for the first-time requires their capacity verification to be carried out by the quality engineers of the purchasing organisation. The choice can then be made on economic considerations.

Capacity verification generally takes the form of survey of venders production resources and his ability for their proper utilisation. Preliminary study of the data furnished by the vendor may indicate firms capability of executing the order and then rule out further consideration of the firm. For accurate assessment of vendors capability, special expertise wherever available should be utilised. Considerable tact will be required to elict necessary information.

4.2. PROCUREMENT PROCESS

Procurement process is known as purchasing through the Purchasing Department. It starts with a requisition from an employee and ends with check/pay order as shown in Fig. 4.1.

The purchasing process involve several steps which get initiated when an employee realizes the need for an item or service. Mostly, it is a reactive operational or clerical job that will follow once a need is realized. The employee submits a material or service requisition containing, the details of required item or service to the immediate supervisor for approval of the requisition. The requisition contains description of the item and the quantity required. A standard office supplies for an individual may not need any approval. However, if there is training for which a set of office supplies are required, it needs an approval from the concerned higher authority. A desktop computer or flight ticket or hotel accommodation normally needs an approval.

The material requisition (MR) is given to the inventory/store/purchasing department. If that is a stocked item then the item is issued to the requester. For a service, if items are out of stock, the purchasing team will take initiative for the buying process. A purchased requisition (PR) is the document which is usually generated by the purchasing department and approved by product manager and finance team. This contains details like in a material request. It may have same or multiple items required by various groups or individuals in the organization. The purchase starts with finding the suppliers or service providers who can deliver. The company issues a tender notification seeking proposal from qualified suppliers. Tenders are usually notified through newspapers and other mass communication media as well as through direct communication with known or incumbent suppliers.

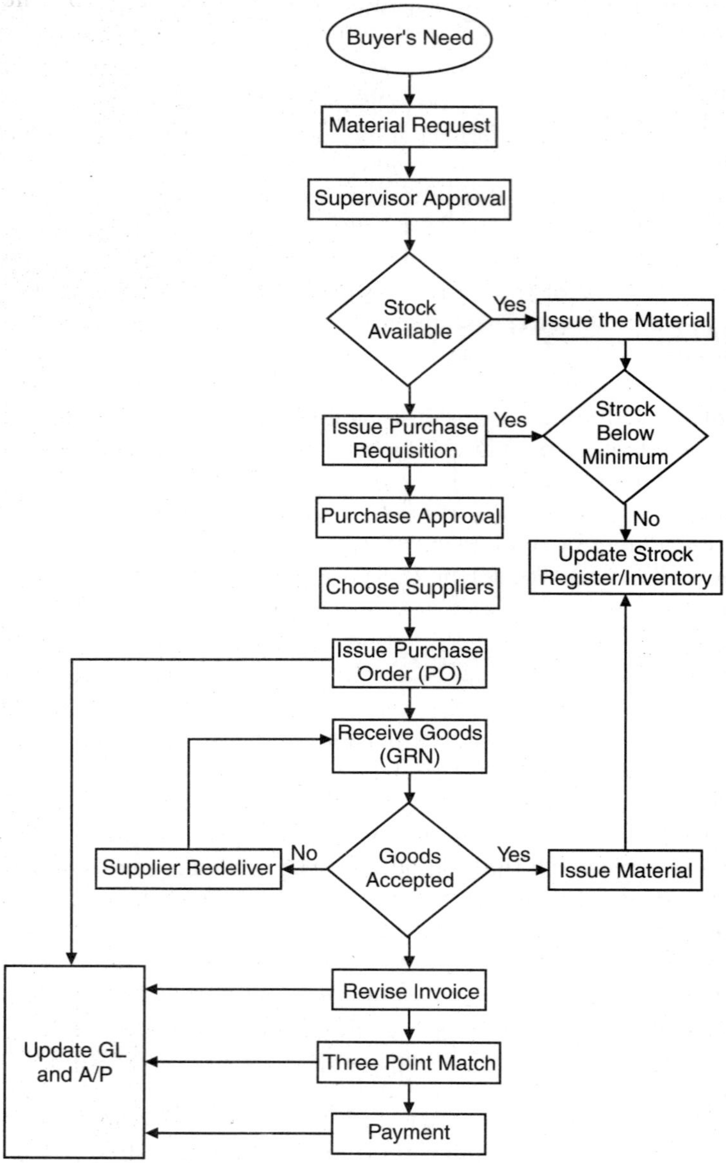

Fig. 4.1. Flow chart of Procurement Process

The suppliers are selected based on the specific requirement to deliver the good or services. A purchase order (PO) is sent to the chosen supplier. Suppliers send a acknowledgment note for the POs received as an acceptance of the order.

Supplier will deliver goods or services based on the agreed up on timeline. This will be notified through a Goods Receipt Note (GRN). The goods go through the inward quality and acceptance process. The passed goods are taken into store and accounted in inventory. If the goods are rejected for any reasons, it is sent back to supplier for replacement.

If the goods accepted then the supplier sends invoice for payment for the goods or services delivered. This is entered in the Account Payable (AP) system as a payable entry. After the necessary finance approvals, the payment is made to the supplier through the agreed means. The AP entry is updated with actual payment details.

The purchasing department of an average company spends an estimated 50 to 70 percent of every revenue on items ranging from raw materials to services. Purchasing has become a strategic function that can be used to control bottom-line costs. Companies are improving purchasing processes for improving customer satisfaction.

4.3. STRATEGIC SOURCING

The normal purchasing process involves steps of requisition, soliciting bids, purchase order, shipping advice, invoice and payment. These have come slow, expensive, and labour intensive. Each transaction generates own paper and the same process is followed for each item even a box of pencil or new truck.

The purchasing is essential. It focused on getting the right quantity and quality of goods to the right place at the right time at a low cost. A shrewd negotiator was needed to obtain the best possible price from suppliers and ensure minimum quality standards. Normal approach is small amounts from many suppliers and playing one against the other to gain price concessions.

In recent years, due to specialized bulk purchases, companies have achieved stunning bottom-line gains. The competition on both the domestic and global levels has made many companies to recognize and make purchasing as an strategic functions. Thus, new strategic are used in purchasing departments.

Strategic sourcing helps in lowering costs and improve quality. It involves analyzing what products the company buys in the highest volume. Procurement strategies are developed to establish the working relationships with the suppliers which are much more integrated than the past.

This process analysis requires answer to the following questions:

(*i*) Why is this product or service brought?

(*ii*) What are market conditions of suppliers?

(*iii*) What are the profit margins of the suppliers?

(*iv*) What is the total price of purchasing from a particular vendor?

(*v*) Where is the good or service produced?

(*vi*) What is the production process?

The products purchased in the highest volume have scope of high cost reductions. The company can then justify the expenses needed to closely study the industry that supplies that product. It needs knowledge of the suppliers operations, the business practices to know and where the most cost is added to the final cost of the product. The supplier is made to redesign process to lower production cost leased on the study.

These efforts leads to significant improvements in product quality with target prices. It works with the supplier to improve the manufacturing process to meet quality target. There is reduced inventory levels, faster time to market, significant cost savings, and reduced development cost leading to enormous benefits for buyers.

The attributes of such suppliers are:

(*i*) Commitment to continuous improvement.
(*ii*) Cost-competitive approach.
(*iii*) Cost-conscious.
(*iv*) Customer-oriented.
(*v*) Employee involvement.
(*vi*) Flexible approach.
(*vii*) Financially stable.
(*viii*) Ability to understand technicality.

4.4. EMPOWERING TEAMS AND JIT PURCHASING

Cross-functional teams are created which involves purchasing personnel in every stage of the product design process. The purchasers and suppliers are included from the start of the new product process to ensure that the products are readily available and are not prone to quality problems. Suppliers become experts in their field and bring a large knowledge base to the design process. This helps to prevent poor designs or manufacturing mistakes.

Purchasing teaches other members of the team to deal directly with suppliers. It eliminates much of the time-consuming work that buyers had to deal with such as solicity bids, creating purchase orders, etc. It spares them to concentrate on the part of their job where their expertise most pays off. "Purchasing should be concerned with the strategic planning aspects of procurement process."

Just-in-time (JIT) manufacturing is one of the biggest facets of industry. JIT companies maintain only enough inventory to manufacture the products they need in the very near future. Parts are ordered on a near-continuous basis and often go directly to the assembly line. The benefits are reduced inventory, improved quality, reduced lead-time, reduced scrap and network, and reduced equipment downtime. JIT purchasing demands that buyers narrow their supplier list to a chosen few who can deliver high-quality products on-demand and in a timely fashion JIT purchases require a minimum of three things in suppliers:

(*i*) demonstrated excellent quality.
(*ii*) ability to make frequent, on-time deliveries, and
(*iii*) ability to provide very large volume commitment or single sourcing arrangements.

In JIT, purchaser should deal only with companies that utilize statistical analysis to verify the quality of their output. The suppliers should be located in the same geographic regions as the buyers for frequent and on-time deliveries. It is also easier for the supplier to react to a sudden unexpected demand for its product. This costs far less to make the frequent deliveries that are needed. Those lower costs can in part be passed on to the buyers.

For decreasing administrative costs associated with repetitive contracts for common material, company uses credit cards or purchasing cards or P-cards. All these have internal checks and audits to ensure appropriate use. Most of the time now purchasing managers make available to negotiate major purchases and setting up of other long term contracts which are generally renewable annually. Purchasing mangers become supply chain managers handling additional functions of their organization operation. Logistic managers, material managers, distribution managers, etc all play the broader function.

4.5. EVALUATION OF SUPPLIERS

Supplier evaluation is the process of evaluating and approving potential suppliers by factual and measurable assessment. It ensures a portfolio of best-in class suppliers is available for use. It is also a process which is applied to current suppliers to measure and monitor their performance for reducing costs, reducing risks and continuous improvement.

4.5.1. Evaluation Process

It is continual process within purchasing department. It is a pre-qualification step within the purchasing process. Supplier evaluation is normally by using five-step processes. These processes may take the form of interview and or site visit. This includes approval of various aspects of the suppliers business including capacity, financials, organizational structure and processes and performance. Based on this evaluation a supplier is either approved or not approved. Some times there is an approved supplier list to which a qualified supplier is then added.

A supplier may be revaluated on a periodic basis. This on going process is known as suppliers performance management.

4.5.2. Benefits and Drawbacks

There are various benefits associated with an effective evaluation process. The benefits are that suppliers provide high standards of product and service levels whilst offering sufficient capacity and business stability. Supplier evaluation helps customers and suppliers identify and remove hidden cost drivers in the supply chain. This may also motivate suppliers to improve their performance.

The supplier evaluation include resources and cost commitments in establishing and maintaining a robust and effective system. There are challenges in specifying and gathering meaningful and relevant information with data integrity. Without going through the root cause of suppliers problems any subjective or inconsistent scoring may result in inaccurate assessment. Another challenge is to make sure that evaluation of current suppliers goes beyond measurement to actual performance improvement by providing feedback to suppliers on their performance and working on continuous improvement opportunities.

4.5.3. Cost Effectiveness

Focusing on processes instead of functions is that processes enable the organization to focus on the customer. Cross-functional and cross-company teams can add product and service combinations that add value.

The organizations are reducing the number of suppliers and estabilishing longer-term buyer-seller relationship for achieving a competitive advantage through more effective use of the supply base. Thus, supplier evaluation and selection is very critical in the development of effective longer-term relationships and performance. The measurement and evaluation of a supplier's performance becomes important on account of this.

Methods are required to establish and accurately reflect the total costs of doing business with individual suppliers. A total-cost approach to supplier evaluations considers the purchase price which is only a fraction of the cost associated with material receipt. The purchasing focus needs to be shifted from primarily a unit-price oriented approach to a total cost based performance evaluation of the suppliers. This approach incorporates incoming quality, quality in the manufacturing process, other related. The total cost based selection and retention of appropriate suppliers is key to product consistency and the organizations ability to meet competition. Indexes related to suppliers non-performance costs, the hidden costs associated with inadequate quality, delivery, reliability, and other elements of performance can be identified by establishing supplier performance.

4.6. CAPACITY VERIFICATION

For capacity verification, the organization should have established, documented, implemented, maintained and continually improved quality management system in accordance with the requirements of ISO 9000–2000. The organization should have identified the processes needed for the quality management system, determined the sequence and interaction of these processes. It should also have established criteria and methods required to ensure the effective operation and control of these processes. It should also ensure the availability of information necessary to support the operation and monitoring of these processes, measurements, monitoring and analysis of these processes and implementation of necessary action to achieve planned results and continual improvement. The capacity verification is done by preparing reports on the following aspects.

(*i*) **Management Commitment:** The top management should be committed to the development of the quality management system.

(*ii*) **Customer Focus:** The top management should ensure that customer needs and expectations are determined. These should be consideration of obligations related to product including regulatory and legal requirements, converted into requirements and fulfilled with the aim of achieving customer satisfaction.

(*iii*) **Quality Policy:** The top management should have defined its quality policy. Policy should be appropriate to the purchase of the organization, committed to meeting requirements of customers and to continual improvement. It provides a framework for establishing and reviewing quality objectives. It is communicated and understood at appropriate levels in the organization, reviewed for continuing suitability.

(*iv*) **Provision of Resources:** The organization should have determined and provided in a timely manner the resources needed to implement and improve the processes of the quality management system and to address customer satisfaction.

(*v*) **Assignment of Personnel:** The personnel should have assigned responsibilities defined in the quality management system are competent on the basis of applicable education, training skill and experience.

(vi) **Training, Awareness and Competency:** The organization should have established a system for identifying competency needs of personnel and provides training, evaluate the effectiveness of the training provided, and maintain appropriate records of education, experience, training and qualifications of its personnel.

(vii) **Facilities:** The organization should have identified, provided and maintained facilities such a workspace, equipment, hardware and software supporting services it needed to achieve the conformity of product.

(viii) **Work Environment:** The organization should have a system for identification and management of human and physical factors of the work environment needed to achieve conformity of product.

(ix) **Planning of Realization Processes:** The organization should have determined quality objectives for the product, project or comfort processes and documentation, resources and facilities specific to the product verification and validation activities, the criteria for acceptability and records that are necessary to provide confidence of conformity in the process planning for product realization.

4.7. DEVELOPMENT OF SOURCES

The capacity verification of potential suppliers who show interest in the product is carried out. However, main thing to be ensured is whether the firm has the minimum essential plant equipment and skilled manpower. Process engineers of the purchasing company can impart the manufacturing know how to the staff of vendor company. To ensure that the quality of the product can be maintained at the aceptable level the quality engineers can set up necessary control.

The purchase department places the order on the vendor who offers the most advantageous terms. Condition of supply must be put down in writing and clearly explained to the vendor and as submission of preproduction sample or a particular type of the packaging etc. Following details should be considered.

(*i*) Any reference to the institutional, national or international standard specification should clearly mentioned in a particular year or the latest specification will apply.

(*ii*) Materials are of critical importance be approved by the purchaser before they are taken into use. The procedure of taking test pieces should be spelt out in detail.

(*iii*) The size of the sample, made of dispatch and the tests to be conducted on the sample should be made known to the vendor if preproduction sample is required.

(*iv*) Sanctioning of deviations from the drawings and specifications or concessions on certain quality requirements should be clearly defined.

(*v*) Method of acceptance inspection should be clarified to the vendors with regard to sampling inspection.

(*vi*) Product nomenclauture, part number and others details should be specified.

(*vii*) Warranty clauses and procedure for claims should be clarified in detail.

After the order is placed with the vendor company, in the initial stages the vendor may require let of clarifications about the drawings and the specification. Minor change in the design to facilitate bulk production in the plant may be required. Requests for changes in design should not be viewed as criticism of the original design but taken as constructive suggestions to facilitate production. Samples should be carefully examined in accordance with store specification. Detailed report of the defects should be made available to the vendor. Quality engineers or process engineers of the purchasing company may advise the vendor how to prevent the defects and improve the quality of the product. In such cases fresh preproduction samples will be called for and clearance of bulk production orders after full quality checks.

SUMMARY

1. Procurement of new product should be done through the quality control organisation which can make comparative evaluation of the products which can meet the general requirements of the company by analysis of the technical specifications of competing products.
2. Purchasing has become a strategic function that can be used to control bottom-line costs. Companies are improving purchasing processes for improving customer satisfaction.
3. Strategic sourcing helps in lowering costs and improve quality. It involves analyzing what products the company buys in the highest volume. Procurement strategies are developed to establish the working relationships with the suppliers which are much more integrated than the past.
4. Cross-functional teams are created which involves purchasing personnel in every stage of the product design process. The purchasers and suppliers are included from the start of the new product process to ensure that the products are readily available and are not prone to quality problems. Suppliers become experts in their field and bring a large knowledge base to the design process.
5. Just-in-time (JIT) manufacturing is one of the biggest facets of industry. JIT companies maintain only enough inventory to manufacture the products they need in the very near future. Parts are ordered on a near-continuous basis and often go directly to the assembly line. The benefits are reduced inventory, improved quality, reduced lead-time, reduced scrap and network, and reduced equipment downtime.
6. Supplier evaluation is the process of evaluating and approving potential suppliers by factual and measurable assessment. It ensures a portfolio of best-in class suppliers is available for use. It is also a process which is applied to current suppliers to measure and monitor their performance for reducing costs, reducing risks and continuous improvement.
7. For capacity verification, the organization should have established, documented, implemented, maintained and continually improved quality management system in accordance with the requirements of ISO 9000–2000.
8. It should also ensure the availability of information necessary to support the operation and monitoring of these processes, measurements, monitoring and analysis of these processes and implementation of necessary action to achieve planned results and continual improvement.

9. The purchase department places the order on the vendor who offers the most advantageous terms. Condition of supply must be put down in writing and clearly explained to the vendor and as submission of preproduction sample or a particular type of the packaging etc.

EXERCISES

1. Why is it necessary to have control on purchased product?
2. Explain procurement process.
3. What is strategic sourcing? Explain.
4. Discuss empowering teams and JIT purchasing.
5. How is the evaluation of suppliers done? Explain.
6. Explain detail of the capacity verification of a supplier.
7. What is the need to develop sources of production? Discuss.

□□□

5 CHAPTER Five

MANUFACTURING QUALITY

CHAPTER OUTLINE

5.1. INTRODUCTION

There is variation in the production system as per need of factory and the product. However, production volume is the most important issue. Types of production systems generally are:

(*i*) Job shop production such as project production.

(*ii*) Batch production for small consumptions

(*iii*) Mass production such as line flow production and continuous production.

5.1.1. Project Production

Examples are several, civil engineering projects for construction or military related activities are project production. Here, large size manufacturing tasks are undertaken. Normally, work is carried out at the site of the work rather than at factory. The resources such as tools, materials, labour etc., are made available at the site themselves. Another example is ship building activity. A definite position plant layout is needed for project production.

5.1.2. Job Shop Production

The job shop production is for low production volume and has the following characteristics:

(*i*) to meet a particular customer's needs.

(*ii*) lot size is small.

(*iii*) variety is very high.

(*iv*) equipments are general purpose and flexible to meet specific customer needs.

(*v*) the labour needed has to be highly skilled to handle the machine/ equipment.

A plot of operation versus time is shown in Fig. 5.1 for the job shop production. Please note that:

Machine time for one component = $\sum$ (Setup time + Process time)

Total elapsed time = $\sum$ (Set up time + Processing time + Waiting time)

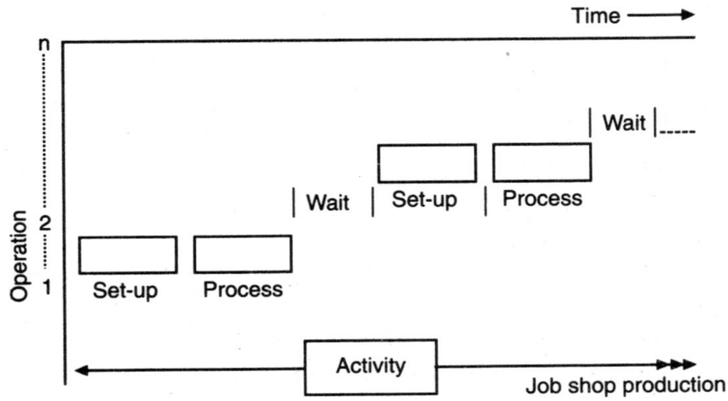

Fig. 5.1. Operation versus time for job shop production

5.1.3. Batch Production

Medium volume lot of same variety is done using batch production. The production order is repeated at regular intervals. More than 75% parts manufactured worldwide are in batches of 50 or less. Batch production characteristics are:

(*i*) for repeated customer orders.

(*ii*) lot size is medium and in batches.

(*iii*) for moderate variety.

(*iv*) machines and equipments are general purpose, but suitable for higher volume production also.

(*v*) jigs and fixtures are used to reduce set up time and increase the production rate.

(*vi*) labour in batch production should have reasonably high skill level, but generally less as compared to job shop production.

A plot of operation versus time is shown in Fig. 5.2. for the batch shop production. Please note that:

Machine time for one component = $\frac{1}{n} \sum$ (Setup time + Process time per part)

Total time = Total set-up time + $n \sum$ Processing time per part.

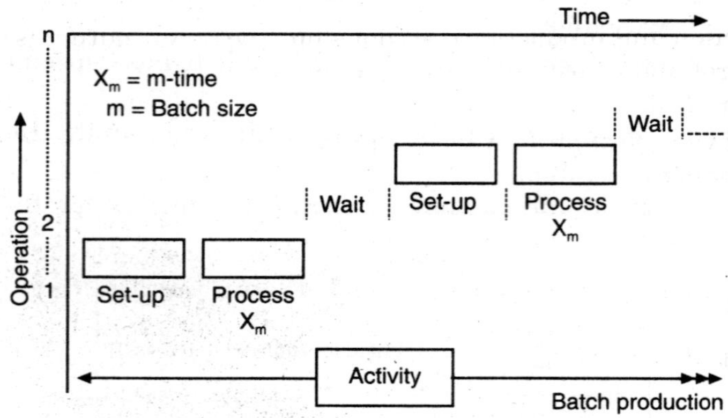

Fig. 5.2. Operation versus time for batch production

5.1.4. Mass Production

Mass production is suited for the high volume production. Mass production characteristics are:

(*i*) for high demand items.
(*ii*) lot size is very high.
(*iii*) variety may be one of its kind.
(*iv*) special machine tools and equipments are used.
(*v*) workers skill level is moderate or same machine is used all the time.
(*vi*) entire plant is designed to cater to a few special varieties of products.
(*vii*) investment in machine is high due to special purpose tools, machines/ equipments.

A plot of operation versus time for mass production is shown in Fig. 5.3.

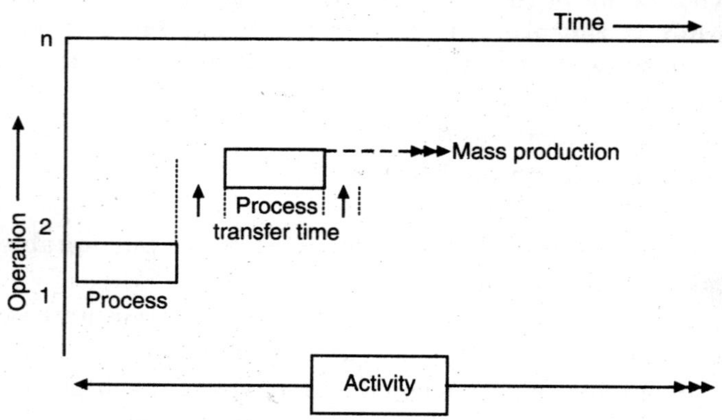

Fig. 5.3. Operation versus time for mass production

Please note that:

Machine time for one component = processing time + transfer time

Total time = processing time + transfer time.

5.1.5. Continuous Production

Continuous production is used in processing industry such as food, oil refinery, drugs and pharmaceutical unit, chemical processing unit etc. In a mass production unit production does not stop. The flow of output is continuous, on-line control and continuous system monitoring is needed. All controls are automated and computer controlled.

A plot of variety of product versus product volume is shown in Fig. 5.4. for different production system.

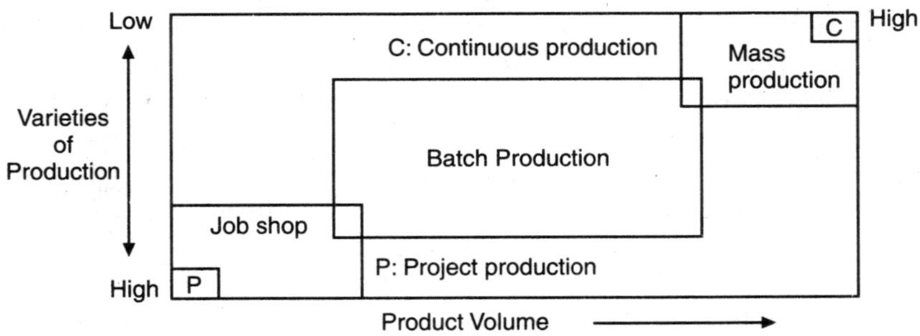

Fig. 5.4. Variety of product versus product volume

5.2. MANUFACTURING

Manufacturing involves making products from raw material by various processes or operations. It is generally a complex activity, involving people who have a broad range of disciplines and skills and a wide variety of machinery, equipment and feeling with various levels of automation, including computers, robots and material handling equipment, see Fig. 5.5.

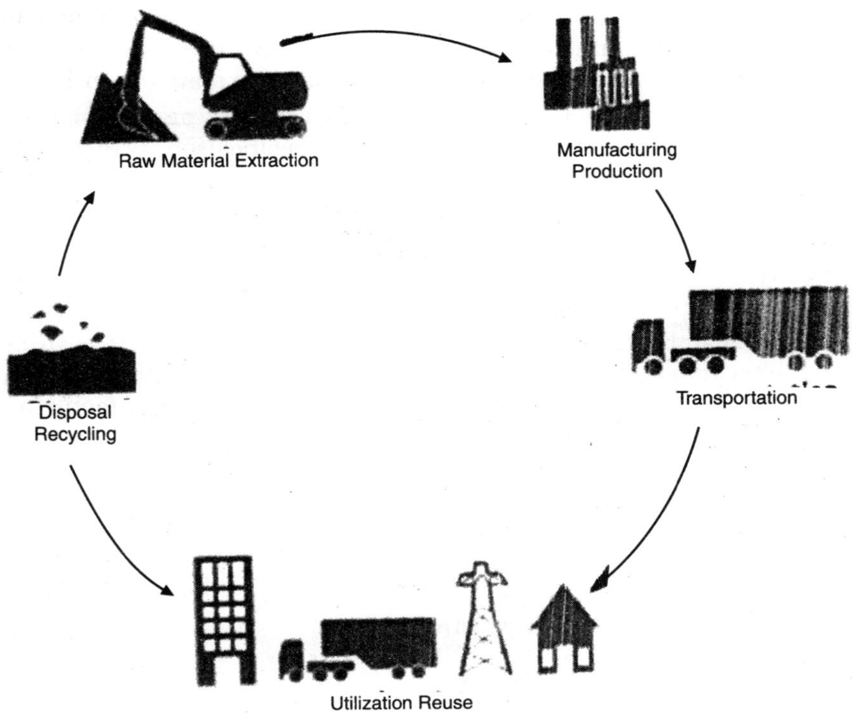

Fig. 5.5. Defining Manufacturing

Manufacturing activities have following considerations:

(*i*) Design must fully meet requirements and specifications of the product.

(*ii*) Manufacturing must be by the most economical methods in order to minimize costs.

(*iii*) Instead of relying on quality testing after the product is made, quality must be built into the product at each stage from design to assembly.

(*iv*) Production methods must be flexible to changing market demands, types of products, production rates, production quantities, and on-time delivery to the customer.

(*v*) A manufacturing organization must constantly be evaluated with a view to their timely and economic implementation considering new developments in materials, production methods, and computer integration of both technological and managerial activities.

(*vi*) Manufacturing activities must be considered as a large system wherein each part is interrelated. Modelling is done to study the effect of factors such as production methods on product quality and cost, changes in market demands, product design, material and various other.

(*vii*) Strive for higher productivity by optimum use of all its resources that is material, machines, energy, capital, labour and technology. In all phases, output per employee per hour must be maximized

5.3. MANUFACTURING STEPS

Manufacturing steps involve pre-production activity, pilot run, production run and delivery to customer. Pre-production activity involves selection of suppliers, develop pilot run plan, and develop manufacturing strategy with production plan, quality plan, test plan, material plan, and supplier plan.

The pilot run is done to validate manufacturing process against objectives set forth in manufacturing strategy and product specification, standard cost, product quality, documentation, tooling, training, process control, validates supplier plan and contracts, and internal failure analysis and corrective action.

Production run is done to produce high quality product on time, while continuing to fine tune the process, first order manufacture, and verification of product cost.

Delivery to customer is done to deliver first production units to customer, refine manufacturing process based on first build and monitor field unit performance to correct any problems. The detailed steps in manufacturing is shown in Fig. 5.6.

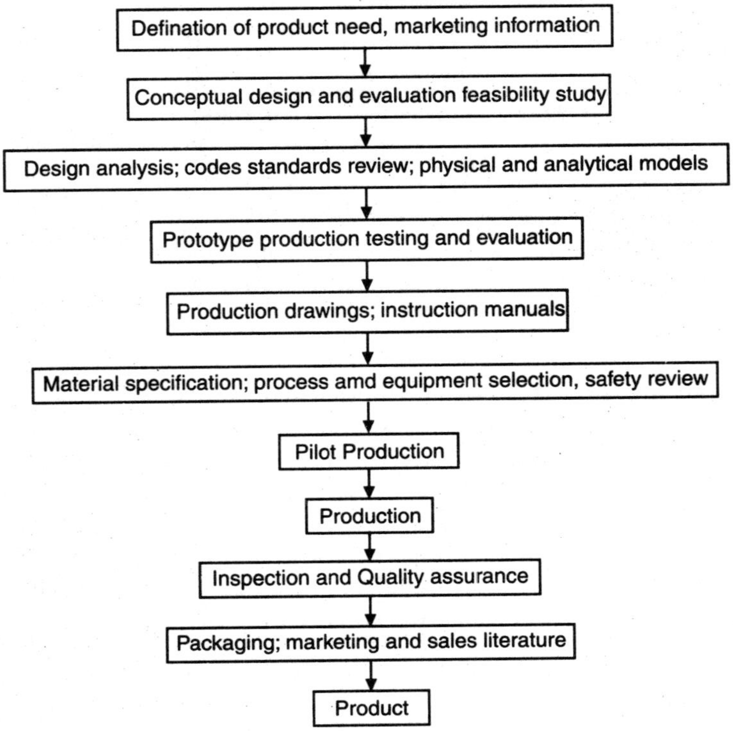

Fig. 5.6. Steps in Modern Manufacturing

5.4. METHODS OF MANUFACTURING

Methods of manufacturing have changed from the rigid, push-oriented production principles of the early 20th century to more flexible, pull principles. Assembly line manufacturers used to push mass production ahead based on estimates of future demand. This often resulted in waste of effort and resources if short of estimated demanded. To avoid waste one needs to wait for actual demand for manufacturing products as per customer requests or take previous batch of products sale. The modern manufacturing method have three main considerations:

(*i*) Waste reduction is done by following several techniques:

(*a*) **Just-in-Time (JIT) Manufacturers:** By just enough supplies to keep the process moving and schedule them to arrive at the factory "just-in-time" for them to be used in production.

(*b*) An automatic request for new supplies are sent to their suppliers when supplies are running low. These alerts are called **Kanban**, and they are generally computerized.

(*c*) **Just-in-Sequences (JIS):** Supplies arrive at the factory at the exact moment. They are needed within the manufacturing sequence. This means production may come to a standstill if the supplies are just a few minutes late.

(*d*) **Total Productive Maintenance (TPM)** is followed to avoid stopping production, some factoring wait until the end of the day to repair minor issues with their machines.

(*e*) **Quick Response Manufacturing (QRM)** in making the time period between a customer's request for a product and the final delivery of that product as brief as possible.

(*f*) **Cellular Manufacturing** separates the factory floor into different sections or cells. Machines are placed in a certain order so that materials flow naturally towards the competition of a product.

(*g*) **Single-minute Exchange of Die (SMED)** is to reduce waste time. Even between projects SMED focuses on changing the factory process from working on an old product to a new product as quickly as possible.

(*ii*) Quality of a Product has to be Maintained

A manufacturing process is given a sigma rating based on the percentage of its product yield being defect free. A one sigma rating designates a process with a yields of 31% and a six-sigma rating is a processes which is nearly perfect that is 99.99966% defect free. Thus, every company wants to achieve six sigmas for all of its manufacturing processes. The Six Sigma method focuses on measuring and analyzing process data in order to find and remove defects. During, lean manufacturing, it is concerned with improving the flow between processes to reduce waste.

Define, Measure, Analyze, Design, Verify **(DMADV)** methodology is followed which is a more anticipatory approach. Based on an analysis of customer demand, manufacturers plan ahead and try to design ways to avoid defects in the first place.

Critical to Quality **(CTQ)** and Quality Function Deployment **(QFD)** are two concepts which try to pinpoint the elements of a product that are most important to customers and translate those needs into a manufacturing strategy that focuses on getting the critical elements exactly right.

Suppliers, Inputs, Process, Outputs customers **(SIPOC)** is followed to allow manufacturers to trace the life cycle of their products from supplier to customer and identify problem areas.

Taguchi Loss Function is considered which suggests that production quality decreases as variation increases.

(*iii*) Accelerate Production

Production is accelerated to make sure products are made available and demand is met. Machine flexibility should exist for change to manufacture different products. Routing flexibility concept is used to focus on the adaptability of the manufacturing process as a whole.

Computer-aided Design (CAD) and Computer-aided Manufacturing (CAM) should be adapted. If a factory has machine and routing flexibility, CAD is used for design of production processes and CAM is to guide parts through those processes with robots and Computer-controlled Machines (CCM).

Computer-integrated Manufacturing (CIM) creates a network of computers that observe and operate the manufacturing process. Computer can monitor every step of production for defects and change the actions of a machine almost immediately leading to flexibility.

Finally, the three major goals of reducing cost, maintaining quality, and accelerating production are achieved through the combined efforts of lean manufacturing theories, six sigma methodologies, and flexible processes.

5.5. STEPS FOR QUALITY MANUFACTURING

Improving quality manufacturing processes leads to decreased waste, better quality products, and an overall improvement in customer satisfaction. To begin with, a committee is formed to ensure the following steps are completed and taken from beginning to end:

(*i*) **Actual Process:** Name the process and its purchase, as well as its starting and ending points, inputs and outputs, and overall requirements. Identify the customers and suppliers who will be affected by this process.

(*ii*) **Areas of Improvement:** A random sampling of the product is done. Product is tested for a variety of things that will have an impact on the end user and customer, which may include durability, materials, toxicity and so forth. Improvement in manufacturing can include disintegration of parts, loose fastness and so forth and should be a main focus.

(*iii*) **Solution for the Problems:** Brainstorming is done in committee, or specialists are consulted. Additionally, get feedback from those who work on or with the process on a daily basis.

(*iv*) **Detailed Solution:** A budget is included to determine what personnel are necessary for making the improvements, conduct a projected cost analysis, and time frame for completing the overall improvements. There may also be a need to determine how the rest of the manufacturing plant will be affected by this and wheather it will slow production at any level.

(*v*) **Put plan into action:** Involve everyone, from the highest levels of management in the manufacturing company down to the workers who utilize the process and implement the action plan.

(*vi*) **Evaluate:** Now, there will be a need to evaluate the improvement process as a whole. The process should have desired effect, the process should be successful, the problem should be fixed, waste should be eliminated. The improvements should be on time and within budget.

(*vii*) The steps two to six should be continuously repeated as often as necessary to achieve improvement within the manufacturing plant. Final goal is to decrease the need for a committee, and instead involve all members of the plant continually working to improve. Six sigma approach is used for maintaining quality in manufacturing which means identifying errors or defects in the production flow and eliminating them to maximize productivity.

5.6. INSPECTION AND CONTROL OF PRODUCT

For purchase of any material or store is there is always the question of its quality. It is desirable to know that the components or materials conform to specifications to avoid production, difficulties in subsequent operations. The quality appraisal may take the form of physical inspection of entire product, or only a small part of it. The inspection of a small portion, is termed as 'Sampling Inspection'.

5.6.1. 100% Inspection

The inspection of each and every piece of store or material is known as 100% inspection. This type of inspection gives complete assurance that all the defective material has been eliminated from the inspected lot. However, as each piece is to be inspected, it proves very expensive and time consuming, particularly where large quantities are involved. There are risks involved in cent percent inspection due to human errors of failing to detect defective articles due to the boredom and fatigue of repetitive inspection operations. 100% inspection diverts the responsibility for the quality from the producer to the inspector. The producer thinks that his responsibility regarding in quality is over even if any defects are found in the product later on.

5.6.2. Sampling Inspection

It is a technique for assessing the quality of a lot comprising of a large quantity by inspecting only some of them. The group of materials chosen for inspection is called a 'Sample'. For example, if a sample of 100 articles out of a lot of 1500 is found to have 2 articles defective, it is reasonable to expect that remaining 1400 will also contain 2% defective articles, provided all 1500 pieces were manufactured under similar conditions. Acceptance of lots on the basis of sample results will involve a certain amount of risk. Sampling inspection provide a strong incentive for the producer to improve the quality of his goods.

5.6.3. Sampling Plan

It is a brief sampling inspection procedure. It decides the number of samples to be taken, the size of each sample and the acceptance criteria. A typical sampling plan is given below:

Single sample size: 100
Acceptance number: 4
Rejection number: 5

This indicates then one sample consisting of 100 units is to be inspected from the lot. If there are 4 or less number of defective units in the sample, the lot is to be accepted. If the defective units are 5 or more, the lot is to be rejected.

5.7. QUALITY IN SALES

Purchasing decisions are based on the price and the quality of a product. The product quality is the first consideration. The sale of a product may suffer if it is priced too high. "It is possible to fail in business even though the quality of the product is good. However, it is not possible to stay in business if the quality is poor, unless one has a monopoly" as per J.M Juran, an authority on quality control. Thus, the importance of product quality in creating and maintaining sales is realized by everyone in the company. Everyone should make conscious efforts to achieve a high standard of quality to enhance the sales appeal of a product. Quality is an aid in competition which is even more effective than price. By improving the quality of its product, company may even take it out of the competing product class but there should be a distinct improvement in quality. In such case, the product should be visibly be superior to the competing brand and can justify a higher price. The use of quality may be in the form, such as better after sales service and longer guarantee periods.

The quality reputation of a company is a great asset which makes selling very much easier and also helps in growth and diversification. A quality image can initially be created by a well-directed sales promotion and the words must be matched by deeds. A lasting quality image can be created only if everyone in the company make it his business to please the customer. The best advertisement for the quality image of any company is a satisfied customer.

The quality should be reflected in the policy statement outlining company objectives. Seriousness of the top management about maintaining the quality of its products creates the right climate for the quality assurance. Every person in the company will try to achieves this objective. Methods of establishing a quality reputation are:

(*i*) Only those products should be sold which fully meet the customer's requirement.

(*ii*) Prompt and courteous adjustment of warranty claims, in case the product fail.

(*iii*) Effective after-sales service.

5.8. PRODUCTS LIABILITY

Competition among manufacturers is a significant motivating factor for providing new and modified products, unproved technology creates hazards that are unknown prior to product use. Highly diverse and technically complicated product design has also increased the severity and frequency of injuries. There is a need to reduce the risk of injuries and to compensate injured consumers. Reducing the risk to injury is best accomplished by stimulating manufacturing creativity. Manufacturers are in the best position to know what are the safest designs, materials, construction methods, and modes of use. State and central laws regulate many products. States have enacted various laws to regulate items such as fireworks, toys, life preservers, etc. Products liability law controls the private litigation of product accidents, which provides compensation to the injured party. The liability may arise as a result of a defect in design or manufacturing, improper service, etc.

The manufacturer of a consumer product must protect itself against the risk of product liability litigation or at least reduce the risk to a level that allow a reasonable profit and continued growth. A formal product safety committee should be established, with either a safety engineer or an outside consultant as a member. Members of the committee should be from the design, manufacturing, marketing and quality areas.

New product are more likely to be involved products liability litigation than well established products; therefore, a special review is required before the product can be released to manufacturing. Customer-oriented tests are performed to predict the misuse of the products. A customer-oriented test tries to determine what happens when the product is misused.

The American National Standards Institute has issued guidelines for the design of product safety. According to the American National Standards Institute, a product safety sign must alert persons to the type of hazards, the degree of hazard seriousness, the consequences of involvement with the hazard and how to avoid the hazard.

A complaint or a claim is a communication between the market place and the organization concerning the performance of the manufacturer or seller. The insurance company reviews the situation and determines the cause of the claim, nature of seriousness, failure mode, negligence of the parties, etc. An early investigation is needed for a quick settlement of reasonable claims or the preparation of defenses. The complaint and claim procedure needs a provision for modifying the appropriate departments. It may even be necessary to implement the product-recall plan. The cost of recalling product varies significantly with the type of product and the quantities involved.

5.9. AFTER-SALES SERVICE

The reliability and maintainability are important quality parameters. A serviceable condition during its intended life is very important. Effective after-sales service is an essential element in ensuring reliability and maintainability. Major mechanical and electrical products such as major automobiles, industrial machinery and domestic appliances requires large service facilities.

After-sales service covers not only maintenance and repair facilities, but also all post-sale efforts in ensuring maximum customer satisfaction. The product may not meet the expectations of the customer which may be due to the ignorance of the customer regarding product capability, incorrect usage, or maintenance, unduly harsh environmental conditions etc. The blame may come on the quality of the product unless the real cause is explained.

In all fairness a customer who paid for the product, should get his money's worth in the form of product service. Important aspects of after-sales service are:

(*i*) Customer Education

(*ii*) Maintenance and repairs service whenever needed.

(*iii*) Importance to customer complaints.

Sales personnel have to do the job of educating the customer. Customer must know about the characteristics and the capabilities of the product, to understand whether product will meet his requirement. Demonstration of the product and the issue of free samples, fulfils some such needs. For major equipment and machinery the education of the customer requires brochures, booklets or technical manuals giving complete details about the product, its characteristics and capabilities. Instructions for installation, operation and maintenance are also incorporated in a manual. Organized training of the customer or assistance in operation and maintenance, may be needed to get the maximum benefit. The design department prepares the documents and manuals in consultation with the quality engineering department.

The machines and pieces of major equipment, require some attention in the form of prevential maintenance and occasional repairs. A customer expects the supplier to provide maintenance, advice, and servicing and repair facilities. Machinery is normally covered by a warranty, as such the supplier is legally based to provide these service upto a stipulated period. A service organization is created which is a branch of the sales department. It has technical staff who has an intimate knowledge of the subject machines or equipment who are capable to provide technical advice regarding installation, operation, maintenance and repair.

Repair facilities are set up by the servicing and repair centres managed by the company. Alternatively, it may issue licence to other repair agencies to provide those services on its behalf.

However, the repair manuals, the training of the repair agency staff, special tools, equipments, spare parts and other accessories required efficient maintenance and repair remains company's responsibility. Periodic quality audits of the repair agencies should also be arranged by the company to assure that the quality of repair standard is and as per requirement.

Feedback on repair problems should be maintained by quality engineers in close liasion with the repair agencies, to know the inadequencies in the quality of the product. The unusual defects could possibly be attributed to faulty design or defective manufacture. Improvement in design, process of manufacturing or repair technique is done by the analysis of these defect reports.

5.10. GUARANTEE

The guarantee of quality is for the benefit of the customer, if the product purchased turns out to be defective. It is an undertaking by the selling company that it will replace the product free of cost or refund the money if the defect is due to the improper materials or faulty manufacture. The guarantee convinces the customer about the quality of the product.

The guarantee is a powerful sales tool. However, it can harm the company, if the company may not have any serious intention to honour its guarantees. In such cases, sales boost will be for short duration. It is not necessary that every single part of the equipment should be guaranteed by the company. The items which can easily be damaged by mishandling may be excluded from the guarantee because it is difficult to know whether the damage is due to manufacturing defect or mishandling. For example, the manufacturers of electrical appliances guarantee does not cover parts made of glass, porcelain and bakelite.

The guarantee terms and conditions should be clear and unambiguous. The following informations should be included in a guarantee card.

(*i*) Validity period.

(*ii*) Manufacturer's liability. The manufacturer decides whether an article is to be repaired or replaced and accordingly free repair or replacement it must be stated.

(*iii*) Guarantee claims procedure. The responsibility for the despatch of a defective article should be clear whether it is to be sent to the company or the accredited service station.

(*iv*) Guarantee invalidation conditions. Accidents, mishandling, improper use, or repair by an unauthorized agency may be the conditions to invalidate a guarantee.

5.11. ANALYSIS OF CLAIMS

The investigation of complaints involves thorough technical knowledge of the product, hence job may be entrusted to the service engineers who may be of the sales department. Guarantee claims may be investigated by the quality control department if decided. If it takes longer for the claims to be finalised, it causes further annoyance to the customers. The minor claims being settled by the regional service centres, and major ones may be referred to the company. Procedure should be simple and ensure that the claims are decided fast.

The advisers should play up the design features of the product, but quality of conformance can also be powerful plank for any sales promotion campaign. The advertisement of quality of conformance must be based on faith. There is substitute for the genuine quality of the product which alone can sustain sales in the ultimate analysis.

Design, production and distribution aim to service the satisfaction of the customer needs. Hence, the feedback to any product plays a vital part in building quality of a product. The sales organization provides a feedback of the customer's needs and performances. The life-cycle of a product generally based on the customer's need. Prospective customers may like the general features of new product, but may not like term of presentation or may required certain additional features. Customer reaction is normally obtained by a market survey which is generally done through the sales organization. Quality improvement is a continuous process. Hence, the sales organization should ensure to get a continuous feed-back of needs and problems. This feedback gives ideas for the subsequent models of the product.

SUMMARY

1. There is variation in the production system as per need of factory and the product. However, production volume is the most important issue. Types of production systems generally are:
   (*i*) Job shop production such as project production.
   (*ii*) Batch production for small consumptions
   (*iii*) Mass production such as line flow production and continuous production.
2. Manufacturing involves making products from raw material by various processes or operations. It is generally a complex activity, involving people who have a broad range of disciplines and skills and a wide variety of machinery, equipment and feeling with various levels of automation, including computers, robots and material handling equipment.
3. Manufacturing steps involve pre-production activity, pilot run, production run and delivery to customer.
4. To avoid waste one needs to wait for actual demand for manufacturing products as per customer requests or take previous batch of products sale. The modern manufacturing method have three main considerations.
5. Improving quality manufacturing processes leads to decreased waste, better quality products, and an overall improvement in customer satisfaction.
6. It is desirable to know that the components or materials conform to specifications to avoid production, difficulties in subsequent operations.
7. Purchasing decisions are based on the price and the quality of a product. The product quality is the first consideration. The sale of a product may suffer if it is priced too high.
8. The reliability and maintainability are important quality parameters. A serviceable condition during its intended life is very important. Effective after-sales service is an essential element in ensuring reliability and maintainability.

Major mechanical and electrical products such as major automobiles, industrial machinery and domestic appliances requires large service facilities.

9. The guarantee of quality is for the benefit of the customer, if the product purchased turns out to be defective. It is an undertaking by the selling company that it will replace the product free of cost or refund the money if the defect is due to the improper materials or faulty manufacture. The guarantee convinces the customer about the quality of the product.

10. The investigation of complaints involves thorough technical knowledge of the product, hence job may be entrusted to the service engineers who may be of the sales department. Guarantee claims may be investigated by the quality control department if decided. If it takes longer for the claims to be finalised, it causes further annoyance to the customers. The minor claims being settled by the regional service centres, and major ones may be referred to the company. Procedure should be simple and ensure that the claims are decided fast.

EXERCISES

1. Discuss methods and techniques of manufacturing.
2. What are the steps of manufacturing?
3. Explain quality aspects in sales.
4. What do you understand by guarantee of a product.
5. What do you understand by manufacturing? Explain.
6. Discuss inspection and control of a product.
7. What is after-sales service? Explain.
8. How is claims analysis is done?

6 Chapter Six

Quality Management

CHAPTER OUTLINE

- Organisation Structure
- Organisation Design
- Quality Functions
- Decentralization of Quality Functions
- Quality as a Staff Function
- Organisation Design for Quality Product
- Quality Cost
- Quality Cost Optimization
- Cost Reduction Programmes
- Process Control
- Human Factor
- Corrective Methods

6.1. ORGANISATION STRUCTURE

Organisational structures depend on business functions and the management. The organisation structure for the management of quality requires, the knowledge of the factors that influence this activity. These factors are briefly discussed as follows:

***(i)* Company Policies and Objectives:** Every aspect of the organisation should lead towards achievement of the stipulated objectives. The philosophy and attitude of a company towards quality management has a great influence on the organisation. Importance of quality, emphasis on type of products, and the market conditions decide the type of organisation. An elaborate organisation for quality engineering and control is essential.

***(ii)* Type of Business:** The business can be traditional with fixed methods of manufacturing for specific customers where the sales are generaly assured by a long term contracts. In such case the products are well established and almost no quality problems may arise and quality organisation can be small. Or else, the business can be diversified having many product types needing flexibility and constant innovations to meet changing market conditions. In such case, considerable effort towards new design and development may be needed. In such cases, several quality problems may exist. Hence, a specialised and large quality organisation may be needed.

***(iii)* Type of Product:** Different products have different quality requirements and parameters. Precision machine tools, automobile components and sophisticated electronic equipments need design evaluation, process controls and after sales services.

***(iv)* Company size and Organisation:** The quality management is a staff and service function for providing service to the other sections of a company. It must suit the organisation of the company. The organisational structure of the company itself decides the quality organisation. Most of the quality functions may be centrally directed by the quality manager. In a divisional type of company structure, quality function may be deligated to the various divisions. The effective channels of communication need be established to ensure that specialised quality services are available whatever they are needed.

***(v)* Market and Customers:** The circle of quality begins and ends with the customer. The customers are well informed that goods of better quality are being offered due to advances in technology. Product specifications are stringenet and all important parameters laid down by customers.

The manufacturer have to have the quality program and organisations to meet these requirements. Due to intense competition, company may not be able to compete in terms of selling prices. It may improve the quality of its products to obtain an edge over its competitors.

6.2. ORGANISATION DESIGN

There are three basic steps in designing any organisation.

***(i)* Identification and grouping of jobs:** First various tasks to be carried out for the achievement of objectives are identified. Then, grouping of these tasks is done into sets of similar jobs. These jobs can be carried out by individuals having similar training and skills.

Subsequently, these individual sets are grouped into allied sets. These can be put under one supervision of the organisation.

***(ii)* Allocation of Authority and Responsibility:** Specific sets of jobs to the individuals with clearly defined authority and responsibility are allocated. The passage of instructions and advice from the management and feedback from the lower ranges of the organisation should be devised in an effective method.

***(iii)* Establishment of Co-operative Relationships:** Various groups are ensured to work as a team as their work is complimentary. Co-operative relationships, are established between various individuals and groups. This aspect is of great importance as quality cannot be built into the product by the quality organisation alone or such the joint efforts of all segments of the company organisation is needed. A specific quality organisation is required to ensure that everyone concerned with quality contributes towards quality. Activities of various groups are co-ordinated and integrated into a unified quality management system.

6.3. QUALITY FUNCTIONS

The quality functions have two distinct groups which are Quality Engineering and Quality Control.

***(a)* Quality Engineering:** Specialist staff functions, activities connected with the development, defining, and planning of quality during the pre-production stage are the components of quality engineering.

Important works are:

(*i*) Laying down realistic quality objectives and advice to the management on the quality policy of the company.

(*ii*) Formulation of design specifications and the analysis of the customers quality requirements.

(*iii*) Improving quality and reducing quality costs through the review and evoluation of product design.

(*iv*) Preparation of product specification and defining quality standards.

(*v*) Formulation of procedures to ensure quality conformance and planning quality standards.

(*vi*) Inspection methods including the design of special test equipment and development of quality control techniques.

(*vii*) Process capability studies work.

(*viii*) Quality cost analysis.

(*ix*) Planning and organising quality control for incoming stores, materials and the capacity verification of the vendors.

(*x*) Company quality audit.

(*xi*) Quality training programme arrangement.

***(b)* Quality Control:** In-process and post-production testing quality plans are aimed at ensuring the quality conformance of products. Important elements are:

(*i*) Help in establishing quality controls in manufacturing process at various points.

(*ii*) Process control equipment maintenance and calibration.

(*iii*) Defect investigation and assistance in solving quality problems during production.

(*iv*) Quality control measures implementation for incoming materials and stores.

(*v*) Stage or interstage inspection or spot checks whenever necessary.

(*vi*) Assessment of the quality of the end-product and the effectiveness of quality control measures through final inspection.

(*vii*) To ensure that the product is able to withstand transportation hazard by checking the quality of the packaging.

(*viii*) Testing for the measurement and analysis of goods based customer complaints.

(*ix*) Defect data and customer complaints feedback to the quality engineering section.

6.4. DECENTRALIZATION OF QUALITY FUNCTIONS

All the quality work elements need not be performed directly by the quality department. These may be performed by the appropriate individual or group to ensure effective quality management in decentalized form as shown in Fig. 6.1.

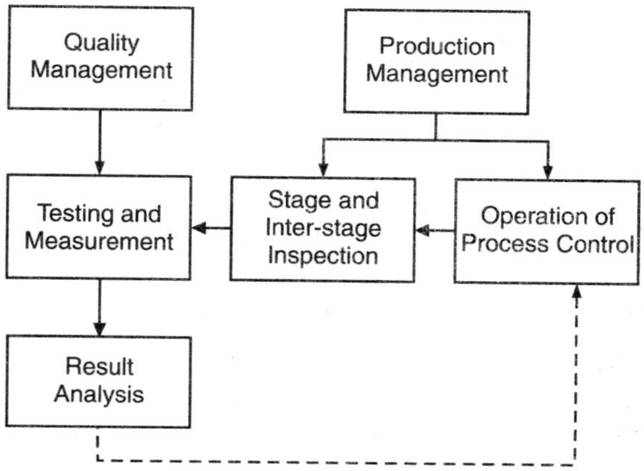

Fig. 6.1. Decentralization of quality functions

The allocation of various quality tasks may be given to the department or sections which can perform these with efficiency for economy. The operation of process controls, stage and interstage inspection may be delegated to the production, if this helps in the smooth flow of production. Inward material control may be delegated to the purchasing department and technically competent persons are made available.

There are advantages from the decentralisation of quality functions. There is the smooth flow of production and less interdepartment friction. However, the longer feedback loop gives slower reactions. That is, the defects in the product take longer time to get rectified and in meantime substantial number of defective products may be produced.

A rush to meet the target may overtabe unsatisfactory state of the process and it actually may go out of control. A large scale decentralisation requires established effective quality control programmes of process.

Delegation of specific responsibilities is an essential quality component an ultimate responsibility of quality management. It should have adequate control mechanism to monitor. The effectiveness of delegated functions require suitable corrective measures.

6.5. QUALITY AS A STAFF FUNCTION

According to Louis A. Allen, organisation staff refers to those positions and elements of an organisation which have responsibility and authority which are accountable for the accomplishment of primary objectives. Staff elements have responsbility and authority for providing advice and service.

In a line organisation, orders and instructions are transmitted from top to bottom and the compliance or progress report is conveyed to the top. The relationship is that of superior and subordinate. A staff position has no direct authority. The staff elements provide advice and service to assist the line element to perform its functions better. The line supervisor has to consider the advice given by the staff department. Normally, the staff members are delegated the authority of issuing certain types of instruction on behalf of the supervisor. In case such instructions can not be complied with due to any reason, the instructions are referred to be modified or cancelled.

The detailed analysis of the work elements of quality management would show that staff functions is essentially an advisory and service function. Staff function itself can not built quality into any product.

6.6. ORGANISATION DESIGN FOR QUALITY PRODUCT

The design of organisational structure starts from the top downwards. The quality manager has to provide the necessary staff advice and service to the company. Manager has to take a decision whether he will be able to take full responsibility of the needs additional manpower delegation of portion or full responsibility to the subordinates. The grouping of the work elements, the quantity of work, and the areas of deployment decide the number of subordinates.

The quality component of internal structure is normally a line organisation. Specialised staff may also be included in the organisation.

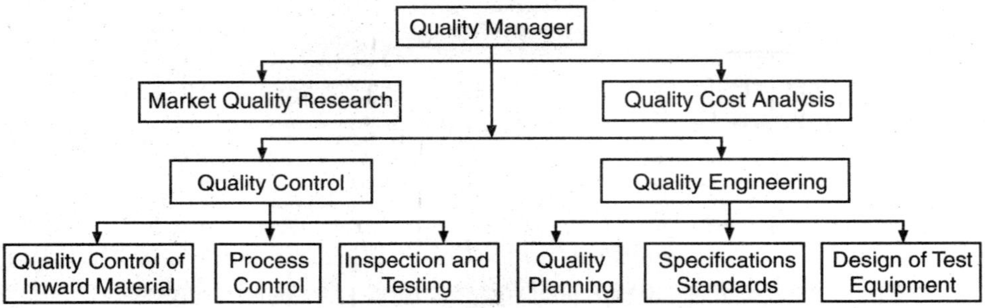

Fig. 6.2. Quality management organisation

In the quality Engineering and Quality Control Organisation, the subgroups are connected to the quality manager by a line relationship, the accountability function is upto the lowest operating level. The responsibilities and authority should be clearly defined while delegating responsibility to various subordinates. Any overlapping of responsibilities and functions should be minimised.

The internal structure of the quality component needs proper relationships with other departments. The quality component or its sub-divisions location and places should be on performance basis to achieve the maximum effectiveness, economy and minimum of friction.

Some typical quality organisation can be categorized as following:

(*i*) Single product organisation
(*ii*) Multiproduct company
(*iii*) Single product with multiplant
(*iv*) Large corporation with different product
(*v*) Jobbing factory
(*vi*) Cottage industry.

6.7. QUALITY COST

Industry wants to secure economic gains in the form of profit. The soundness of an industry is judged by its financial viability. Any industrial function or activity generally assessed by the economic yardstick. Quality improvement helps sales and long term profitability. Any quality programme must be justified by the returns it brings to the company. The demand for higher quality goes on increasing and improvement in quality generally involves increased cost of manufacture leading to higher product price. If higher quality product is costly, then the customer chooses a cheaper product of lower quality. However, there is a minimum limit for the quality level. Quality limits keep on changing generally in upward direction due to social and technological development. The general relationship between quality value or quality cost of a product with increasing quality level is shown in Fig. 6.3.

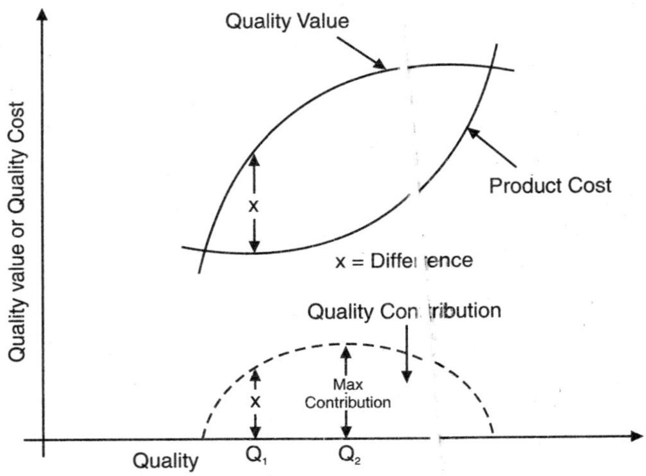

Fig. 6.3. Quality value or quality cost versus quality

If the quality of a product is improved, then the cost rises at an increasing rate. The difference between the value and cost of a product at any particular quality level is known as quality contribution. The contribution can be increased by improving the product quality upto point Q_2 and subsequent improvement in quality decreases the contribution. It is the middle quality level which gives the maximum benefit. Here, the quality means the quality of design or the grade of the product.

Quality of conformance is the product conformance to the design. As quality conformance also costs money, quality control programme needs to ensure minimum overall quality costs. Cost incurred to ensure that the product is of requisite quality is known as quality cost.

Quality affects all aspects of the organization and has dramatic cost implications. Poor quality creates dissatisfied customers and eventually leads to loss of business.

Detecting and correcting defects during product design and product production is considerably less expensive than when the defects are found at the customer sites as shown in Fig. 6.4.

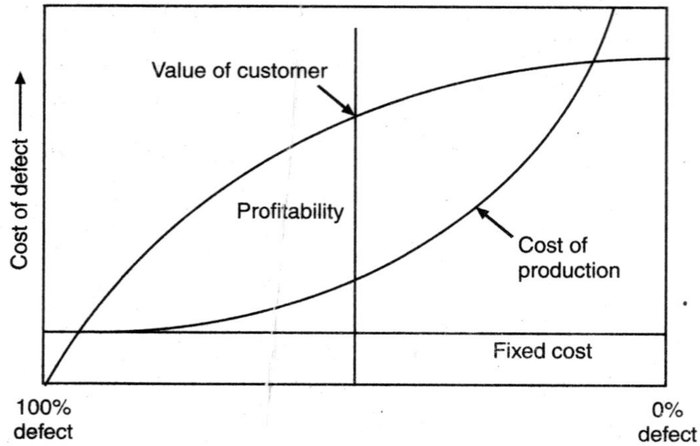

Fig. 6.4. Cost of Defects

There is a need of cost balancing. This is the problem of balancing the value of given result against the cost of a given result. It becomes very expensive if the quality specifications narrow the limits. This expense is very greatly closer to perfection as shown in Fig. 6.5.

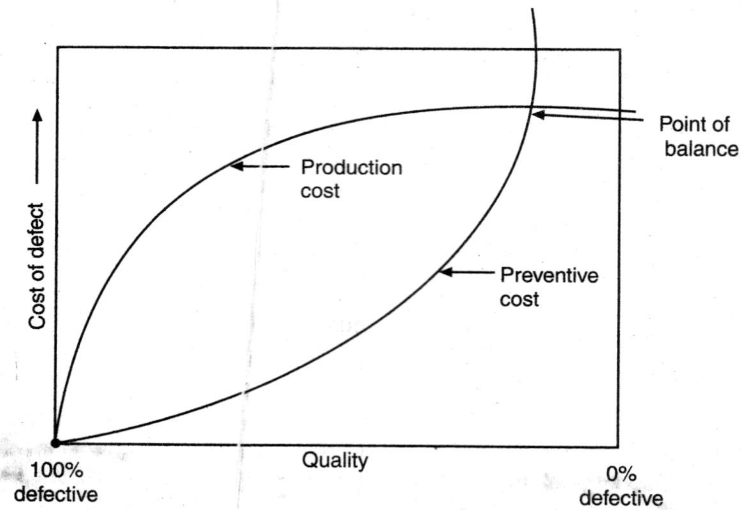

Fig. 6.5. Cost Balancing

Refer Fig. 6.6 for demand of even great precisions. Note that quality at level Y is the most economical. A quality drop to level X reduces the cost by M. It reduces, the value as N is greater than M. A rise of quality to level Z increase the quality cost value by O while the cost is increased by P which is greater than O.

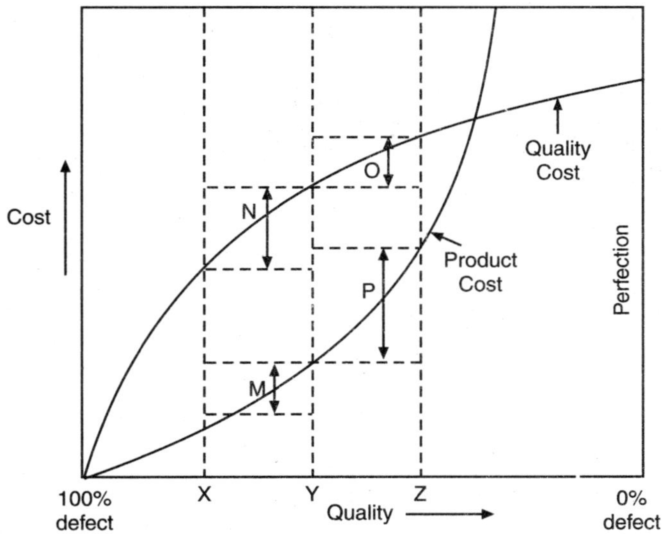

Fig. 6.6. Cost of Quality Versus Quality

6.8. QUALITY COST OPTIMIZATION

The cost of a product and quality are linked. Higher the quality of the item produced, greater would be its cost of production. This maybe due to several reasons, such as, better quality of materials used in production, higher degree of skills required and greater cost incurred in improving quality. All the cost elements which affect the total cost of maintaining and controlling quality can be grouped together as cost of quality or 'quality cost'. Quality costs broadly comprise of Failure costs, Appraisal costs, Prevention costs.

(*a*) Failure Costs: The direct and indirect costs incurred on those products or services which fail to comply with their prescribed specifications are the failure costs. Such costs may be internal or external.

(*b*) Internal Failure Costs

(*i*) Cost of rejections or scrap produced which will not be accepted by the customer and which will become a total loss. It includes the cost of power and various inprocess materials spent in producing the rejections.

(*ii*) Cost of corrective operations, or rework : This is case of such items which have not been passed during inspection but can be made acceptable after certain rework or repair. For example, brazing, welding, straightening, pressing, filling, re-heat treatment etc.

(*iii*) Cost incurred in fault investigation, trouble-shooting, defect-analysis, etc.

(*iv*) Cost incurred in affecting design modification, if necessary, in order to overcome defects in future.

(*v*) Loss in capacity of production due to the rejections produced.

(*c*) External Failure Costs

1. Loss of future orders due to poor performance in service.

2. Cost incurred in attendig to customer complaints and providing customer service, including cost of returned material processing, tansport, loss of taxes.

(*d*) Appraisal Costs: Costs involved in actual checking of the quality *viz.*, the cost of carrying out actul inspection are known as appraisal costs. These costs include:

(*i*) Cost of inspection or testing of raw materials, including the cost of their movement for the purpose of inspection.

(*ii*) Cost of maintaining and providing laboratory services for the purpose of inspection.

(*iii*) Cost of calibration and maintenance of test and inspection equipment at regular intervals.

(*iv*) Cost of product inspection.

(*e*) Prevention Cost: Prevention costs are those which are involved in ensuring quality. Thus, faulty or defective work or rejections are not produced in the very first instance. The main constituents of prevention costs are listed as follows:

(*i*) Cost involved with training and retraining of operators, supervisors and other staff so that they can all carry out their functions most effectively and can maintain desired quality levels.

(*ii*) Cost of repairing and maintaining gauges, dies, machines, tools and equipment to be able to achive desired quality.

(*iii*) Cost of R & D efforts put-into continuous work for developing better materials, processes and techniques, so as to maintain high quality of product.

(*iv*) Cost involved to motivate the staff and develop high quality of work life.

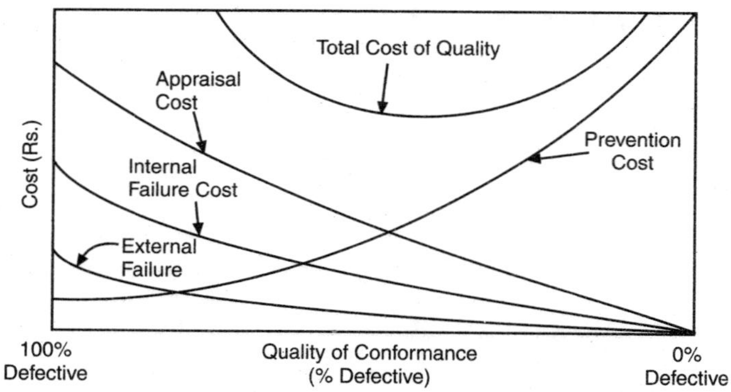

Fig. 6.7. Quality cost versus defects

Appraisal costs and failure costs decreases as the quality improves, whereas, prevention costs increase with improvement in quality as shown in Fig. 6.7. In case of failure costs, both internal and external costs obviously have the same trend. When failure rates are high, there are more rejections, greater need for repairs and so failure costs go up. As the quality level go up, the amount of inspection can be safely reduced. Hence appraisal costs come down. The trend is reverse incase of prevention costs because quality will improve and it will be maintaind only when sufficient efforts are put in and concepts of total quality management and human resource development are introduced.

The total cost of quality curve is obtained by taking the cumulative sum of all the components of quality costs.

6.9. COST REDUCTION PROGRAMMES

Preventive quality control programmes should be utilised in such a manner as to yield maximum results. This gives the confidence to the company management and generates enthusiam for quality programmes. The frequency of failures in the process result in the production of defective jobs. It could begin with the isolation of the product having a high rate of rejection as shown in Fig. 6.8.

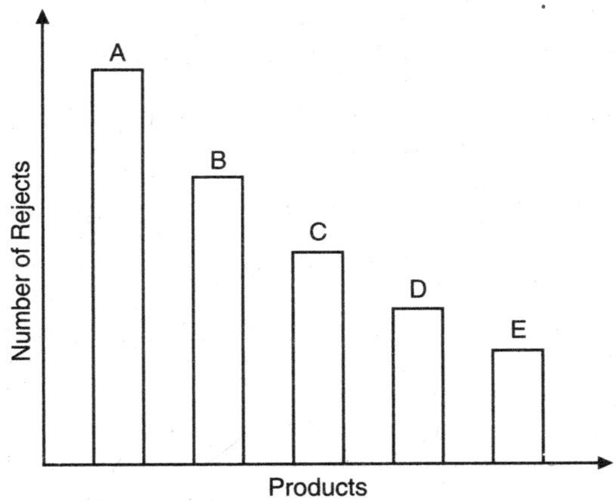

Fig. 6.8. Histogram of rejects

On the basis of the cost of failures of products, the graph may look like Fig. 6.9. The cost of failure due to each major cause should be estimated and tabulated. The cause showing the highest failure cost deserves first attention.

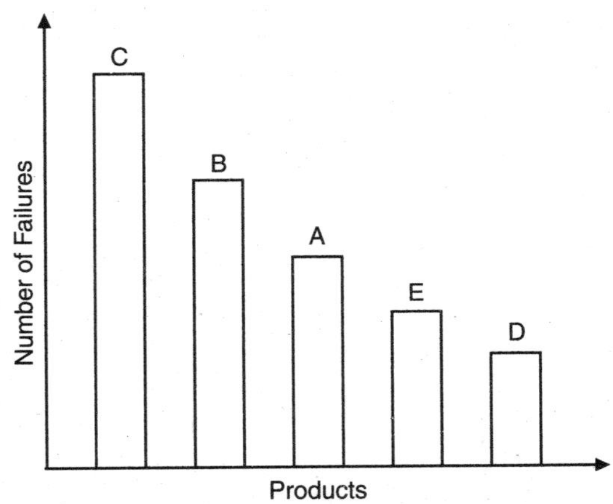

Fig. 6.9. Cost of failures

A quality study may reveal a design weakness or unsuitability of a particular equipment or machine which has been selected. Proper planning should be done especially in the case of continuous or repeated production process. Maintenance aspects must be considered (which is discussed in subsequent chapters). Maintenance aspects give maximum cost reduction, therefore, it is an essential element.

6.10. PROCESS CONTROL

Process control deals with architectures, mechanisms and algorithms for controlling the output of a specific process. It is a statistics and engineering discipline. Process control can be divided in the following categories:

(*i*) Discrete Process

These are motion based such as automotive production, robotic assembly, packaging applications etc. These are characterized as discrete process control. The discrete manufacturing involves the production of discrete components. The component itself can be a final product of a company or assembly of components leading to product like a car or refrigerator etc.

Discrete manufacturing process is mostly automated. It uses Computer Numeric Control (CNC), Computer Aided Design (CAD), Computer Aided Manufacturing (CAM), Copying machines and robotics etc.

(*ii*) Continuous Process

Continuous process is a physical system represented through variables which are smooth and uninterrupted in time. These require that specific quantities of raw materials be combined in specific ways for particulars durations to produce an intermediate or end result such as chemicals, fuels, plastics, food items, beverages and medicine etc. Now-a-days, continuous process control use Programmable Logic Controller (PLC), Distributed Control System (DCS). Supervisory Control And Data Acquisition (SCADA) system, etc.

Production process control and quality assurance are elements of a quality management system. This is the set of policies, procedures, and processes used to ensure the quality of a product or a service. It is well known that quality management systems improve the quality of the product and services produced. This improves market share, sales growth, sales margins, and competitive advantage and helping to avoid litigation.

Quality control techniques in production were initially developed by statistician W. Edward Deming. These ideas were adopted in post-World War II.

The international organization for standardization prepared quality principles and the procedures for implementing a quality management system in ISO 9000: 2000, ISO 9001: 2000 and other documents. These documents have become the best practices for ensuring quality and in many fields including service as the basis for regulation. ISO 9000 guidelines provides a comprehensive model for quality management systems that can make any company competitive.

The collection and analysis of data that measure the quality of the raw materials, components, products, and assembly processes is an important element of quality assurance. Data collection and analysis techniques during the design, development and production stages have become statistics based.

Statistics are used for:

(*i*) Acceptance sampling

(*ii*) Statistical process control (SPC)

(*iii*) Troubleshooting studies.

6.10.1. Acceptance Sampling

Acceptance sampling determines whether a batch of product, supplier components or finished units is of acceptable quality. Instead of testing 100% of the batch, a random sample of the batch is done, and a decision about the whole batch is taken from the sample test results. Acceptance sampling was developed during World War II to test bullets. Subsequently, standards have been developed to include various types of quality measurements, and testing and sampling methods. These standards involve sampling by variables Vs attributes, use of simple Vs double or multiple sampling, rectifying Vs non-rectifying with respect to non-conforming items. These help to determine an appropriate sampling plan.

6.10.2. Statistical Process Control (SPC)

SPC is a technique of monitoring a production process through the use of control charts. In process data or random samples of the output at various stages of the production processes are collected. One can defect variations or trends in the quality of the materials or processes which may affect the quality of the end product. As data are gathered during the production process, problems can be detected and prevented earlier than techniques used to determine quality of the end product. Production processes are streamlined through the identification of bottlenecks, delays by use of SPC techniques.

6.10.3. Troubleshooting Studies

Identification of a problem in the end-of-the-line product leads to troubleshooting study to determine whether changes in certain inputs such as raw materials or process characteristics are responsible with the output variables. The analysis of data recorded on production inputs and outputs help in such studies.

Statistical techniques can detect associations between raw materials or process attributes and end-of-the-line product outcomes. These observational studies may not prove the existence of a cause-and-effect mechanism, but results of troubleshooting analysis may indicate potential targets for corrective actions, off-line testing or further measurements and analysis to establish the root cause of the manufacturing problem may also be achieved by troubleshooting studies.

6.11. HUMAN FACTOR

The human is the most important and considered as a critical component system in any industry and it is most difficult to control. It matters at all level from the lowest worker to the chief executive. There is a increased probability of accidents and errors due to human factor. There is lower productive output, increases in lost time, higher medical costs, higher material costs, increased absenteeism, low quality work, injuries, strains, etc.

"Human beings by their very nature make mistakes, therefore, it is unreasonable to expect error-free human performance" as per Shappell and Weigmann.

Human nature is to blame an operator when something goes wrong. The determination of responsibility, fault, and blame are the games people play.

Blaming the worker is easier for management. Acceptance of the fact that the workplace, procedure, environments or system needs improving is to be ensured. The behaviour of the human, their abilities, limitations, and their characteristics to the design of tools, machines, system, tasks, jobs and environment for productive, safe, comfortable and effective human use in a productive and safe manner are based on human factors. Human factor is just not applying checklists and guidelines. It is just also not using oneself as the model for designing things or just commonsense human use. Various human factors are considered here.

(*a*) Cycles of human factor: These are the goals of humans which are achieved through several steps in the human cycle which depicts the human operator brain and body; and the system with which he or she is interacting. The problems and deficiencies in the human-system interaction of an existing system are necessary to be diagnosed.

There are five different approaches that can be used:

(*i*) Equipment design: This changes the natue of the physical equipment with which human work.

(*ii*) Task design: This focuses more on changing what operators do instead of changing the devices.

(*iii*) Environmental design: This provides improved lighting, temperature control and reduced noise in the physical environment.

(*iv*) Training of individuals: It is done for preparing the worker for the job conditions.

(*v*) Selection of individuals: It is based on physical and mental dimension for good system performance. Selecting operators with best profile of characteristics for the job is optimized.

Human Factors Engineering (HFE) is used to determine human capabilities and limitation in the design of the products, processes, systems, and work environments. It improves the usage, system performance and reliability, and user satisfaction. It reduces operational errors, operator stress, training requirements, user fatigue and product liability.

(*b*) Attitude of top management: The attitude of the top management is reflected at all lower levels, and the departmental heads of the quality department. Organizations must not only posses information but also acquire and create new information and knowledge. The quality manager is required to convince the top management about the necessity of introducing quality control program.

The quality manager has to make a plan to meet the quality needs of the company. A positive top management attitude leads to higher internalization intent, higher receptivity for training and knowledge base building.

(c) Co-operation of functional groups: The active co-operation of functional groups is required. They should be involved at the planning stage itself during the formulation of the quality program. A committee approach is followed for the management of wide spread quality activities.

Various committees comprising of executive from all the concerned departments is formed for the planning and coordination of specific quality activities such as product planning and design review, quality planning and process control, defect investigation of customer complaints and quality motivation.

Functional layout consists of functional grouping of machines. All turning machines may be grouped together at one place and all milling machines at another. Parts may be moved in different sequences from one machine to another depending on their processing requirements. Functional and process layout dominates batch production facilities. This leads to a large number of setups for different parts, high material handling costs, and high work-in processs inventory.

This group technology is for identifying and bringing together related or similar components in a production process which gives advantage of similarities in making use of the inherent economies of production flow methods.

(*d*) Operators attitude: The operator work is a decisive factor in determining the quality of the product. Operator contributes the maximum to the quality of product. However, the obsession for higher production has affected the attitude of operator towards the quality. The quality is attahced by organizational culture which describes the psychology, attitudes, experiences, beliefs, personal and cultural values of an organization. These are the specific collection of values and norms that are shared by people and groups in an organization which control the way they interact with each other and with stakeholders outside the organization.

Organizational values develop organizational norms, guidelines, or expectations that prescribe appropriate kinds of behaviour by an employee in particular situation. These control the behaviour of organizational members towards one another. Strong culture exists where staff respond to situations based on their alignment to organizational values. Strong cultures help firms operate like well oiled machines.

(*e*) Operators responsibility: The management has to provide the necessary means to the operator. These means are used by management to exercise effective control over the process. For this, the operator must know what he is supposed to do, must have means by which he can know the result of his action and must be able to regulate satisfactory operation of the process. Quality can be achieved only if the process is kept-under a state of control by the operator. Normally, the management takes the credit to maintain good quality of product. However, if the quality of the product is poor, then the worker is held responsible.

(*f*) Causes of operator errors: There are three main causes of operator errors which are as follows:

- **(*i*) Incompetence:** The operator training should be done to acquire necessary proficiency in the job.
- **(*ii*) Lack of Awareness:** Errors can be avoided by making the process fool proof. Appropriate devices may be incorporated in the process to provide a visual and sonic alarm in the event of an error. Proper lighting,

clean working conditions, avoidincy under fatigue, and long working conditions, proper incentive, provision of health regarding policies, and rewards for extra ordinary work also reduce the opertor error.

(iii) **Carelessness:** Unsuitable working conditions cause the operator do not take interest in the quality of the product. The main cause of disinterest is the lack of knowledge about the product. Ignorance and lack of understanding leads to poor quality.

Nowadays, the product is broken down into small discrete elements. A worker may be producing thousands of small parts every day without knowing their fit and function. Then he cannot be blamed if he is not very much concerned about its quality.

6.12. CORRECTIVE METHODS

The main corrective methods are motivation, education and financial assistance.

(*a*) Motivation of Worker: Quantitative output is more important to worker because it determine his wage. The workers needs to be convinced that he should attain required quality. Thus, the motivation of the operators and the workers are necessary to achieve the good quality products.

(*b*) Education of Workers: Workers should be well informed about the products. The workers should be able to comfortably understand the needs of the company, their plans, the process of the manufacturing the products. The worker should be educated about company's product, company's quality policy, and the effects of quality on company's sales and profitability.

(*c*) Financial Incentive: Generally financial incentives are based on price rate system which are output oriented. Financial inventives are the major motivating factor. Some forms of financial inventives linked with quality of the products such as quality bonus or merit payment provides a strong motivation for the operator to improve the quality.

SUMMARY

1. Quality control gives the causes for variation in the characteristics of products. It also gives solutions by which these variations can be controlled. It achieves economy and objective. It meets procedure and dynamics in its operation.
2. Quality is closely related to cost and customer needs. Quality is the degree to which it fulfils the requirements of the customer. Quality means degree of perfection. Quality can be determined by characteristics such as design; size; material; chemical composition; mechanical functioning; workmanship; finish and other properties.
3. Quality affects all aspects of the organization and has dramatic cost implications. Poor quality creates dissatisfied customers and eventually leads to loss of business. All the cost elements which affect the total cost of maintaining and controlling quality can be grouped together as cost of quality or 'quality cost'.
4. Process control deals with architectures, mechanisms and algorithms for controlling the output of a specific process. It is a statistics and engineering discipline.
5. Discrete manufacturing process is mostly automated. It uses Computer Numeric Control (CNC), Computer Aided Design (CAD), Computer Aided Manufacturing (CAM), Copying machines and robotics etc.

6. Continuous process control use Programmable Logic Controller (PLC), Distributed Control System (DCS). Supervisory Control And Data Acquisition (SCADA) system, etc.
7. Acceptance sampling determines whether a batch of product, supplier components or finished units is of acceptable quality. Instead of testing 100% of the batch, a random sample of the batch is done, and a decision about the whole batch is taken from the sample test results.
8. SPC is a technique of monitoring a production process through the use of control charts. In process data or random samples of the output at various stages of the production processes are collected. One can defect variations or trends in the quality of the materials or processes which may affect the quality of the end product.

EXERCISES

1. Discuss the factors to be considered for organisational structure for quality management.
2. Explain the basic steps to design an organisation structure for quality management.
3. What do you understand by quality functions. Explain in detail.
4. How is decentralization of quality function done? Discuss.
5. Quality is also a staff function. Justify.
6. How is designing and fitting of an organisation done for different types of products and company? Explain.
7. What is a quality value and how does it corelate with quality cost?
8. Explain quality of conformance.
9. How is quality cost optimization done?
10. How are the cost reductions done for quality product?
11. Human factor is a most important element in quality of a product. Justify.
12. What corrective methods can be adapted to improve and ensure quality due to human factor?
13. What do you understand by quality control? Explain.
14. What are the dimensions of quality?
15. What is the meaning of quality control?
16. Define quality control.
17. Explain cost of quality.
18. Discuss cost of quality and its relationship using a diagram.
19. Discuss process control in detail.
20. Explain acceptance sampling.
21. What is SPC? Explain.
22. Discuss troubleshooting studies in relation with quality control.
23. Explain effect of top management towards quality.
24. How does co-operation of groups affect quality?
25. Does operator attitute affect quality? Explain.
26. What relationship responsibility has with quality?
27. What are the causes of apparatus errors?
28. What correhives measures can be taken to eliminate apparatus errors?

□□□

7 CHAPTER Seven

CONTROL CHARTS

CHAPTER OUTLINE

- Introduction
- Statistical Process Control Methods
- Control Charts for Variables
- Control Charts for Attributes
- C-Charts
- Sampling Plans
- Operating Characteristics (OC) Curves

7.1. INTRODUCTION

One of the best technical tools for improving product and service quality is statistical process control (SPC).

Statistics is defined as the science that deals with the collection, tabulation, analysis, interpretation and presentation of quantitative data. Each division is dependent on the accuracy and completeness of the preceding one. Data may be collected by a technician measuring the tensile strength of a plastic part or by an operator using a check sheet.

Statistical quality control describes the set of statistical tools used by quality professionals. There are three broad categories of statistical quality control.

7.1.1. Measures of Central Tendency

A measure of central tendency of a distribution is a numerical value that describes the cultural position of the data or how the data tend to build up in the center. There are three measures commonly use in quality.

(*i*) average

(*ii*) median

(*iii*) mode.

The arithmetic average, or the mean is a statistic that measures the central tendency of a set of data. Mean is sum of all the observations and divided by the total number of observations. The mean calculation is done by:

Acerage or Mean, $\overline{X} = \frac{\sum_{i=1}^{n} X_i}{n}$

where, $\overline{X}$ = the mean

X_i = observation i, $i = 1, ..., n$

n = number of observations

Σ = sum of

Another measure of cultural tendency is the medians which is defined as the value that divides a series of ordered observations so that the number of items above it is equal to the number below. When the number in the series is odd, the median is the midpoint of the values, provided the data are ordered. When the number in the series is even, the median is the average of the two middle numbers. The mode, is the value that occurs with the greatest frequency. The average is the most commonly used measure of ventral tendency. It is used when the distribution is symmetrical. The median is used when an exact midpoint of a distribution is desired. When a distribution has extreme values, the average will be adversely affected, whereas the median will remain unchanged.

7.1.2. Range and Standard Deviation

A second tool of statistics is composed of the measures of dispersion, which describe how the data are spread out or scattered on each side of the central value. One of the measures of dispersion is the range, which is the difference between the largest and smallest values of observation. It is given by

$$R = X_l = X_s$$

where R = range

X_l = largest observation

X_s = smallest observation

There are two measures which can be used to determine the amount of variation in the data. The first measure is the range, which is the difference between the largest and smallest observations. Another measure of variation is the standard deviation. The standard deviation is given by:

Standard deviation, $\sigma = \sqrt{\frac{\sum_{i=1}^{n} (X_i - \overline{X})^2}{n-1}}$

where, σ = standard deviation of a sample

$\overline{\overline{X}}$ or $\overline{X}$ = average of averages

X_i = observation i, $i = 1, ... n$

n = the number of observations in the sample

The standard deviation is a reference value that measures the dispersion in the date. The smaller the value of the standard deviation, the better the quality, because the distribution is more closely compacted around the central value. In quality control the range is a very common measure of the dispersion. It is used in one of the principal control charts.

The primary advantage of the range is in providing a knowledge of the total spread of the data. As the number of observations increases, the accuracy of the range decreases. The standard deviation is used when a more precise measure is desired.

When a distribution is symmetric there are the same number of observations below and above the mean. This is common observation when only normal variation is present in the data. When a disproportionate number of observations are either above or below the mean the data has a skewed distribution. Figure 7.1 illustrates symmetric and skewed distributions.

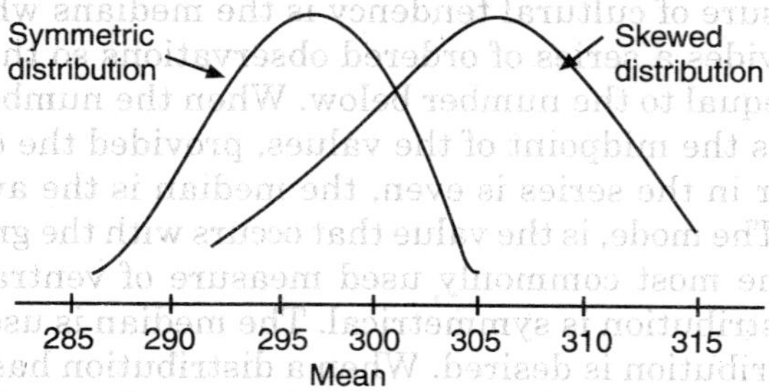

Fig. 7.1. Differences Between Symmetric and Skewed Symmetric

7.2. STATISTICAL PROCESS CONTROL METHODS

The most commonly used tool for monitoring the production process is a control chart. Different types of control charts are used to monitor different aspects of the production process.

A control chart or process chart or quality control chart is a graph which shows whether a sample of data falls within the common or normal range of variation. It has upper and lower control limits that separate common from assignable causes of variation. The common range of variation is defined by the use of control chart limits. A process is out of control when a plot of data reveals that one or more samples fall outside the control limits.

Figure 7.2 shows a control chart. The x-axis represents samples taken from the process over time. The y-axis represents the quality characteristics that is being monitored (ml. of liquid). The center line (CL) of the control chart is the mean, or average, of the quality characteristics that is being measured. The mean is 300 ml. The upper control limit (UCL) is the maximum acceptable variation from the mean for a process that is in a state of control.

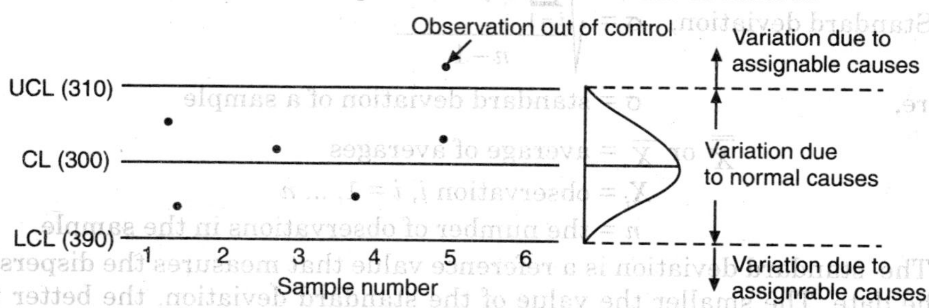

Fig. 7.2. Quality Control Chart

Similarly, the lower control limit (LCL) is the minimum acceptable variation from the mean for a process that is in a state of control.

The upper and lower control limits are 310 ml and 290 ml. respectively. If a sample of observation fall outside the control limits, assignable causes are to be found out.

Control charts are used as tools in statistical process control. They can be used to measure any characteristics of a product, such as the weight of a chocolate box, the number of product, or the volume of bottled softdrink. There are different characteristics that can be measured by control charts can be divided into two groups: variables and attributes. A control chart for variables monitoring characteristics can be measured and have a continuum of values, such as height, weight or volume. A water bottling operation is an example of a variable measure, since the amount of liquid in the bottles is measured and can take on a number of different values. The height of a bag of a sugar, the temperature of a baking oven, or the diameter of plastic tubing are other examples.

A control chart for attributes monitors characteristics having discrete values which can be counted. Colour, taste, or smell are examples. An attribute requires only a single decision, such as yes or no, good or bad, acceptable or unaceptable therefore, attributes normally takes less time than that of variables because a variable needs to be measured.

7.3. CONTROL CHARTS FOR VARIABLES

It monitors characteristics which can be measured and have a continuous scale, such as height, weight, volume or width. Two of the most commonly used control charts are the mean and the standard deviation or range. The process is assumed not to be in control when observed values go outside the control limits.

7.3.1. Mean (X-Bar) Charts

A mean control chart is also known as an X-bar chart. It monitors changes in the mean of a process. Multiple samples are taken and their means are calculated. These samples are small, about four or five observations. Each sample has its own mean, $\overline{X}$. The center line of the chart is then calculated which is the mean of k samples, where k is the number of samples. Figure 7.4 shows a mean (X-bar) chart.

$$\text{Central Line, } CL = \overline{\overline{X}} = \frac{\overline{X}_1 + \overline{X}_2 + \overline{X}_3 + \ldots + \overline{X}_k}{k}$$

The following formulas are used to construct the upper and lower control limits of the chart.

$$\text{Upper control limit, UCL} = \overline{\overline{X}} + z\sigma_{\overline{X}}$$

$$\text{Lower control limit, LCL} = \overline{\overline{X}} - z\sigma_{\overline{X}}$$

where, $\overline{\overline{X}}$ = the average of the sample means

z = standard deviation which is 2 for 95.44% confidence and 3 for 99.74% confidence.

$\sigma_{\overline{X}}$ = standard deviation of sample means, taken as $\dfrac{\sigma}{\sqrt{n}}$

σ = population or process standard deviation

n = sample size or number of observations per sample

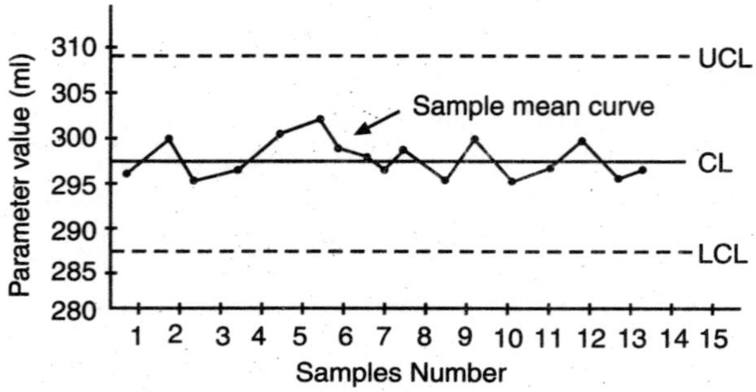

Fig. 7.3. A mean (X-bar) Control chart

Another method to construct the control limits is by using the sample observation values variation range as an estimate of the variability of the process. It is known that the range is simply the difference between the largest and smallest values in each sample. The spread of the range gives the variability of the data. The control limits will be calculated as follows:

Upper control limit, UCL = $\overline{\overline{X}} + A_2R$

Lower control limit, LCL = $\overline{\overline{X}} - A_2R$

where, $\overline{\overline{X}}$ = average of the sample means

R = average of range of the samples

A_2 = factor is made available from ASTM quality control manual. It includes three standard deviations of ranges and is dependent on the sample size being considered.

7.3.2. Range (R) Charts

Range charts monitor the dispersion or variability of the process. The method of developing and using R-charts is the same as that for X-bar charts. The center line of the control chart is the average range, and the upper and lower control limits are calculated as follows:

$$\text{CL} = \overline{R}$$

$$\text{UCL} = D_4\,\overline{R}$$

$$\text{LCL} = D_3\,\overline{R}$$

where, values for D_4 and D_3 are from Table 7.1. Factors of control limits. The values are made available from ASTM quality control manual. An example of control chart is shown in Fig. 7.4.

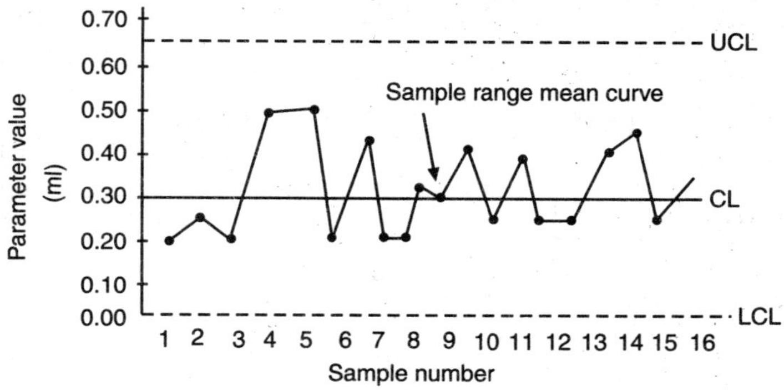

Fig. 7.4. Range (R) Charts

7.3.3. Using Mean and Range Charts Together

The mean and range charts are used to monitor different variables. The mean or X-bar chart measures the central tendency of the process. The range chart measures the dispersion or variance of the process. Both mean and range chart can be used to monitor a process. There can be a shift in the mean of the product, but there is no change in the dispersion. The dispersion is the same, but shift detected in X-bar chart as shown in Fig. 7.5. (*a*). In another case, there could be a shift in the dispersion of the product without a change in the means *X*. This condition is detected by a range chart but not by an $\overline{X}$-bar chart as shown in Fig. 7.5. (*b*). A shift in either the mean or the range means that the process is out of control. Thus, it is important to use both charts to monitor the process.

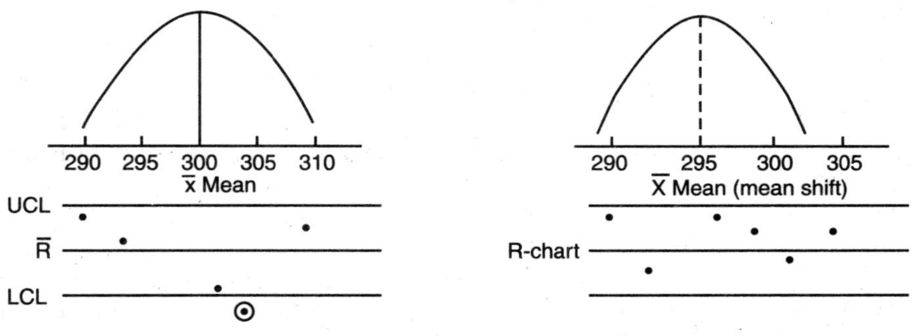

Fig. 7.5. (a) Shift in mean detected by $\overline{X}$-chart but not by R-chart

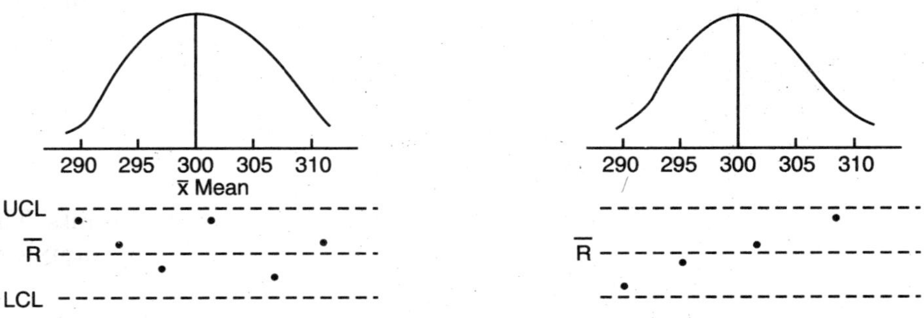

Fig. 7.5. (b) Shift in dispersion detected by R-chart but not by $\overline{X}$-chart

7.4 CONTROL CHARTS FOR ATTRIBUTES

Control charts for attributes are used for counted products rather than measured products. Attributes are discrete in nature. The products are simple yes-or-no decisions. Two of the most common types of control charts for attributes are P-charts and C-charts.

7.4.1. P-charts

P-charts are used to measure the defective proportion in a sample. The calculations of the center line as well as the upper and lower control limits are similar to the calculations for the other kinds of control charts. The center line value which is calculated as the average proportion defective in the population. This is like obtained by taking a number of samples of observations at random and computing the average value of P across all samples.

For constructing the upper and lower control limits for a P-chart, the following calculation formula is used:

$$UCL = \bar{P} + z\sigma_P$$

$$LCL = \bar{P} - z\sigma_P$$

where, z = variable of multiplier of standard deviation

$\bar{P}$ = sample proportion defective

σ_P = standard deviation of the average proportion defective.

Similar to the other charts, z is taken as 2 or 3 value, depending on the amount of data to be captured in our control limits. Normally, value of z is taken as 3.

The sample standard deviation is computed as follows:

$$\sigma_P = \sqrt{\frac{\bar{P}(1-\bar{P})}{n}}$$

where, n is the sample size.

The centre line of the chart is given as:

$$CL = \bar{P} = \frac{\text{total number of defective bottles}}{\text{total number of observations}}$$

EXAMPLE 7.1.

Consider an example of a bottle manufacturing plant. The number of defective bottles in twenty random samples with forty observation each is found by inspection. The defective bottle found in each sample are:

Sample number	Number of Defective bottles	Number of observations sampled	Fraction Defective
1.	6	40	0.15
2.	4	40	0.10
3.	2	40	0.05
4.	4	40	0.10

5.	2	40	0.05
6.	6	40	0.15
7.	6	40	0.15
8.	4	40	0.10
9.	2	40	0.05
10	4	40	0.10
11.	6	40	0.15
12.	4	40	0.10
13.	4	40	0.10
14.	2	40	0.05
15.	2	40	0.05
16.	4	40	0.10
17.	8	40	0.20
18.	6	40	0.15
19.	2	40	0.05
20.	2	40	0.05
Total	80	800	

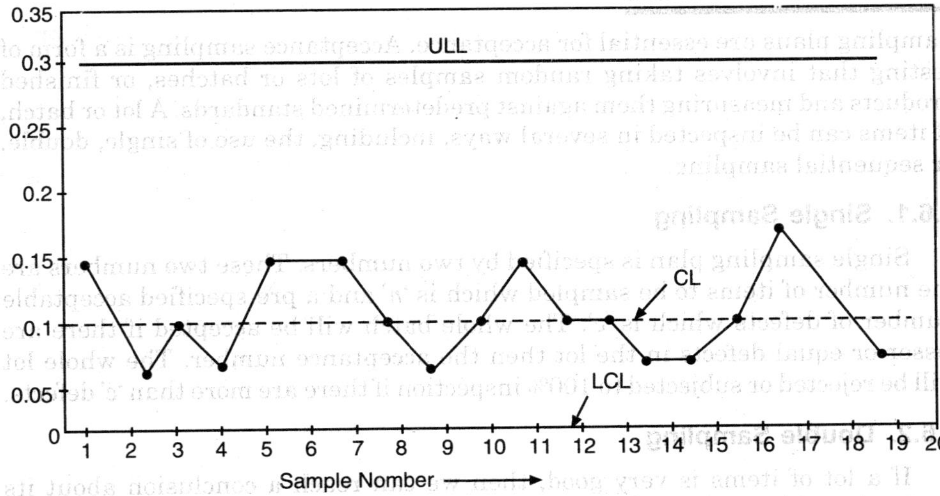

Fig. 7.6. P-Control Chart Plot

Solution: $\bar{P} = \dfrac{80}{800} = 0.10$

and $\sigma_P = \sqrt{\dfrac{\bar{P}(1-\bar{P})}{n}}$

$$= \sqrt{\frac{(0.10)(0.90)}{40}} \simeq 0.05, \text{ where } n = 20$$

$$\text{UCL} = \bar{P} + z(\sigma_P) = 0.10 + 3 \times 0.05 = 0.25$$

$$\text{LCL} = \bar{P} - z(\sigma_P) = 0.10 - 3 \times 0.05 = -0.05 \simeq 0$$

If LCL becomes negative in the calculation the LCL is tables as zero because a negative control limit cannot exist. Finally, the *P*-control is plotted.

7.5. C-CHARTS

C-chart is a control chart used to monitor the number of "counter" the number of defects occuring in a given unit of time.

It can also monitor the total number ratio (c) of defects to number inspected (n) for each. This total number of defects (c) divided by number of units inspected. May vary as subgroup size in C-chart. A value for the defects per sample (c) is calculated for each sub group. The total number of defects must at always be less than the number of units inspected.

The average member of defects, $\bar{c}$, is the centre line of the control chart. The upper and lower control limits are given by:

$$\text{UCL} = \bar{c} + 3\sqrt{\bar{c}}$$

$$\text{LCL} = \bar{c} - 3\sqrt{\bar{c}}$$

7.6. SAMPLING PLANS

Sampling plans are essential for acceptance. Acceptance sampling is a form of testing that involves taking random samples of lots or batches, or finished products and measuring them against predetermined standards. A lot or batch, of items can be inspected in several ways, including, the use of single, double, or sequential sampling.

7.6.1. Single Sampling

Single sampling plan is specified by two numbers. These two numbers are the number of items to be sampled which is 'n' and a pre-specified acceptable number of defects which is 'c'. The whole batch will be accepted if there are lesser or equal defects in the lot then the acceptance number. The whole lot will be rejected or subjected to 100% inspection if there are more than 'c' defects.

7.6.2. Double Sampling

If a lot of items is very good, then we can reach a conclusion about its quality by taking a smaller sampler (n_1) then would have been used in a single sampling plan (n). The lot can be accepted if the number of defects in this smaller sample (of size n_1) is less than or equal to some lower limit (c_1). The whole lot can be rejected if the number of defects exceeds an upper limit (c_2). However, if the number of defects in the n_1 sample is between c_1 and c_2, a second sample (of size n_2) is drawn. The lot accepted or rejected based on this cumulative results. This concept is known as double sampling, as samples have been drawn twice.

7.6.3. Sequential Sampling

An extension of double sampling is multiple sampling. The samples are used sequentially until a clear decision can be made. The units are randomly selected from a lot and tested one-by-one. Such cumulative member of inspected pieces and defects are recorded. This process is known as sequential sampling.

The whole lot will be rejected if the cumulative number of defects esceeds an upper limit specified for that sample. The lot will be accepted if the cumulative member of rejects is less than or equal to the lower limit. However, if the number of defects falls within these two boundaries, we continue to sample unit from the lot. It may happen in some sequential plans that the whole lot is tested, unit by unit, before the lot is accepted or rejected.

The types of products being inspected and their expected quality lend decides the selection of the best sampling approach single, double or sequential.

In many cases, a single sampling plan is easier and simpler for workers to conduct, although the number sampled may be greater than under double or sequential sampling plans.

7.7. OPERATING CHARACTERISTICS (OC) CURVES

The Operating Characteristics (OC) curve establishes appropriateness of acceptance plan which discriminated between good and bad lots. A curve is a combination of n (sample size) and percent defect c (acceptance level). It shows the probability that the plan will accept lots of various quality levels.

A highly discriminating sampling plan and OC curve would be preferable. If the complete supply of parts has an very high level of defects which is unacceptable level, then there is very high probability of rejecting the complete supply.

Figure 7.7. shows a perfect discrimination plan for a company which has plan to reject all lots more than 3% defective. Best way to assure 100% acceptance of good lots is through full inspection which is often very costly method.

Figure 7.8 indicates, that for the same sample size, a smaller value of c yield steeper curve than does a larger value of c. Hence, method to increase the probability of accepting only good lots is by setting very tight acceptance levels.

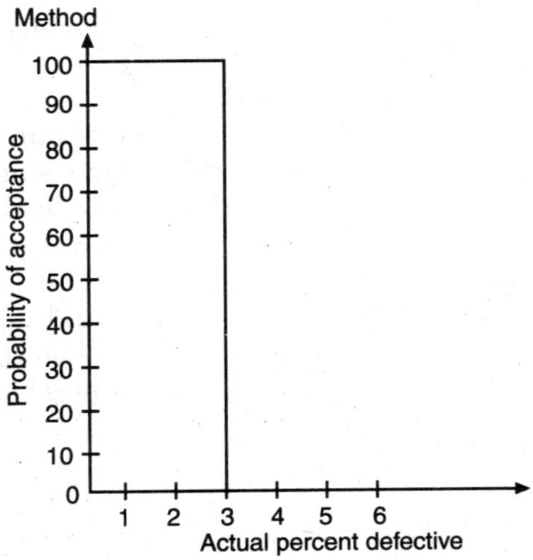

Fig. 7.7. Perfect Discrimination for Inspection Plan

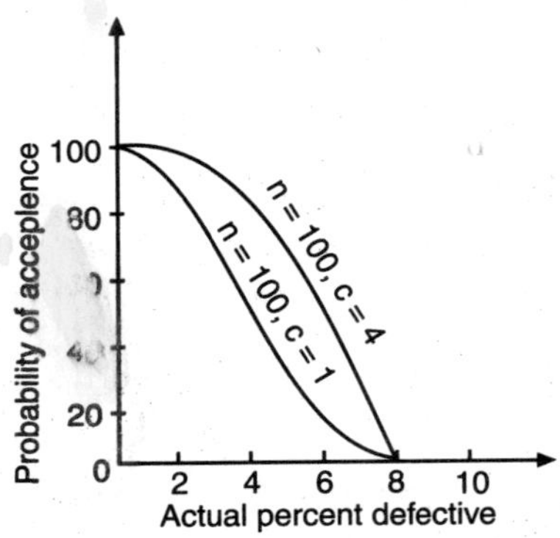

Fig. 7.8. OC Curves for Two Different Acceptable Levels of Defects for the Sample Size

Another method to develop a steeper and better OC wire is to increase the sample size. Figure 7.9. shows that even when the acceptance number is the same proportion of the sample size, a larger value of ***n*** will increase the probability of accurately measuring the lot's quality. In this case, both curves use a maximum defect rate of 4%. The OC curve for ***n*** = 25, ***c*** = 1 has more defective parts then the second plan. The following table of data is an example of this:

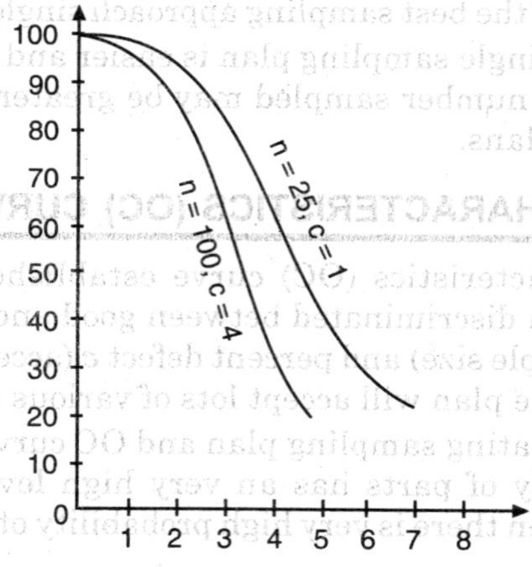

Fig. 7.9. OC Curve for two Different Sample Sizes but Some Acceplame

S.No.	Actual Defect in the lot (%)	Probability of Accepting the lot (%)	
		n = 100, ***c*** = 4	***n*** = 25, ***c*** = 1
1.	1%	98%	96%
2.	3%	80%	82%
3.	5%	43%	63%
4.	7%	16%	47%

It can be seen that, the probability of acceptance of a lot with 1% defects is 98% for n = 100 whereas only 96% for n = 25. Similarly, the probability of acceptance a lot with 5% defects, is only 43% for n = 100. But, it is 63% using the smaller sample size at n = 25.

It is important to understand that improper sampling has the danger of leading to a wrong conclusion.

Consider that there is a lot of 1000 components. Say, in actuality only 30 or 3% are defective. Thus, the lot should be accepted of 4% defect rate is acceptable. But if a random sample of n = 50 components were drawn, we may end up with zero defects and accept that lot or there could be 30 defects in the sample. In the latter case, it will look that 60% are defective and as such on the lot goes for rejection.

7.8. ATTRIBUTES OF CONTROL CHARTS

A statistical technique for controlling the quality of a product is known as control charts. These charts are also known as Shewhart Charts. A control chart can predict the defects in the pieces. A corrective action is taken accordingly before production starts.

Certain variations in the quality of any product will always occur. These variations are due to several causes acting in a random manner which are an inherent feature of any process. Few variations are due to one or two unusual causes. These causes can be identified and eliminated. The variations due to random causes follow a normal curve. If the pattern of variations are disturbed, then the process is out of control and needs fixing.

The maximum variation in the quality characteristics is +3 σ of the normal distribution for keeping the process under control. Figure 7.10 the principle of control charts. X-axis represents the quantity of the product and Y-axis represents the variation in the quality characteristics. The middle line is the mean value, the upper line is +3 σ above and the lower line is –3 σ below. The upper and lower lines are used as limits of variation such that process is under control. Each piece is drawn from production run at regular intervals. The data so obtained is plotted. The process is considers under control if the points remain within the control limits.

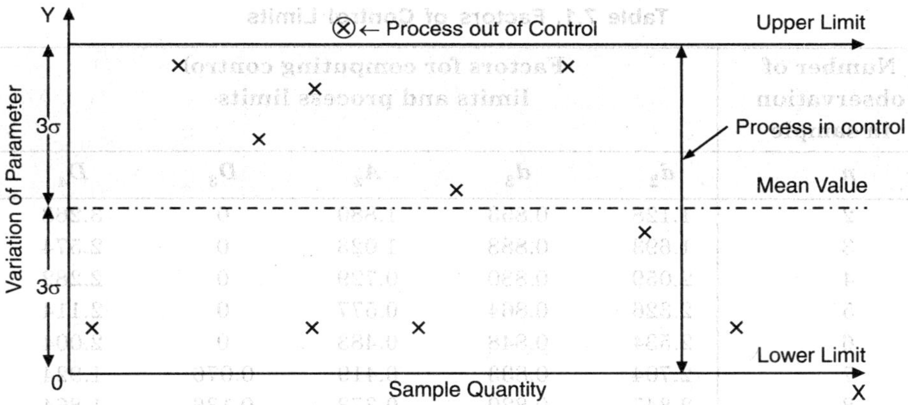

Fig 7.10. Principle of control charts

X-control Chart

The control line represents the grand mean $\overline{\overline{X}}$ of the sample means $\overline{X}$. The upper and lower control lines are 3 σ limits for sample means.

$$\text{Upper control limit (UCL)} = \overline{\overline{X}} + 3\sigma_{\overline{X}}$$

$$\text{Lower control limit (LCL)} = \overline{\overline{X}} - 3\sigma_{\overline{X}}$$

Where, $\overline{\overline{X}}$ = grand mean of the sample means

$\overline{X}$ = sample means.

$\sigma_{\overline{X}}$ = standard deviation of the sample means

The $\sigma_{\overline{x}}$ of the sample means is not the same as σ of individual measurements.

$$\sigma_{\overline{X}} = \frac{\overline{R}}{dz\sqrt{n}}$$

Where, n = sample size

$\bar{R}$ = mean range

d_2 = value from statistics table

$$\therefore \quad UCL = \bar{\bar{X}} + \frac{3\bar{R}}{d_2\sqrt{n}}$$

and

$$LCL = \bar{\bar{X}} - \frac{3\bar{R}}{d_2\sqrt{n}}$$

If

$$A_2 = \frac{3}{d_2\sqrt{n}},$$

then

$$UCL = \bar{\bar{X}} + A_2\bar{R}$$

and

$$LCL = \bar{\bar{X}} - A_2\bar{R}$$

Again A_2 can be picked up from a statistical table which is as follows

Table 7.1. Factors of Control Limits

Number of observation in sample	Factors for computing control limits and process limits				
n	d_2	d_3	A_2	D_3	D_4
2	1.128	0.853	1.880	0	3.268
3	1.693	0.888	1.023	0	2.574
4	2.059	0.880	0.729	0	2.282
5	2.326	0.864	0.577	0	2.114
6	2.534	0.848	0.483	0	2.004
7	2.704	0.833	0.419	0.076	1.924
8	2.847	0.820	0.373	0.136	1.864
9	2.970	0.808	0.337	0.184	1.816
10	3.078	0.797	0.308	0.223	1.777

R-charts

The basic data for the construction of $\bar{x}$ – R charts is the variability of the process. Actual construction can be demonstrated by an example. Consider that a final machining operation of pins of diameter 6.00 + 0.6 mm is to be done. The measurements of 25 samples comprising of 5 pins each are given in the table. The range of measurement for each sample is computed. The reading of sample 1 are: 5.99 mm, 5.98 mm, 6.01 mm, 5.97 mm and 600 mm.

$$\therefore \quad \text{Range (R)} = \text{highest value} - \text{lowest value}$$

$$= 6.01 - 5.97 = 0.04 \text{ mm.}$$

The range for all samples is computed in similar way and tabulated as given in the table.

$$\text{The average or mean range } (\bar{R}) = \frac{\sum_{i=1}^{25} Ri}{25}$$

$$\therefore \qquad \bar{R} = \frac{0.88}{25} = 0.035 \text{ mm}$$

that is the centre line of R-chart is 0.035 mm,

therefore, $\text{LCL} = D_3 \bar{R}$

and $\text{UCL} = D_4 \bar{R}$

Table 7.2. Measurement of Pin diameter $\bar{X}$ and $\bar{R}$

Sample No	Pin Diameter in mm					Sum in mm	Mean value	Range variation
	1	2	3	4	5	Total	$\bar{X}$	$\bar{R}$
1	6.00	6.01	5.98	5.97	5.99	29.95	5.990	0.04
2	6.00	5.97	6.00	5.99	6.01	29.97	5.994	0.04
3	5.98	6.00	5.99	5.99	6.00	29.96	5.992	0.02
4	6.01	6.00	5.99	5.99	5.98	29.97	5.994	.03
5	5.99	6.01	6.01	6.00	6.02	30.03	6.006	0.03
6	6.03	6.00	6.01	6.02	5.99	30.05	6.010	0.04
7	6.03	6.02	6.01	6.01	6.00	30.07	6.014	0.03
8	6.02	6.04	6.00	6.02	6.01	30.09	6.018	0.04
9	6.01	6.00	5.98	5.97	5.99	29.95	5.990	0.04
10	6.02	6.01	6.01	6.00	6.02	30.06	6.012	0.02
11	5.98	6.02	5.99	6.00	6.01	30.00	6.000	0.04
12	6.00	5.99	5.98	6.02	5.99	29.98	5.996	0.04
13	5.99	5.98	6.02	6.00	5.97	29.96	5.992	0.05
14	6.02	6.01	6.00	5.99	5.98	30.00	6.000	0.04
15	6.01	6.00	6.03	6.02	6.01	30.02	6.014	0.03
16	6.04	6.02	6.02	6.01	6.00	10.09	6.018	0.04
17	6.01	6.01	6.00	5.99	5.98	29.99	5.998	0.03
18	6.02	6.00	6.01	6.00	6.00	6.00	6.006	0.02
19	6.02	5.99	6.01	6.01	6.02	30.05	6.010	0.03
20	5.99	6.00	6.04	6.01	6.01	30.05	6.010	0.05
21	6.01	6.03	6.01	6.04	6.00	30.09	6.018	0.04
22	6.02	6.02	5.99	6.01	6.02	30.06	6.012	0.03
23	6.03	6.02	6.03	6.04	6.01	30.13	6.026	0.03
24	6.01	6.02	6.03	6.04	6.01	30.11	6.022	0.03
25	6.02	6.03	6.01	5.99	6.04	30.09	6.018	0.05
						Total	150.160	0.88

The sample size is 5. d_3 and d_4 values from table are zero and 2.114 respectively. Hence, LCL will also be zero and UCL = 2.114 × 0.035 = 0.074 mm.

Construction of R-chart is basically plot range of variations in mm various sample number which is shown in Fig. 7.11.

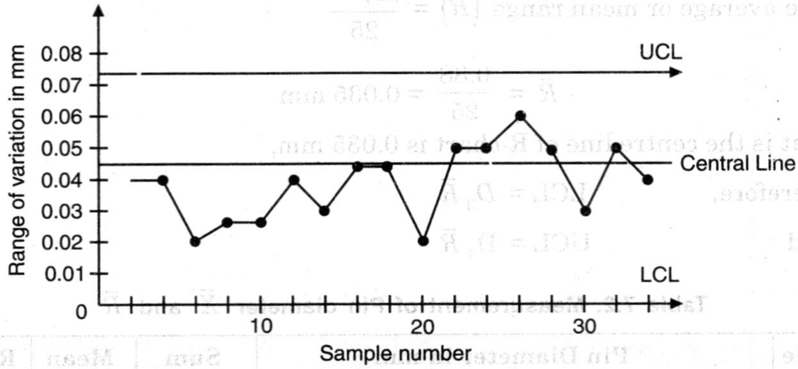

Fig. 7.11. Construction of R-chart

Analysis of R-charts

In the plot, all the points within control limits. Cause for any point lying outside is found and eliminated. These may one or two stray points outside the limit for which no reasons are assignable and these as chance points. Area under the normal curve between + 3 σ of the mean value is 0.997. That is 99.7% of all readings are likely to fall within + 3σ limits. A lower range means less variability which is good. A genuine reduction in range may be due to assignable causes such new tool or new material.

Construction of $\bar{X}$-charts

If R chart shows that the process is under control, then $\bar{X}$-charts are constructed. For sample number-1 values are:

5.99 mm, 5.98 mm, 6.01 mm, 5.97 mm, 6.00 mm,

$$\bar{X} = \frac{\sum_{i=1}^{i=5} X_i}{5} = \frac{29.95}{5} = 5.99\,\text{mm}$$

$\bar{X}$ for all samples is computed as shown the table. The overall mean or the grand mean is given by:

$$\bar{\bar{X}} = \frac{150.160}{25} = 6.006\,\text{mm}$$

The centre line of $\bar{X}$ chart is 6.006 mm,

Hence $\quad UCL = \bar{\bar{X}} + A_2 R$

and $\quad LCL = \bar{\bar{X}} - A_2 R$

where, A_2 is obtained from table depending on the sample size.

The sample n is 5 and the value of A_2 from table is 0.577.

$\therefore \quad UCL = 6.006 + 0.577 \times 0.035 = 6.026$ mm

$\quad LCL = 6.006 - 0.577 \times 0.035 = 5.986$ mm.

The plot of pin diameter versus sample number is shown in the Fig. 7.12.

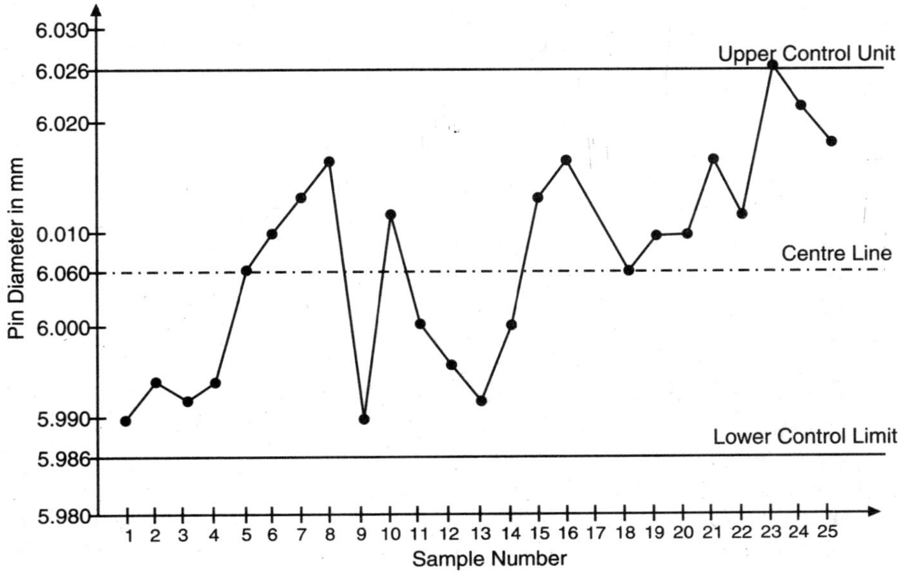

Fig. 7.12. Construction of $\bar{X}$ control chart

Analysis of Control Charts

Examination of $\bar{X}$ -chart shows that the manufactured pins will be under control process. It the majority of points are on one side of mean line, then the process needs setting.

The process setting may be shifted towards UCL or LCL to have better quality control. Any gradual ship. of majority points may be due to tool wear or similar reasons.

Process Capability Study

It is under after $\bar{X}$ and R control charts give satisfactory indication. It is done to ensure that tolerance is possible through process control. Frequency distribution is done for a new process or after considerable time process may need settling. It is better to study process capability before production. Process study capacity is defined as 6 σ deviations. This is for individual observations

$$\sigma_I = \frac{\bar{R}}{d_2}$$

Where, σ_I = standard diviation for individual observations.

R = Average range of samples size from table

d_2 = Factor as per sample

Process Capability = $6\,\sigma_I$

and $\sigma_I = \dfrac{\bar{R}}{d_2}$

$$\therefore \quad \text{Process capability} = 6\,\sigma_I = 6\,\frac{\overline{R}}{d_2}$$

$$R = 0.035, d_2$$

for $n = 5$ is 2.326.

$$\therefore \quad \text{Process capability} = 6 \times \frac{0.035}{2.326} = 0.090 \text{ mm.}$$

Use of Control Charts

If the initial construction of control chart indicates that process is under control w.r.t. range and also the central value, the control charts can be used. The control charts with centre line and control limits should be near the machine such that operator is able to see for reference while working. This way, process control becomes interesting to the operator who gets involved in quality improvement. The nature of process and criticality decide the periodicity of making sample checks. The sample checks will be made at shorter intervals for a process with volume production having fast process drift. After decision on periodicity intervals for sample checks, these are executed by supervisor and operator. The observations are plotted in the form of control charts. If the range or mean value does not change reasonably for long period of time, the control chart characteristics showing variable need be used to control the quality.

SUMMARY

1. Statistical quality control describes the set of statistical tools used by quality professionals.
2. The arithmetic average, or the mean is a statistic that measures the central tendency of a set of data. Mean is sum of all the observations and divided by the total number of observations.
3. The standard deviation is given by:

   Standard deviation, $$\sigma = \sqrt{\frac{\sum_{i=1}^{n}(X_i - \overline{X})^2}{n-1}}$$

   where, σ = standard deviation of a sample

   $\overline{X}$ = the mean

   X_1 = observation i, $i = 1, \ldots n$

   n = the number of observations in the sample
4. Control charts are used as tools in statistical process control. They can be used to measure any characteristics of a product, such as the weight of a chocolate box, the number of product, or the volume of bottled softdrink. There are different characteristics that can be measured by control charts can be divided into two groups: variables and attributes.
5. A mean control chart is also known as an X-bar chart. It monitors changes in the mean of a process. Multiple samples are taken and their means are calculated. These samples are small, about four or five observations.
6. Range charts monitor the dispersion or variability of the process. The method of developing and using R-charts is the same as that for X-bar charts.

7. P-charts are used to measure the defective proportion in a sample. The calculation of the center line as well as the upper and lower control limits is similar to the calculation for the other kinds of control charts.
8. C-chart is a control chart used to monitor the number of defects per unit. The types of units of measurement are a period of time, a surface area, or a volume of liquid.
9. Sampling plans are essential for acceptance. Acceptance sampling is a form of testing that involves taking random samples of lots or batches, or finished products and measuring them against predetermined standards. A lot or batch, of items can be inspected in several ways, including, the use of single, double, or sequential sampling.

EXERCISES

1. What is descriptive statistics? Explain.
2. Explain SQC.
3. Discuss statistical process control methods in brief.
4. Explain CL, UCL and LCL.
5. What is mean and range chart? Explain.
6. Discuss range charts.
7. What is single and double sampling?
8. Explain sequential sampling.
9. Discuss application of various charts.
10. Explain operating characteristics of quality curves.
11. What is process capability study? Explain.
12. Explain construction and analysis of c-charts.
13. What is P-chart? Explain.
14. Explain defect analysing by construction and analysis of charts.
15. Explain improvements by control charts.
16. What is variable sampling? Eplains.
17. Explain operating characteristics and its curves.

8 CHAPTER Eight

DEFECTS DIAGONSIS AND PREVENTION

CHAPTER OUTLINE

- Defect Study
- Product Quality
- Maintainability
- Failures
- Definition of Reliability
- Product Reliability Aspect
- Advantages of Reliability Engineering Program
- Availability
- Factors of Reliability
- Collection of Reliability Data from the Beginning
- Reliability Measurement
- Derivation of Reliability Function
- Zero Defect
- Removing Impediment to Quality
- Quality Circles
- 5S-Concept

8.1. DEFECT STUDY

Home appliances are made of few components, automobiles are made of several hundred components and aerospace vehicles are made of several thousand components. Any component failure makes the equipment or machinery non-functional. Just imagine during a critical war a missile fails due to a component failure, it would be disastrous. In short, failure of a component can lead to loss of time, money, inconvenience.

Modern systems consist of several sophisticated components which are connected intricately. Even breakdown of a component, may bring down complete system. It can be seen that reliability of any equipment or system is inversely proportional to its complexity.

Decision whether a system will accomplish its mission successfully, it becomes essential that an evaluation of system reliability is done. Earlier very high safety factors were used which added tremendously to weight of an equipment. Another approach followed was, to learn from failures of previous designs. However, these approaches are impractical for the new types of airborne and electronic equipments.

Nowadays, every design its drastically different from the previous one within a short span of time, thereby, not allowing enough time to learn from past mistakes. The re-design with institutive approach, gave a new approach to reliability. It was statistically defined, calculated and designed into pieces of equipment.

Inspite of proper planning and process control some defects may be found during assembly or final inspection. These defects have to be investigated to find the cause, for its elimination. Normally any process control analysis is a form of defect investigation Based on the defects, the cause is identified. This gives a clue to the corrective action. However, some defects have many possible causes. Thus, formal defect studies are carried out to known the real cause to take corrective action.

Defect studies cost money as it requires the time of engineers and technicians and test facilities. In some cases the products destroyed in testing. Hence, the defect studies are done an a selective basis. The defects can be sporadic and chronic. A sporadic defect generally signifies that some new factors has entered into the process. Such defects have to be investigated on priority. Chronic defects keep on occurring in a certain percentage of the product due to some unknown causes. Incidence of such defect are to be minimised if possible chronic defects are a large number of different types of defects varying magnitude. Some vital defects may be evaluated. The defects having highest losses need first attention. Defects resulting in serious failures of equipment in service are vital defects. There is loss of customer goodwill due to failures. The cost of the study should be compared with the expected savings due to the elimination of defects. An approval is taken for the sanction of necessary funds and the active collaboration of other departments like engineering and production.

A defect study requires proper planning. Its responsibility should be clearly defined. Generally a team is given responsibility. The team may have personal term quality control engineering production, etc. A study coordination is also identified. The investigation team must from thorough knowledge of the product. An indepth study of the product, its function, quality characteristics, operation, method of manufacture and the salient features of the process involved is done. During defect study of the product, final inspection results, process control data, final inspection results, process control data, material inspection results and material or process changes are recorded.

8.1.1. Identification of Defects

A defect is generally an apparent departure from the specified quality standard. In a complex product, the defective process stage has to be identified to find the basic cause. Study of symptom and their correlation with the defects can be used. In another approach, the assemblies or the system may be replaced one by one until the defect is removed. If the defective assembly has been identified, same process of elimination is followed, until the actual defective component is identified. Similarly, the defect in a product involving several processes, the process causing the defects has to be identified. Defects may be due to faulty casting, defective material, poor metal finishing or improper heat treatment. Finally, the process responsible for the defect is identified.

8.1.2. Analysis of Defect

The basic cause of the defect needs to be identified for remedial actions. The process is analysed for variable factors. If there are several variables, extensive measurement/testing is needed. Special tests may be needed in complex processes. The cause can be linked to non-conformance of certain quality characteristics. The analysis should lead to the ultimate reason of nonconformance. The ultimate reason may be the inadequate process capability of the machine, poor setting, a worn jig. Analysis of defect includes several factors. It may be improper design and specifications such as illegible drawing prints. It can be impractical design or incompatible component. It can be improper materials such as untested substandard materials. It can be improper operating and supervisory staff such as inadequate skill, unable to understand manufacturing drawings, or inadequate supervision. It can be improper process control and inspection such as inadequate process, improper test equipment, dated calibration, unskilled inspections.

8.1.3. Corrective Measures

Normally, the cause itself indicate the remedy a such as replacement of an obsolete drawing or calibration of equipment. In other cases, remedial measures may require considerable analysis. The process may be unable to achieve the specified tolerance. Remedy can be alternative method, relative assembly, or 100% inspection of the affected component, or provide wider tolerance. Consultation with all the concerned departments is done on the various possible methods of avoiding defects. The remedial measures may consider economics of various alternatives.

After corrective measures, it must be carefully monitored to see wheather the defect has been eliminated or minimised. This process of the adjustment of remedial measures and study of their effect is continued until the problem is completely resolved.

8.2. PRODUCT QUALITY

8.2.1. Definition

The quality of a product may have greater or lesser significance depending on the need and requirement of the user. The easiest way to define quality would be "The degree to which a product meets the requirement of a customer." Quality control includes all efforts to manage quality and maintain assurance of continued high quality of product or service. Then quality control conveys an idea about determining and maintaining that quality of the product or service which will satisfy the customer by its performance, cost and delivery. An item, even if produced to a level of quality higher than that demanded by the customer will not be appreciated if its cost is too high or if it is delivered too late.

"Customer satisfaction is the key to effective quality assurance."

In the year 1987, the ISO 9000 series was brought out as a guideline for developing a quality system. This help a company in designing, building and maintaining quality of a product thus satisfying the requirements of the customer on a consistent basis. A large number of companies, not only in the manufacturing sector but also in the service, education and training sectors all over the world have accepted these standards and have got the necessary accrediation from authorised agencies.

8.2.2. Variable Parameters

Inspite of the fast evolution of quality requirements, the business of industry are facing a severe challenge, which has to be faced with confidence.

The challenges are identified as follows:

(*i*) Increasing demand by the customer for quality.
(*ii*) More stringent demand for quality (Very strict that must be obeyed).
(*iii*) Increasing competition in the market, squeezing profit margins.
(*iv*) Changes in technology and materials used.
(*v*) Increasing wages and labour costs.
(*vi*) More vigorous laws for environment control.

8.2.3. Quality Control

Quality control has to be systematically introduced from the very beginning of production, or rather at the design stage and followed up till the finished product has been put into performance. Quality control deals with the whole system of production or service and methods which are employed to establish and achieve desired standards of quality and performance. Such a system of quality control is commonly termed as total quality control or TQC. Quality can affect all the vital functions of a company, such as cost, productivity, delivery, skills and expertise of the employees and management.

8.3. MAINTAINABILITY

The time taken to repair a system depends on how it has been designed. Further, the design and installation characteristics will also dictate the maintenance policy.

"Maintainability (M) is the probability that a system will be restored to operational (specified conditions) effectiveness within a given period of time when the maintenance action is performed in accordance with prescribed procedure."

It is a characteristic of the design and installation of the unit or system.

Maintenance is of two types:

(*i*) Preventive
(*ii*) Corrective

Both types of maintenance activity result in down time. In case of preventive (or scheduled maintenance). The down time is called scheduled interruption of service, whereas in the case of corrective (or unscheduled maintenance) it is called unscheduled or emergency interruption of service.

Meantime of maintenance action ϕ

$$\phi = \frac{\text{Total maintenance action time in hours}}{\text{Number of maintenance actions}}$$

The maintenance action rate is calculated as the reciprocal of the mean time of maintenance action and is expressed as:

$$\mu = \frac{1}{\phi}$$

Maintainability Equation

The maintainability equation is expressed as:

$$M = [1 - \exp(-\mu t)]$$

Since $\mu = \frac{1}{\phi}$,

This may also be written as:

$$M = [1 - \exp(-t/\phi)]$$

The second term in parenthesis represents the probability of not performing a maintenance action within the maximum available time t, therefore, $[1 - \exp(-\mu t)]$ is equal to the probability of completing the action in this time. The maximum allowable time, t is called the maintenance time constant.

8.3.1. Factors Affecting Maintainability

The factors which affect maintainability may be grouped under two major groups namely Design and Installation.

Some factors related to design are reliability, compatibility, visibility, repairability, accessibility and configuration.

The installation factors generally relate to the human being who is charged with the equipment, as well as with the associated environment. These may be listed as the experience, training, skill and supervision level of maintenance personnel as well as techniques used for maintenance and logistic support and other factors including environment, manuals, equipment overhaul, modification schedule, available test and calibration techniques, etc.

Factors which can be incorporated in the design of a system to achieve optimum maintainability are:

(*i*) Design for minimum maintenance skills.

(*ii*) Design for minimum tools.

(*iii*) Design for minimum adjustments.

(*iv*) Use of standard interchangeable parts/components.

(*v*) Grouping of sub-systems which can be easily located and identified.

(*vi*) Provision for visual inspection.

(*vii*) Use of colour coding for wires to facilitate tracing faults.

(*viii*) Use of overload indicators, alarms and lighted fuses.

(*ix*) Making parts and components easily accessible.

(*x*) Designing for safety using interlocks, safety covers and guarded switches.

8.3.2. Measure of Maintainability

The first step in measuring maintainability is to define its constituent namely down time. It is the interval during which the systems is not in an acceptable operating conditions. Down time is divided into three components:

(*i*) Active repair time

(*ii*) Logistics time

(*iii*) Administrative time

Active repair time is the time during which one or more technicians are actually working on repair, replacements and adjustments, etc. This time is sensitive to environment, technician skill level, procedure, etc. Logistic time is the down time in hours lost in waiting for parts or components needed to do the repair/placement.

Administrative time includes all down time which is not specifically allocated to either active or logistic time. This time is due to administrative procedure such as filling forms accounting and storage, etc.

Mean down time takes into account the down time due to preventive and corrective maintenance. The mean down time $\overline{M_t}$ is given by the relation

$$\overline{M_t} = \frac{F_c M_{c_t} + F_p M_{p_t}}{(F_c + F_p)}$$

where, F_p, F_c = No. of preventive and corrective maintenance tasks per 1000 hours respectively.

M_{ct}, M_{pt} = Mean corrective and preventive maintenance down time, respectively.

8.4. FAILURES

Continually measuring the reliability of a machine, product or process is an essential part of total Quality Management. When acquiring new machines, creating a new product, or even modifying an existing product, it is always necessary to determine the reliability of the product or process. One of the most powerful methods available for measuring the reliability of the product or process is FMEA. FMEA attempts to detect the potential product-related failure modes.

Any deviation in the properties of an item from the prescribed conditions is considered a fault and an item is considered to have failed under following three conditions.

(*i*) When it becomes completely inoperable.

(*ii*) When it is still operable but is no longer able to perform its intended function satisfactorily.

(*iii*) When serious deterioration has made it unreliable or unsafe for continued use.

Frequency of failures is a measure of reliability. If failure frequency is high, then the system is considered unreliable.

8.4.1. Causes of Failures

The sources of failures have many aspects of the product namely design, material selection, material interfection, fabrication and processing, assembly, inspection, testing, quality control, storage and shipment, service conditions, maintenance, unanticipated exposure to overload or mechanically chemical damage in service etc. The causes of failure are as follows:

(a) Deficiencies in Design

(i) Engineering efforts and changes in design:

This is due to making very little engineering efforts to avoid design features known to be conductive to failure for example.

Intersection of two mechanical notches. When changes in design are made without adequate consideration of possible introduction of stress raisers.

(ii) Upgrading of a Part

A new application in which a part must withstand service conditions more sever than its former use, sometimes may result in early failure.

(iii) Insufficient Design Criteria

Deficiency in design may result from impossibility of making reliable stress calculations for complex parts.

(b) Deficiencies in Material

Many failure originate due to imperfection in material as given under:

(*a*) Material imperfection reduces the overall strength of the material.

(*b*) It provide preferential path for propagation of cracks.

(*c*) Act as notches or provide paths for intergranular corrosion.

(c) Deficiencies in Processing

Few of processes that contribute to failure are cold forming, machining and grinding, identification marking, improper heat treatment, acid picking, electroplating, welding, re-working etc.

(d) Errors in Assembly

Failure in service sometimes result from error in assembly, that were not detected in inspection by the manufacturer or purchaser. Failure may also occur due to an operator's negligence.

(e) Improper Service Conditions

Abnormal severe conditions of speed, loading temperature and chemical environment is often a major contribution to occurrence of service failures.

(f) Inadequate Maintenance

Maintenance procedures should be thoroughly revaluated when failures recur despite scheduled maintenance.

8.4.2. Types of Failures

Any system which is well designed, well engineered, thoroughly tested and properly maintained should in operation. It has been observed that in practice, a well designed, well engineered, thoroughly tested and properly maintained systems, also fail sometimes. There are three failures which may occur during life cycle of a system. These are early failures, random failures and wear out failure. But failures caused by damage due to careless handling, storage, improper operation by the user or change in operating conditions.

(*a*) Early Failures: Variation in manufacturing process and poor quality control techniques during production cause early failures and these failures occur in the early life of system operation. In electronics and software based systems, early failures may be eliminated by debugging or burn in-process. Debugging is done before releasing the system for actual use. The system is run continuously for number of hours under stimulated operating conditions and corrections are made to overcome the failures.

Normally, the production units test the system for a pre-determined period, and release the system make necessary modifications. Then, release the system with a understanding that of system failures, it will be under warranty for an estimated time. The warranty period estimation is done on the basis of statistical distribution.

(*b*) Chance Failures: During the actual working of the system, the chance failures are pre-dominant. These failures are random, irregular and unexpected. Sudden stress accumulations beyond the design strength of the component can be one of the reasons of chance failures. Although, prediction of chance failure is not possible, but it follows certain rules of collective behaviour such that the frequency of their occurrence during sufficiently long periods is nearly constant. Useful life of the system is the phase during which only chance failures occur. Reliability techniques like duplicating the components, incorporating the safety margins in design, etc. reduce the occurrence of chance failures.

(*c*) Wear Out Failures: Aging or wearing out of components cause wear out failures. Such failures are due to improper maintenance or no maintenance, and the failure became frequent. These failures can be arrested for sometime by preventive maintenance of the system at particular intervals. The period between two preventive maintenance should be shorter than the mean wear out life of the system.

(*d*) Life Cycle Curve: The studies by manufacturers, research institutes and other agencies have provided reliability data. Based on these, life cycle consists of three distinct failures. There is high frequency of failure during the testing, *i.e.*, first phase. It drops off rapidly as the failed components are replaced by good and tested component. The second phase is during actual operation or the useful life of the system where in chance failures take place. Next period is wear out or aging when failures starts increasing. The life cycle curve is shown in Fig. 8.1.

The past data from experimentation of reliability or observation during in-service operation are used in establishing the probability distribution of these failure patterns. Then, it become possible to measure and predict the frequency of each type of failure from the sample data.

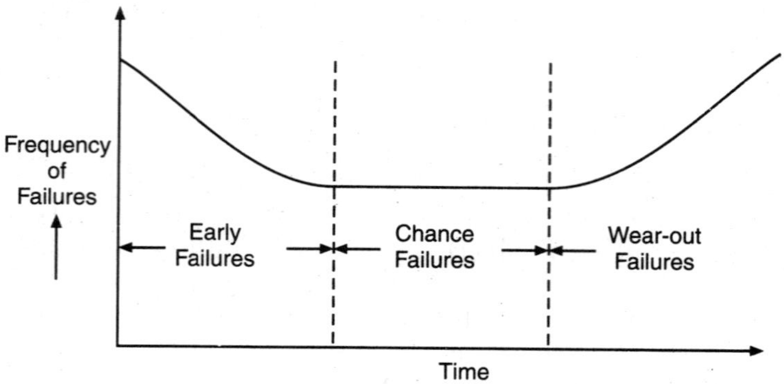

Fig. 8.1. Life cycle curve of system

8.4.3. Modes of Failures

(*a*) **Catastrophic Failures:** System becomes inoperative suddenly due to catastrophic failures. It can be like a fuse blowing in a circuit after several hundred hours of operation. Such failures are random. If the failure rate is known, then one can statistically predict the probability of occurrence of one or more catastrophic failures for any time period.

(*b*) **Degradation or Creeping Failures:** These failures occur gradually due to change of some parameter with time, *e.g.*, a drop in power output can occur due to a decrease in the value of a transconductance in a radio tube. An excessive humour frequency shift in a receiver can occur due to a change in the value of the resistance of a resistor. Proper inspection can detect such failures and subsequently eliminated.

(*c*) **Independent Failures:** Such failures are not generated by other failures. It can be catastrophic, degradation or wear out type. Say, failure of a fan belt of an automobile is no way related to the puncturing of the tyre of the wheel.

(*d*) **Secondary Failures:** Some primary failure can cause secondary failure. Say, a tyre burst can cause breaking the spokes in a motorcycle travelling at high speed. Thus, breaking of spoke is a secondary failure. If a resistor in an electronic circuit is shorted, then it may cause an excessive drain on the tube. Consequently, tube failure is a secondary failure.

8.5. DEFINITION OF RELIABILITY

There are numerous definitions of reliability. The important definitions of reliability which are comes across quite often are:

(*i*) It is the integral of the distribution of probabilities of failure-free operation from the instant of switch on to the first failure.

(*ii*) Component or system reliability is the probability that the component or system will not fail for a time *t*.

(*iii*) It is probability that a device will operate without failure for a given period of time under given operating conditions.

(*iv*) It is a mean operating time for a given specimen between two failures.

(v) It is system capacity for failure free operation for a definite of time under given operating conditions and minimum time lost for repair and preventive maintenance.

It is arbitrarily assumed to be the equipment capacity to maintain given properties under specified operating conditions and a given period of time.

However the reliability definition given by the Electronic Industries Association (EIA) USA states that:

"Reliability is the probability of an item performing its intended function over a given period of time under the operating conditions encountered."

The above definition has following important elements:

(a) Probability

(b) Intended Function

(c) Time

(d) Operating Conditions

These are being considered one-by-one.

(a) Probability

Some items may have very short life, some very long life and a group of items may have a certain average life. Frequency distribution is used to identify life of each item. Suppose the probability of survival of an item operating for 15 hours is 0.80, it indicates that we could expect the item would be operating only 80 times out of 100 after a 15 hour operating period.

(b) Intended Function

There are several occasions where in one component failure does affect the main function considerably. But there are cases where a component failure may lead to complete breakdown of the system or equipment or machinery and intended function cannot be achieved. Simplest example can be an automobile where in if a spark plug fails, the automobile is still functional. In case, engine fails, automobile becomes non-functional.

(c) Time

It is period during which the system or machinery is expected to function without failures. It can be seconds, minutes, hours, days, months or years depending on the system or machinery.

(d) Operating Conditions

It can be temperature, humidity, pressure, shock, vibration, gravity. All these parameters have to be in a specific range for normal functioning.

8.6. PRODUCT RELIABILITY ASPECT

Reliability engineering covers all aspects of a product's life, from its conception, subsequent design and production processes, through its practical use lifetime, with maintenance support and availability. Reliability engineering covers:

- Reliability
- Maintainability
- Availability

(a) The Reliability Bathtub Curve: Most products exhibit failure characteristics as shown in the bathtub curve of Fig. 8.2.

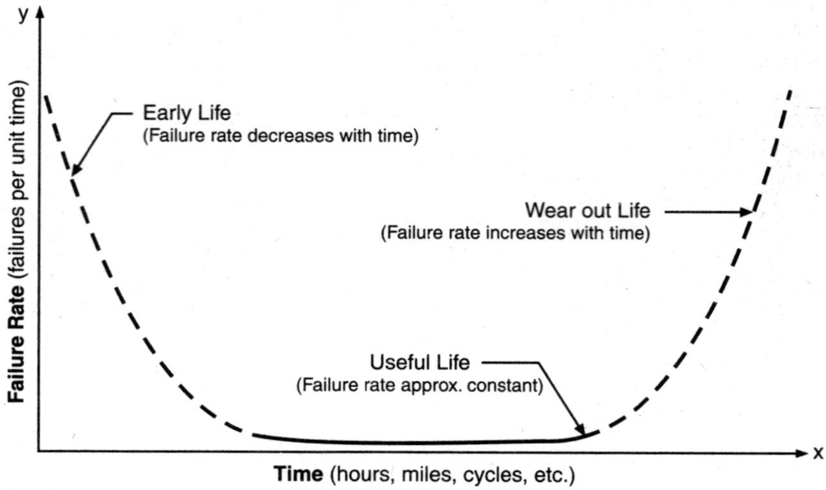

Fig. 8.2. An idealised reliability bathtub curve with the three major life regions: early, useful and wear out

This curve is plotted with the product life on the x-axis and with the failure rate on the y-axis. The life can be in minutes, hours, years, cycles, actuations or any other quantifiable unit of time or use. The failure rate is given as failures among surviving units per time unit. As can be seen from many products will begin their lives with a higher failure rate which can be due to manufacturing defects, poor workmanship, poor quality control of incoming parts, etc., but exhibit a decreasing failure rate as shown in the plot. The failure rate then usually stabilises to an approximately constant rate in the useful life region, where the failures observed are chance failures. As the products experience more use and wear, the failure rate begins to rise as the population begins to experience failures related to wear-out. In the case of human mortality, the mortality rate, is higher during the first year or so of life, then drops to a low constant level during our teens and early adult life and then rises as we progress in years.

Burn-In

Looking at this particular bathtub curve, it should be fairly obvious that it would be best to ship a product at the beginning of the useful life region, rather than write-off the production line; thus preventing the customer from experiencing early failures. This practice is commonly referred to as burn-in, and is frequently performed for electronic components. The determination of the correct burn-in time requires the use of reliability methodologies, as well as optimisation of costs involved. That is costs of early failures versus the cost of burn-in is used to determine the optimum failure rate at shipment.

(b) Minimising the Manufacturer's Cost: Figure 8.3 shows the product reliability on the x-axis and the producer's cost on the *y*-axis.

If the producer increases the reliability of his product, he will increase the cost of the design and/or production of the product. However, a low production and design cost does not imply a low overall product cost. The overall product cost should not be calculated as merely the cost of the product when it leaves the shipping dock, but as the total cost of the product through its lifetime. This includes warranty and replacement costs for defective products, costs incurred by loss of customers due to defective products, loss of subsequent sales, etc. By increasing product reliability, one may increase the initial product costs, but decrease the support costs. An optimum minimal total product cost can be determined and implemented by calculating the optimum reliability for such a product. The total product cost is the sum of the production and design costs as well as the other post-shipment costs. It can be seen that at an optimum reliability level, the total product cost is at a minimum. The *optimum reliability level* is the one that coincides with the minimum total cost over the entire lifetime of the product.

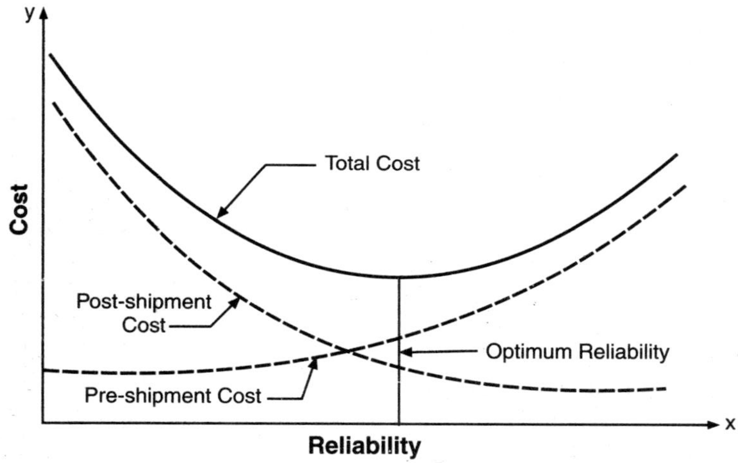

Fig. 8.3. Total product cost vs. product reliability

8.7. ADVANTAGES OF RELIABILITY ENGINEERING PROGRAM

The following list presents advantages that can be obtained with the implementation of a sound reliability program:

- Optimum break-even period.
- Optimum warranty period and estimated warranty costs.
- Optimum preventive replacement time for components in a repairable system.
- Spare parts requirements and production rate, resulting in improved inventory control through correct prediction of spare parts requirements.
- Better information about the types of failures experienced for parts and systems that aid design, research and development efforts to minimise these failures.
- Establishment of system to know failure occurance time in the life of a product and better preparation to cope with them.
- Studies of the effects of age, service duration and application and operation stress levels on reliability.

- A basis for comparing two or more designs and choosing the best design from the reliability point of view.
- Evaluation of the amount of redundancy present in the design.
- Estimations of the required redundancy to achieve the specified reliability.
- Guidance regarding corrective action decisions to minimise failures and reduce maintenance and repair times, which will eliminate overdesign as well as underdesign.
- Help provide guidelines for quality control practices.
- Optimisation of the reliability goal that should be designed into products and systems for minimum total cost to own, operate and maintain for their lifetime.
- The ability to conduct trade-off studies among parameters such as reliability, maintainability, availability, cost, weight, volume, operability, serviceability and safety to obtain the optimum design.
- Reduction of warranty costs or for the same cost, increase in the length and the coverage of warranty.
- Establishment of guidelines for evaluating suppliers from the point of view of their product reliability.
- Promotion of sales on the basis of reliability indexes and metrics through sales and marketing departments.
- Increase of customer satisfaction and an increase of sales as a result of customer satisfaction.
- Increase of profits or for the same profit, provision of even more reliable products and systems.
- Promotion of positive image and company reputation.

8.8. AVAILABILITY

The availability of a system is the probability that it is operating satisfactorily when called upon to perform its function. It takes into account both failures and repair of system.

The term reliability is concerned with the system's capability of survival, whereas maintainability is related to the system repair by combining, there two factors, a family of parameters is obtained which indicates the system availability.

Example: A telephone subscriber would judge the availability of telephone service on the basis of the performance of the hand set. If it operated well with little delay most of the time, he would conclude that he had a high telephone availability. There might have been failures of some switches in the system, but when there failures occur, the switches take over and complete the call due to the automatic switchover redundancy without affecting the system's operation. On the other hand, the subscriber's success in completing his call instead of being due to redundancy alone, might also be due to a very high equipment reliability which results in a reduced probability of failure. Thus, it may be seen that availability actually consists of two components. Maintainability and reliability. This means that poor reliability can be offset by correspondingly improved maintainability. The faster the maintenance action rate, the better is the resulting availability.

Given below various definitions of availability according to their applicability.

(*i*) Intrinsic Availability (A_i) (Inherent Availability)

$$A_i = \frac{\text{Operating time}}{\text{Operation time + Active down time}}$$

$$A_i = \frac{\text{MTBF (Mean Time Between Failures)}}{\text{MTBF + Mean time waiting for spares + Mean time for repair + Mean administrative time}}$$

$$A_i = \frac{\text{MTBF}}{\text{MTBF + MTTR}}$$

(*ii*) Equipment Availability

$$A_a = 1 - \exp(-\mu t_m)\,[1 - \exp(-\lambda T)]$$

where t_m = Maintenance action time,
T = Mission time,
μ = Maintenance action rate,
λ = Failure rate

$$A_a = 1 - \exp[-\mu/\text{MTTR}]\,[1 - \exp(-T/\text{MTBF}]$$

where MTTR = Mean time to repair
MTBF = Mean time between failures.

EXAMPLE 8.1.

For an four stroke car, the following data was obtained

MTBF = 600 hr.

Mean time waiting for spares = 10 hr.

Mean time for repair = 50 hr.

Mean administrative time = 5 hr.

Calculate the availability of the automobile

Solution: We have availability

$$= \frac{\text{(Mean Time Between Failures) MTBF}}{\text{MTBF + Mean time waiting for spares + Mean time for repair + Mean administrative time}}$$

$$= \frac{600}{600 + 10 + 50 + 5} = \frac{600}{665} = 0.9022$$

90.22% is the availability of the car.

8.9. FACTORS OF RELIABILITY

Life characteristics operating conditions and the failure distribution help in prediction of system reliability. Hence, the first step in reliability prediction is life characteristics determination.

A number of samples will fail successively in time, if a random sample of items are taken from a population and are put to test or use under a set of fixed or given environmental or operating conditions. The life length will be represented by the data so obtained. Repairability or non-repairability decides measurability of life length. The life of a repairable item can be measured by failure rate or Mean Time Between Failures (MTBF).

***(a)* Failure Rate:** Failures per unit time is the failure rate. It can be expressed in failures per hours or failures per 100 or 1000 hours. It is defined as:

$$\text{Failure rate, } \lambda = \frac{f}{T}$$

where f = Number of failures during the test interval

T = Total test time

Failure is almost constant if the design is stabilised. The reliability of the system is higher if the failure rate is smaller.

***(b)* Mean Time Between Failures (MTBF):** When failure rate is almost constant, MTBF is reciprocal of the failure rate or the ratio of the total test time T and the number of failures.

$$\text{MTBF, } m = \frac{1}{\lambda} = \frac{T}{f}$$

MTBF is also known as the average time of satisfactory operations of the system, Reliability of the system is higher if the MTBF is larger.

***(c)* Mean Time to Failure (MTTF):** If $t_1, t_2, t_3, ..., t_n$ are failure times of n items in a life test, then the mean time failure is given as:

$$\text{MTTF} = \frac{1}{n} \sum_{i=1}^{n} t_i$$

If the item is non-repairable, MTTF is used.

There is a basic advantage of collecting failure data. These help to predict or calculate the reliability of a system whenever it is made to function under circumstances which these data represent. Enhancement of product reliability is another reason.

Failure analysis helps in determining the basic reason of failure, and based on its determination and subsequent corrective action.

At the time of planning data, objective of information should be first define. This helps such as technicians, inspectors, machinist etc., who are involved with reliability data to know what information is needed.

One must be specified sources of data to make ensure that data is unbiased.

Evaluation procedure must be decided before hand, test programme must also be identified.

8.10. COLLECTION OF RELIABILITY DATA

Failure analysis should be directed from the begining towards finding out specific information on the manufacturing, processing and service histories and photographic records of failed components or system, etc.

Background data collection on manufacturing and fabricating history of component needs to begin with getting specification and drawing. It should cover all design aspects of the components.

Manufacturing and fabricating related data may be Mechanical Processing, Thermal Processing, Chemical Processing.

Data can be based on service history. Special attention need to be given to environmental details. For example, normal and abnormal loading, accidental overloads, cyclic loads, variations in temperature, pressure, etc.

Analyst needs to decide whether photograph of the failed component or system is required or not. It is better to have all photographs (4″ × 5″ minimum) taken by a large camera to permit detailed enlargement.

Analyst needs to ensure that the samples are suitable for intended purpose and that they sufficiently represent the characteristic of failure.

Realiability Data Usage

Reliability data usage are as follows:

(*i*) The information regarding the reliability requirements w.r.t maintainability, availability and mean life.

(*ii*) The causes of failure and relegate them to their proper perspective can be highlight whether these failures are due to design, usage or other factors.

(*iii*) The areas needing improvement, modifications, effect of sub-systems and interconnecting links can also be determined.

(*iv*) A means of comparing results with those obtained by simulated tests in factory and establish correlation between them can also be provided by field data.

(*v*) Deviation, if any, between simulated field tests can be found out using this information.

(*vi*) The information about logistics, maintenance, operations, administrative delays and spares requirements can also be provided by using reliability data.

(*vii*) Estimation of the degradation and wear out characteristics of components can done using reliability data.

Reliability Data Sources

There are many sources for collecting data. Some of sources at various levels are as follows:

(*i*) Report on raw material

(*ii*) Report of standard part supplies

(*iii*) Inspection record receiving

(*iv*) Shop order and inspection record

(*v*) Report of welding and *X*-ray

(*vi*) Log sheets of test

(*vii*) Report of failure

(*viii*) Build ups and assembly logs

(*ix*) Report on field failure

(*x*) Report of service trouble

(xi) Records of maintenance
(xii) Department of quality control
(xiii) Market research
(xiv) Experience of customers

Some agencies have created/collected failure/reliability data on specific component/equipments. Such data can be used for future design, development and production of similar type of components.

8.11. RELIABILITY MEASUREMENT

Determination of life characteristics is the initial step in reliability prediction.

Measurement of the life length depends on whether the item is repairable or non-repairable.

In case of repairable items, measurement of life can be done by failure rate. (λ and MTBF)

In case of non-repairable items, measurement of life can be done by mean time to failure (MTTF).

(a) Failure Rate (l)

Failure rate is computed as simple ratio of the number of failures, F, during a specified test interval to the total test of the items under going test.

$$\lambda = \frac{F}{T} \qquad ...(i)$$

where, λ = Failure rate
F = No. of failures during the test interval
T = Total test time.

If the failure rate value is smaller then the reliability of system is higher.

(b) Mean Time Between Failure (MTBF)

It is reciprocal of the constant failure rate or ratio of test time due to the number of failures

$$m = \frac{1}{\lambda} = \frac{T}{F} \qquad ...(ii)$$

MTBF is also taken as average time of satisfactory operation of system. Largest the MTBF, higher the reliability.

(c) Mean Time to Failure (MTTF)

If we have life test information on n items with failure times $t_1, t_2 \ldots\ldots\ldots t_n$, then MTTF is defined as:

$$\text{MTTF} = \frac{1}{n}\sum_{t=1}^{n} t_i$$

8.12. DERIVATION OF RELIABILITY FUNCTION

Suppose an event can occur in N mutually exclusive and equally likely ways, n of which has the characteristic A. In such case, the probability of A occurrance is n/N. Mathematically,

$$P(A) = \frac{n}{N} \quad ....(8.1)$$

The reliability or Probability of survival can be expressed as a function at any time t during test.

Consider that m is the number of cases for which event A will not occur.

$$\text{Realiability, } R(t) = \frac{n}{N} = \frac{n}{(n+m)} \quad ...(8.2)$$

where, n and m are counted at specific time t.

Expression for probability of failure at any time t is given by:

$$F(t) = \frac{m}{N} = \frac{m}{(m+n)} \quad ...(8.3)$$

Thus, $\quad R(t) + F(t) = 1 \quad ...(8.4)$

Equation (8.2) can be written as:

$$R(t) = \frac{n}{(n+m)} = \frac{(N-m)}{N} = 1 - \left(\frac{m}{N}\right) \quad ...(8.5)$$

Differentiating equation (8.5) w.r.t t, we get

$$\frac{dR(t)}{dt} = \frac{d}{dt}\left(1 - \frac{m}{N}\right) = -\left(\frac{1}{N}\right)\frac{dm}{dt} \quad ...(8.6)$$

Rearranging it, we get:

$$\frac{dm}{dt} = -N\,\frac{dR(t)}{dt} \quad ...(8.7)$$

which is the rate at which component fails. Since, $m = N - n$, after differentiation we get

$$\frac{dm}{dt} = \frac{d}{dt}(N-n) = -\frac{dn}{dt} \quad ...(8.8)$$

Equation (8.8) gives the negative rate at which component survive. Dividing equation (8.7) by n on both sides, we have

$$\frac{1}{n}\frac{dm}{dt} = -\frac{N}{n}\frac{dR(t)}{dt} \quad ...(8.9)$$

Left hand side of equation (8.9) is failure rate $\lambda(t)$ and

$n/N = R(t)$, thus:

$$\lambda(t) = -\frac{1}{R(t)}\frac{dR(t)}{d(t)} \quad ...(8.10)$$

Equation (8.10) is re-arranged as follows:

$$\lambda(t)\,dt = -\frac{dR(t)}{R(t)} \quad ...(8.11)$$

or $$\int_0^t \lambda(t)\,dt = -\int \frac{dR(t)}{R(t)} = -\ln R(t)$$

$$\ln R(t) = -\int_0^t \lambda(t)\,dt$$

$$R(t) = 1 \quad \text{at } t = 0,$$

$$R(t) = \exp\left[-\int_0^t \lambda(t)dt\right] \qquad ...(8.12)$$

Equation (8.12) is a general formula for computing reliability. Here, equation $\lambda(t)$ can be variable. However, if we specify that $\lambda(t)$ is constant over time and $\lambda(t) = \lambda$(say), then reliability function becomes:

$$R(t) = \exp[-\lambda t] \qquad ...(8.13)$$

Also,

$$R(t) = \exp[-t/m] \quad \text{as} \quad \lambda = \frac{1}{m} \qquad ...(8.14)$$

8.13. ZERO DEFECT

Crosby first introduced this concept in his absolutes for quality management. The purpose of the system is to eliminate identified defects by prevention of occurrence to a maximum extent, if not completely. The spirit of zero-defect is prevention of defects by doing it right first time and every time. This leads to minimization of defects to bare minimum and tending to zero with drive for continuous improvement. What this programme achieves is continual lowering of defect level by following different improvement tools. Japanese industries successfully used this technique as a strong business driver in their global competitiveness. An outstanding example is the "Six-Sigma" movement of Motorola, which aims to continually decrease the defect level in the process to three to four parts per million and yet again considerable cost, product and performance superiority over the competitors. The important direction is not what one does but how it does it. The zero-defect programme is to continually bring down the defect levels in the process or a product to near zero value.

Zero defects is a concept which promotes a constant, conscious desire to do a job right the first time. It is management technique where the job is taken as a challenge by the individual worker. He is given all facilities to meet the challenge. Achievers receive public recognition which motivates to excel in respective job leading to pride in the workmanship. This is just integration into a formalised programme. The zero defect programme has been applied in many US companies with good success. General electric has made contribution to the development of the zero defects approach to quality improvement.

"Doing it right the first time" is less expensive than the costs of detecting and correcting nonconformities. Four absolutes of quality management are: quality is conformance to requirements, prevention of nonconformance is the objective not appraisal, the performance standard is zero defects not "that's close enough," and the measurement of quality is the cost of nonconformance.

8.13.1. Zero Defects Programme Organisation

A careful preparation and an efficient organisation is required for the successful execution of the zero defects programme. It embraces all departments and levels from the chief executive to the shopfloor worker. The full faith and support of the chief executive is the pre-requisite for this programme.

The zero defects programme involves motivation of the management supervisors workers, co-operative actions by the management and workers, laying down targets, measurements of result, and recognition of achievers.

A whole time programme administrator along with some staff are required for planning, execution and coordination of the activities. The ability of the administrator is key in the success of the zero defect programme. A dynamic personality enjoying the respect of the management and worker is required. Full faith in the programme and a through knowledge of its administration is needed. The administrator needs be assisted by a committee comprising of representatives of all major departments. This enables the department to participate in the administration of the programme.

8.13.2. Management Motivation

Motivation efforts are first towards the managerial staff. A series of briefings are initiated by the Chief Executive to show his support to the programme. The programme administrator does the actual explaining of the aims and conduct of the programme. The briefing may be arranged at various levels if the managerial staff is too large. The chief executive briefs the top level managers. The assistance of the programme administrator may be taken. There after each manager briefs staff. The briefing session should not be one way affair where the higher management issues just orders and instructions. A relaxed atmosphere should exist for a genuine discussion. Doubts on its efficacy should be removed by logical reasoning and examples of companies should be given.

8.13.3. Supervisor and Workers Indoctrination

Indoctrination and motivation should be for the shop floor staff and workers. First, quality awareness should be done by educating on the company's product and uses. The defect affect the product performance and the customers. The defects can be avoided by conscious action. The use of posters, brochures, informal shop discussions, etc. would be required for the indoctrination campaign.

The campaign should be individual oriented. The individuals contribution to the quality of the final product, dramatising the effect at the user's end, of a seemingly minor error on the production worker's part should be highlighted in every poster. The support of the trade union should also be ensured. The zero defect programmes enhances the dignity of the man and his job, therefore, it have been welcomed by the trade unions. A zero defect programme is used as a reward for superior achievement.

The employee may be asked to volunteer themselves for participation in the programme. Signing zero defect pledge cards to strive for achieving higher quality and superior workmanship is very important. The signing of pledge cards should be voluntary and persuation by explaining the aims and the benefit of the programmes. The programme administrator should not unduly wrong even if some employees do not sign the pledge cards. They are likely to join if they see the programme working. Those who sign up for the zero defect programme should be recognised by giving them zero defects badge or a pin should be worn like a decorative piece of jewellary which employee may want to wear. An additional motivating factor in making employees sign up for the programme, will also happen if the badge is attractive enough.

8.14. REMOVING IMPEDIMENT TO QUALITY

First management do everything to eliminate the came of error controlled by them before asking workers to produce defect free jobs. Still some causes may not be in the awareness of the management. The workers may know some causes of errors, actual or potential. The causes should be eliminated on the basis of feedback from employees.

8.14.1. Error Cause Removal (ECR)

The actual or potential causes of error are identified by the workers and reported to the management for their elimination. Standardised forms are used with three parts. Part-I of the form giving in brief the cause of errors, and suggestions if any is filled by the workers. Supervisor checks the entries for completeness and gets more information if necessary to properly identify the causes. The supervisor signs the part II of the form and returns it to the worker as receipt of suggestion. The supervisor studies the cause of error and takes suitable action. Most of the cases are removed by the supervisor or the shop foreman. The action taken is entered in part III of the form and the worker is informed through a copy of part-III form. The zero defect programme administrator gets the other copy for rewarding the employees. In case the cause of error cannot be removed the ECR form is given to the programmer administrator for appropriate action at higher level. Action should be taken expenditiously and the originator of the suggestion should be informed. ECR helps in the removal of impediments to error-free performance and also serves as a strong motivating factor.

8.14.2. Setting Targets

Quality awareness should be made to every individual as he can contribute. Individual and group quality targets may be presented as a challenge. A dull repetitive job becomes interesting as the worker does the job without any error to prove that he met the challenge. There should be careful solution of the target. Too low targets are not challenging or too high targets will reduce enthusiasm. A target should be realistic and achievable. The quality target should be dynamic based on experience. The revision is normally in the upward direction and the ultimate aim being a zero defect product.

8.14.3. Measurement of Achievements

The zero defects programme motivates workers properly. The measurement of the result of their efforts will be desirable by the employees. The fear of poor performance is removed in zero defets programme due to rewards for superior achievement no punishment for poor performance after necessary efforts. This positive approach ensures success of zero defect programmes. Method of measurement or criteria of checking achievement should be simple and objectives easily understandable. The judging should be fair.

8.14.4. Achievers Recognition

Recognition should be selective. Only really outstanding performance should be recognised. The award of a badge, medal or certificate should be personally given to the employee preferably by the Chief Executive. Achievement of employees should be well published as it serves as a morale booster, and an inspiration.

8.14.5. Application of Zero Defect Programme

Zero Defect programme has a valuable motivational technique. This can be used in any industry for quality improvement. It cannot cure-all as it solves quality problems based on causes factors controlled by management. First management should have provided all necessary means of quality control and the proper state of mind of the worker is achieved through this.

8.15. QUALITY CIRCLES

There is specialisation in the manufacturing job into design, process engineering, tooling and actual production. All creative and innovative ideas for quality improvement come from the designer and the process engineer. Japanese realised that creative power was not limited to the designers and engineers only. The shop floor worker manufacturing the product can contribute in the quality improvement programme. For this proper environment and motivation is needed. The concept of quality circles was introduced by Dr. Ishikawa, a Japanese quality expert. This calls for the participation of maximum number of people engaged in industry in quality control programmes. It discriminated the message of quality down below. Quality consciousness grew bottom spread horizontally with the encourangement from the top. The basic rules are based upon the SQA techniques.

The concept of quality circle was first introduced in Japan in 1962. "Quality Circle" (QC) was defined as: a small group of working people who, under the leadership of their own foreman or supervisor, are trained to identify, analyze, and solve quality related problems on a volunatory basis and present their solutions to their managers. The idea is to train and stimulate the people of a work area for solving their problems on their own by teamwork, brainstorming and learning to be creative in the workplace.

By 1967, there were 10,000 QC registered in Japan; which further grew to the level of 20,000 in 1969. Soon, it become part of the "total quality" movement globally, and drew keen attention of all "Quality Masters" including Dr. Jaran, because of its tremendous power of getting people of the workplace involved in quality improvement programmes.

The steps of work for a quality circle group are to:

(*i*) Select the problem in areas of common interest.

(*ii*) Name and register the circle with the "Quality Circle Steering Committee."

(*iii*) Assign responsibility to each member.

(*iv*) Decide on the schedule of the meeting and communicating to all concerned.

(*v*) Collect data and do brainstorming on the problem and possible solutions.

(*vi*) Train QC members as per the needs.

(*vii*) Find out right solutions.
(*viii*) Fine tune the solution.
(*ix*) Collect data and do analysis of improved methods/solutions.

The important features of quality circles are:

(*i*) The group is comprised of people who do similar or connected work.
(*ii*) The size be limited to preferably 6–10 or 3–12 at most.
(*iii*) "The quality circle" should have an inspiring name, and registered with the company.
(*iv*) Group should meet regularly for atleast about an hour per week.
(*v*) Own supervisor of the team should lead the group.
(*vi*) The group must make presentation of their ideas to management.
(*vii*) The group should take the responsibility for implementation of the improvement ideas.

In effect, quality circle work brings a cultural change in the work environment whereby people at all levels in the organization become a willing partner of the progress. In India, many companies namely BHEL, Tata Group, etc. have successfully introduced regular "quality circles" at the shop-floor levels as self motivated groups for changing the work culture and effecting long-term improvement.

Quality circle concept benefits are not only quality or cost improvement, but it also changes the work culture and can be applied to all areas of business. Successful quality circles members become the real "change agent" of the work environment that TQM quality system demands. Quality circles can potentially lead to quality revolution in an organization.

8.15.1. Definition and Objectives of Quality Circle

Quality circle is small group of employees working together who meet each other for analysing and resolving work related problems. The meeting are informal which are held regularly. There are 5 to 10 men in a group. The meeting is held at least once a week for one hour. At the meetings talks are only about the work and problems. The administrative or personal problems are not discussed in this forum. Quality circle is voluntary by nature and all the members are highly motivated. Objectives of quality circles are:

(*i*) improve quality and productivity,
(*ii*) make promote quality, safety and cost consciousness among workers.
(*iii*) employees learn new techniques of identifying technical problems.
(*iv*) solve the problems in the work area which gives job satisfaction
(*v*) improve employees motivation
(*vi*) inspire for effective team work
(*vii*) improve communications and develop healthy relationships.
(*viii*) development of a complete problem solving methodology.

8.15.2. Quality Circle Functioning

The members of a circle are from the same work area or who perform similar work. As such the problems seluted are familiar to all and it is easy to find solutions. A circle may also comprise of members from different work areas wherein the problem is studied from all angles to arrive at optimum solutions. The membership is voluntary. A circle can have 5 to 10 members. The size must be small so that each member participates and contribute in the meetings. There can be more than one circle in a given work area. Good co-ordination should be maintained between the leaders of these circles and the facilitator to avoid duplication of activities in the same work area.

8.15.3. Quality Circle Structure

Top management, steering committee, co-ordinator, facilitators, leaders, members and non-members constitute the structure of quality circles.

(i) **Top management:** Should provide all the necessary support through personal presence at major activities of quality circles. The top management has to ensure the success of implementation of QC concept. The top management should have understanding and faith in the concept of Quality Circles, provide budget for the expenditure, encourage competition by instituting an award.

(ii) **Steering committee:** The steering committee consists of the chief executive, divisional heads of departments and the coordinator. The chief executive acts as chairman and the coordinator acts as secretary of the steering committee. The meetings are convened at least quarterly. The steering committee promotes quality circles, develops working methodology, gives the guidelines, establishes the programme objectives and resources. It provides policy guidelines and directions, nominates coordinator and facilitators. It obtains feedback from facilitators and reviews activities. It provides necessary budget, decides on awards and also reviews the overall working.

(iii) **Coordinator:** The activities of quality circles are coordinated by coordinator and makes the operation of quality circles smooth and effective. Coordinator function as registers quality circles, has liaison with the facilitators. It convenes steering committee meetings and circulates record notes. It gives assistances to circles, whenever asked for. It organises top-management presentations, training programmes for facilitators and leaders, lecturer, seminars, exhibition, presentation of quality circle case studies in sister units.

(iv) **Facilitator:** Facilitator is a senior officer nominated by the management. He guides and makes quality circles operation a success. Function of a facilitator are to attend quality circle meeting, coordinates for necesssary training programme in SQL techniques, or working procedures etc. Acts as intermediary in resolving quality circle problems. He communicates the steering committee decision, recommendations, searches for new ideas and publicises the programme. He also collects and arrange dissemination of literatures.

(v) **Leader:** Leader is chosen by the members of a quality circle or may decide to have leader by rotation. To begin with, the supervisor may be a leader and another worker as deputy leader. Functions of a leader are conduct circle meetings, maintain registers, train the members, maintain cohesiveness and involve every member. Interaction with other functional area in problem-solving, reviews, progress, arrange for mid term and top management presentation.

(*vi*) **Members:** Members are a small group of people from the same work-area of doing similar type of work. The member contribute to the effective functioning for better performance. Functions of members are to meet regularly at one hour in a week, contribute problem solving ideas. Provides assistance to the leader in data collection, record-keeping, interacting with other areas, catalyse cohesiveness.

8.15.4. Quality Circle Limitations

It is not a substitute for the main functions of the management. Main functions of management are to plan, organise, allocate resources, control, check, review etc. Quality circle is not directly responsible for improvement of quality. Quality circle does not exclude the role of individual suggestions for job or quality improvement. It is not an agency for diverting problems by the management. It is neither pro nor against union or management.

Quality circle improves the work, aims at making the work more meaningful and effective. The persons involved feel satisfied in doing it. It is a movement for improvement of work culture and productivity.

8.16. 5S-CONCEPT

5-S is a concept of work culture for effecting improvement in performance mainly through creating good work environment and work culture. The 5-S programme focuses on having orderliness, cleanliness, discipline and organization in the workplace. 5-S as stands for in Japanese and its literal English translation are:

Seiso – Cleanliness
Seikelsu – Clean-up time
Seiton – Orderliness
Seiri – Tidiness
Shitsuke – Discipline

The system is sort, systematize, sweep, standardize, and sustain. The benefits of 5-S implementation are:

(*i*) Quality and productivity improvement.
(*ii*) Standardization of working methods.
(*iii*) Improvement of safety in the workplace.
(*iv*) Reduction of inventory, storage space and costs.
(*v*) Reduction in cycle times.
(*vi*) Reduction in machine down time.
(*vii*) Higher employee morale.

SUMMARY

1. **Reliability:** It is the probability of an item performing it intended function over a given period of time under the operating condition encountered.
2. **Availability:** The availability of a system is the probability that it is operating satisfactorily at any point in time when used under stated condition. The various types of availabilities are (*a*) Intrinsic availability, (*b*) Equipment availability, (*c*) Mission availability.

3. **Maintainability:** It is the probability that a system will be restored to operational efficiency within a given period of time when the maintenance action is performed in accordance with prescribed procedure.
4. **Failure:** Any deviation in the properties of an item from the prescribed condition is considered a fault.

   Causes of failure: (*a*) Deficiencies in design, (*b*) Deficiencies in Material, (*c*) Deficiencies in Processing, (*d*) Error in assembly, (*e*) Improper service conditions.
5. **Types of Failures:** (*a*) Early failure, (*b*) Chance failures, (*c*) Wear out failures.
6. **Modes of Failures:** (*a*) Catastrophic failures, (*b*) Degradation or creeping failures, (*c*) Independent failures, (*d*) Secondary failures.
7. **Failure Rate:**

$$\lambda = \frac{f}{T}$$

where f = Number of failures during the test interval
T = Total test time

8. **Mean Time between Failures (MTBF):**

$$\text{MTBF, } m = \frac{1}{\lambda} = \frac{T}{f}$$

9. **Mean Time to Failure (MTTF):**

$$\text{MTTF} = \frac{1}{n} \cdot \sum_{i=1}^{n} t_i$$

10. **Advantages of a Reliability Engineering:**
    (*a*) Optimum burn-in-time or breaking-in period.
    (*b*) Optimum warranty period and estimated warranty costs.
    (*c*) Evaluation of the amount of redundancy present in the design.
    (*d*) Help provide guidelines for quality control practices.
    (*e*) Promotion of positive image and company reputation.
    (*f*) A basis for comparing two or more designs and choosing the best desing from the reliability point of view.
11. **Reliability Data Usage:**
    (*a*) The information regarding the reliability requirements w.r.t. maintainability, availability and mean life.
    (*b*) Deviation, if any between simulated field tests can be found out using this information.
    (*c*) Estimation of the degradation and wear out characteristics of components can done using reliability data.
12. **Data sources:** Some of sources are:
    (*a*) Report on raw material
    (*b*) Report of standard part supplies
    (*c*) Market research
    (*d*) Records of maintenance
    (*e*) Experience of customers.
13. **Reliability:**

$$R\,(t) = \exp\left[-\int_0^t \lambda(t)dt\right]$$

14. **Zero defects:** Zero defects is a concept which promotes a constant, conscious desire to do a job right the first time.
15. The zero defects programme involves motivation of the management supervisors workers, co-operative actions by the management and workers, laying down targets, measurements of result, and recognition of achievers.
16. **Quality circle:** Quality circle is small group of employees working together who meet each other for analysing and resolving work related problems.
17. Quality circle is voluntary by nature and all the members are highly motivated.

EXERCISES

1 Define reliability and quality. Explain with examples.
2 What do you understand by availability and maintainability? Explain in detail.
3. What are the factors affecting maintainability?
4. What are the types of failures? Explain each type with an example.
5. Explain various causes of failures.
6. Explain various parameters of the system effectiveness.
7. What are the concepts of failure modes? Explain.
8. Explain the difference between MTTR and MTTF.
9. What is a bathtub curve? Describe in detail.
10. What are various advantages of reliability engineering programme?
11. Draw total product cost versus product reliability curve. Explain in detail.
12. Explain the importance of reliability in everyday life.
13. Explain various causes of poor reliability of a system.
14. Compare the followings:
    (*a*) Reliability and Maintainability
    (*b*) Reliability and Availability.
15. Distinguish between reliability and quality of a product.
16. What is a difference between reliability and durability of product.
17. Derive the reliability function.
18. What are the methods of data collection, storage and recovery?
19. How do you establish component reliability from test data?
20. What is zero defect? Explain.
21. How do you remove impediments to quality? Discuss.
22. Explain quality circles in detail.
23. Discuss identification and analysis of defects.
24. Explain calculation of reliability.
25. Discuss building reliability in the product.
26. How is evaluation of reliability done? How is interpretation of test results done?

□□□

9 CHAPTER Nine

QUALITY MANAGEMENT SYSTEM

CHAPTER OUTLINE

- Definition of Quality and Quality Control
- Responsibility for Quality Control
- Cost and Quality
- Quality Costs
- Quality and Cost Relationship
- Quality Policy
- Quality Assurance
- Quality Mangement System
- Quality System and its Activities
- Quality Audit
- Difficulties in Implementing
- TQM Need for Quality Improvement and Control
- Quality ystem Standards
- ISO 9000
- Definitions of Important Terms as used in ISO: 9000 Publications
- Important Clauses in ISO: 9000 Specifications
- Elements of a Quality System
- Essential Steps in Implementing Quality System for ISO: 9000
- Management Responsibility
- Quality System
- Contractive Review
- Design Control
- Document and Data Control
- Purchasing
- Control of Customer Supplied Product
- Product Identification and Trace Ability
- Process Control
- Control of Inspection, Measuring and Test Equipment
- Inspection and Test Status
- Control of Non-conforming Product
- Corrective and Preventive Action
- Handling Storage, Packaging, Preservation and Delivery
- Quality Records
- Training
- Servicing
- Statistical Techniques
- Content of Level 2 Procedure Manual

9.1. INTRODUCTION

The word quality has different meanings under different circumstances. Customer satisfaction is the key to effective equality assurance. The quality of a product may have greater or lesser significance depending on the need and requirement of the user. Simplest way to define quality would be "the degree to which a product meets the requirements of a customer".

From the very beginning of production, all efforts to manage quality and maintain assurance of continued high quality of product or service Quality control should be included. Quality control has to be systematically introduced and followed up till the finished product has been put-into performance. Prevention rather than detection of defective item to be produced is the main objective of quality control.

Every employee has to directly or indirectly contribute towards the quality of the product. Quality is everybody business. Even the sweeper has some thing to contribute towards the final quality of the company and this must be kept in view by the designer. The capabilities of the plant and the customer's requirements must be considered by the designer. He must design for simplicity, producibility and lowest total cost. Manufacturing must provide satisfactory materials, tools, and working conditions, train operators and plan right methods. Employees must try to produce accurately for the first-time with full care and attention, and report any problems or short comings to their superiors. Sales must not promise deliversies that cannot be met, prices that are low and cannot be achieved and quality which canot be produced.

9.2. QUALITY POLICY

The first step for the top management should be to announce its quality policy at the time of introducing the quality system. The mangement should then ensure that this policy is correctly understood and sincerely adopted by all concerned. Some of the important points to be taken up are:

(*a*) To write down quality objective in terms of factors, such as requirements of quality, quality criteria, performance in service, safety and reliability.

(*b*) To ensure that the design of the product and its components fully meet the custormer's requirements, for example, dimensional accuracy, chemical composition, mechanical properties and any other test criteria, keeping in view economy and ease of manufacturing, adopting correct materials and manufacturing methods.

(*c*) To give due importance to prevention of defect rather than dependence on detection after occurency, attempting to produce 'right first time', using total quality control at all stages of manufacturing.

(*d*) To incorporate continuous review of the product quality and performance utilising results of market survey and feedback from customers and quality audits.

(*e*) To provide training and orientation programme for the customers in order to acquaint them with product usage and maintenance.

(*f*) To do proper vender selection and vender development and to take up evaluation of vender performance, coupled with providing necessary assitance to build and maintain desired quality. There are no definite rules about framing of the quality policy, but it is desirable that this policy reflects the underlying means to be adopted by the company to build and maintain the quality required by the customer and assure timely deliveries, continuous product improvemet, and reliability of the product performance.

9.3. QUALITY ASSURANCE

Definition is as follows:

All the planned and systematic actions necessary to provide adequate confidence such that a product or service will satisfy given requirements for Quality.

Similar to 1057: 1971 described quality assurance in some what fewer words such is progress using the defination as follows:

The provision of evidence or proof that requirement for quality have been met.

Both definitions focus on the need for evidence. Quality, like justice, must not only be achieved but must be seen to be achieved.

(a) Quality Assurance Principles: The basic priciples which enable an organisation to satisfy its customer's requirement in the most effective manner may be written as follows:

- "Quality is company wide"
  — every management function and activity carries a specific quality related responsibility.
- "Right first time"
  — plan and anticipate problems.
- "Communicate and cooperate"
  — every one knows their responsibilities such as, what to do, where they fit within the organisation and with whom they interface. These basic principles can be most effectively put into use by the development and operation of a documented quality system.

(b) Quality System: A quality system is the collection of resources, organisation, equipment, people and procedures, which implement the quality policy.

The necessary points of a good quality system are to:

- Clearly define the customer expectation.
- Ensure that the raw materials used in the processes are of an acceptable and consistent quality.
- Control the production or service operation to get correct and consistent quality.
- Follow procedures to take care of defective raw materials, defective product or service and complaints from customers.
- Documantation to such an extent that in the event of a failure the root cause can be identified and the problem prevented from recurring by changing the documentation.
- Monitor quality levels and should ideally monitor quality costs.

For achieving the requirements of quality, for meeting the expectations of the customer and also to protection of the interest of the company in terms of optimum cost for the desired quality, every company should develop and implement a quality management system. Such a system should be designed to perform the following major activities:

(*i*) Building customers confidence in the quality and reliability of all products manufactured.

(*ii*) Making a reference guide for all concerned in order to understand their role and responsibilities.

(*iii*) Ensuring complete understanding with the suppliers of materials and services and the process to be used for its adoption.

(*iv*) Educating all employees in order to develop a clear understanding about the role and working of the quality assurance department.

(*v*) Providing a reference for quality audit purposes.

The main activities in a quality system are as follows:

(*i*) Marketing: Total quality control starts from the stage of marketing the product itself. The quality aspects in marketing would include determining the exact market demand, the types, grades or compositions required, competitors, prices and prices to be charged.

(*ii*) Design and Development: The design of a component or a product under manufacture must be such that it leads to maximum economy in manufacturing enabling, maximum saving, in the use of materials, labour, equipment, tools, power and other resources, and still fulfilling the requirements of customer in terms of dimensional tolerances, mechanical properties, metallurgical quality, chemical composition and etc.

(*iii*) Material Procurement: The quality of products manufactured would depend mostly on the quality of raw materials and other in-process materials used.

(*iv*) Manufacturing and Process Control: Proper records must be maintained for various production processes adopted, equipment and tooling used and personnel. The monitoring and control methods to be used during production should also be documented as a part of process control.

(*v*) Inspection: The quality of a product would depend on the degree of quality control exercised on the incoming materials, inspection at different stages of production, and final inspection. It is necessary to introduce procedures for each of the requirements.

(*vi*) Post Manufacturing Operations: This includes following:

(*a*) Handling and storage of the products prior to dispatch in a safe manner, avoiding any damage or loss.

(*b*) Packaging to enable easy and safe handling, loading, unloading, transportation and subsequent processing at the customer's premises, avoiding any damage or loss in transit.

(*c*) Safe delivery at the customer's end.

9.4. ISO 9000

ISO is the International Organization for standardization, based in Geneva and Switzerland. ISO 9000 is a set of written standards, laying down a quality system.

The basic elements of system are defined through, ISO-documentation. ISO ensures a uniform system, which is universally recognized. ISO offers a customer focused quality system which can be maintained.

ISO is linked with organization looking for excellence in quality. ISO 9000 is a set of international standards for quality management systems. Many customers demand from their suppliers to achieve ISO 9000 certification. ISO 9000 is not specific to any industry and consists of a series of five internal standards for "quality management".

Table 9.1. ISO Description

S.No.	ISO Series	Description
1.	ISO 9000	It is considered as a road map for use of the other standards in the series.
2.	ISO 9001	It specifies a model when two parties require the demonstration of a supplier's capability to design, produce, install, and service product.
3.	ISO 9002	It specifies a medal for quality assurance in production and installation
4.	ISO 9003	It is a model for quality assurance in final inspection and testing
5.	ISO 9004	It provides quality management guidelines for developing and implementing a quality system.

9.5. QUALITY AUDIT

It is conducted to compare the given aspects of quality performance with a standard for that performance. It is one of the important tools for achieving the objective set out by the organisation. It is an information gathering activity so that the need for improvement or corrective action may be followed. Audits are carried out to verify that idividual elements within a quality system are effective, sufficient and suitable in achieving stated quality objectives. The process of an audit has to be carried out in a systematic and organised manner. Types of audits are being given here.

(*a*) First Party-Internal: This is an audit carried out by a company on its own system, using its own staff, provided that they are independent of the function being audited. It can also be carried out by an outride agency on behalf of the company.

(*b*) Second Party External: This is an audit carried out by one organisation on another, with whom they either have or intend to have, a contract to purchase goods or services. The purpose is to provide the purchasing organisation with a level of confidence in the supplier's ability to provide goods or services to an agreed quality level.

These audits can be carried out by either purchasing organisation or an outside agency on behalf of the purchasing organisation. the problem with second party audit is that. They quite rightly tend to concentrate only on the areas of the auditee's business which directly effect the supplied goods or services and completely ignore those that do not.

Second party assessments schemes are run by such well known operations as the MOD (AQAP), the Ford Moter company (Q101), British Telecom (Delegated Release), and retail conglomenates such as ASDA, Marks and Spencers and Litle Woods.

***(c)* Third Party External:** This is the asessment of an operation by an auditor or a group of auditors who have no involvement with or commercial interest in the outcome. They are conducted by recognised certifification Bodies and, if successful result in some formal listing or registration which the company can very often use to marketing advantage.

There is another formal third party audit operation whcih should be mentioned, and that is the audit conducted by such statutory bodies as the Health and safety Executive (OSHA). Environmental Health Departments (FDA and the EPA) and the Atomic Energy Authority.

The third party assessments and the result can be of considerable importance to a company. Regrittably for the auditee, They don't result in any certificates for being a good boy, only problems for being bad! But evidence of success in such audits might well be relevant to an assessment team involved in certification activity.

***(d)* Difficulties in Implementing TQM:** The main dfficulties are given below:

(*i*) Lack of commitment on the parts of top management.

(*ii*) Lack of employee involvement.

(*iii*) Lack of team work. Without Team work TQM cannot be successful.

(*iv*) Lack of customer oriented approach.

(*v*) Lack of attention paid to customer feedbacks and complaints. The supplier must. Care for the requirements of his customer and must keep him satisfied.

(*vi*) Supplier control. This control includes vendor rating, vendor inspection and evaluation, training of vendors, materials inspection before dispatch and technical suppout for quality improvement.

(*vii*) Reviw of quality Procedures.

The steps of quality improvement and control are as follows:

(*i*) Set Quality Standards

(*ii*) Plan methods, technology, materials tooling and personal to achieve the specified quality.

(*iii*) Manufacture right the first time

(*iv*) Inspect and report any quality shortcomings.

(*v*) Carry out corrections in process, controls, toolings etc.

(*vi*) Re-plan for long term quality control

Steps (*ii*) and (*iii*) can be further explained as follows:

(*i*) Incoming materials control, supplier maintenance and audit, supplier improvement, supplier rating.

(*ii*) Employee, training and development
(*iii*) Process planning, process documentation and product introduction.
(*iv*) Gauge and Fixture control.
(*v*) Equipment maintenance and spare parts control.
(*vi*) Set up and first piece control.
(*vii*) Process capability determination.
(*viii*) Statistical process control.
(*ix*) In-process inspection and test.
(*x*) In-process audits and finished goods audits.
(*xi*) Quality data system and documentation.
(*xii*) Corrective action program.
(*xiii*) Outgoing quality level improvement.

Deaming has evolved fourteen major points for improving quality which are summarised below:

(*i*) Establish the objective of working towards continuous innovation and improvement.
(*ii*) Adopt a new philosophy that we cannot accept the old mistakes and defects.
(*iii*) Cease dependence on mass inspection; require statistical evidence that quality has been built in the process.
(*iv*) End the practice of awarding business on the basis of price.
(*v*) Use statistical techniques to find the trouble spots.
(*vi*) Institute modern methods of training on the job.
(*vii*) Improve quality of supervision, doing what is right for the company not just turning out quantity.
(*viii*) Breakdown warriers between different departments and also with suppliers and customers so that there will be open and effective communication.
(*ix*) Eliminate posters and slogans. they do not help in solving problems. show people how.
(*x*) Eliminate work standards that prescribe a numerical quota and disregard quality.
(*xi*) Remove barriers between workers and their right to pride in workmanship.
(*xii*) Establish a regular re-training program to keep up with changes and new developments.
(*xiii*) Top management must accept the above points with full sincerity.

9.6. ISO 9000 SERIES OF STANDARDS

Quality system standards specify the minimum requirements of quality systems for application to specific products or services. Standards are normally used for following purposes:

- As guidance to an organisation introducing a quality system
- As a basis for evluating an organisation's quality system assessment.
- To specify the quality assurance requirements when involved in a contract.

There are two types of quality system standards:

- General standards.
- Industry related standards.

General standards are those issued by national bodies for use in quality system development, guidance or evaluation.

Regardless of the country of origin on the industrial applications, the differences within these standards are in the main, quite small.

Industry related standards are those developed by purchasing boides for a particular industry. Example include defence, aerospace and nuclear industries. These are normally produced as an aid to Government procurement or used where there are important and over riding safety requirements.

ISO 9000 is a series of International quality standards which tell suppliers and purchasers what is required of a quality management system. They are basically common sense set down on paper and can be used by all types of industrial or service operation. Equally they apply to any size of company, from 5 to 5000.

The standards identify the basic disciplines and specify the procedures and criteria to ensure that products or services meet the customer's requirements. By building in qualty at every stage, through soundly based and efficient procedures, waste and rework will be reduced, resulting in satisfied customers.

The ISO 9000 standard has eight clauses: Scope, Normative references, Definitions, Quality Management system, Management Responsibility, Resource Management, Product and/or Service Realization, and Measurement. Analysis and Improvement. For the five required clauses, the system is shown in Fig. 9.1.

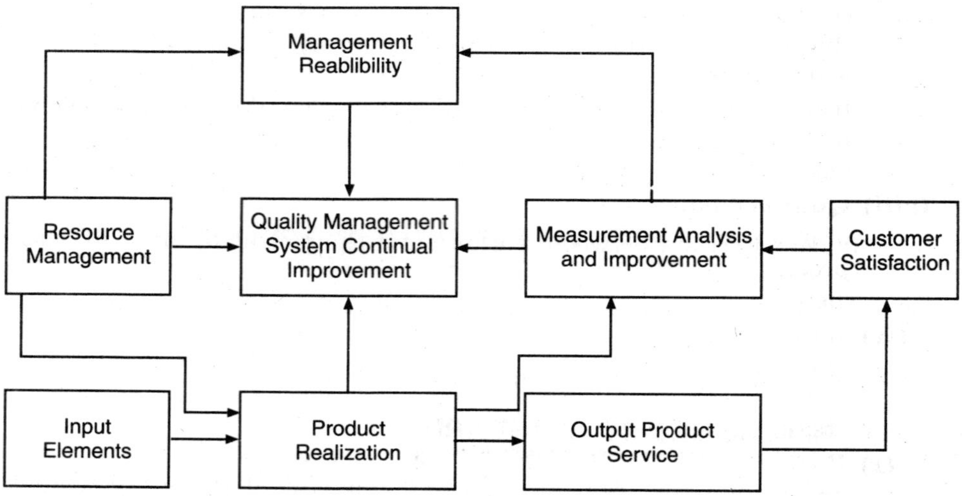

Fig. 9.1. A process based quality management system

International quality system standards contain a minimum series of requirements in the form of procedural controls over critical activities.

9.7. DEFINITIONS OF IMPORTANT TERMS AS USED IN ISO 9000 PUBLICATIONS

(*i*) **Quality Control:** Operational techniques and activities used to fulfil the requirements for quality.

(*ii*) **Quality Assurance:** All these planned and systematic actions necessary to provide adequate confidence that a product or service will satisfy given requirements of quality.

(*iii*) **Quality Managment:** One of the functions of management that determines and implements the quality policy. It covers strategic planning of quality allocation of resources for achieving that quality and systematic activities to be introduced for producing that quality.

(*iv*) **Quality System:** The organisational structure, responsibilities procedures, processes and resources required for implementing total quality management. The management should develop, establish and implement such a quality system by which the specified policies and objectives may be accomplished. It should be adapted to that particular type of business activity.

(*v*) **Quality Policy:** The overall quality intentions and direction of an organisation as regards quality, which form an integral element of the corporate policy of the company.

(*vi*) **Quality Manual:** It is an apex manual or main guiding document that contains the company's policies and procedures which describe the quality management system established by the company's management and servies as a permanent reference for Implementation and maintenance of that system. It tells 'what you do'.

(*vii*) **Quality Plans:** The management should prepare its quality plans, consistant with the quality management system. It should comprise of the quality objectives to be attained, allocation of responsibilities and authority at different stages, procedures, methods and work instructions to be followed testing, inspection and audit procedures to be adopted, and all other measures necessary to meet the quality objectives.

(*viii*) **Quality Procedures:** It is a middle level management manual or code of practice instructing the staff in broad terms, how the policies and procedures outlined in the quality manual are to be achieved. It tells 'how you do.'

(*ix*) **Work Instructions:** This document is for a lower level of managements, *e.g.*, supervisors and workers; describing indetail how a specific activity is undertaken and performed meeting with the specific standards of quality. It tells 'how each of you do.'

(*x*) **Records:** These are the various types of forms. to be regularly prepared and maintained to demonstrate that a product has been developed and manufactured in accordance with th specifications and requirements laid down. These show, evidence of what your have done.

9.8. IMPORTANT CLAUSES IN ISO: 9000 SPECIFICATIONS

1. Management Responsibility: The main elements of this clause are:

(*a*) To state the quality policy of the company

(*b*) To define responsibilities and authorities of its people with regard to quality particularly those who need organisational freedom and authority to initiate action to prevent occurence of non-conformities, identify and record product quality problems recomment solutions and verify implementation of those solutions.

(*c*) To allocate adequate resources to operate the quality system.

(*d*) To provide a mechanism of review its effectiveness and ensure effective implementations.

(*e*) To appoint a 'management representative' who should be assigned to ensure implementation and should have defined resposniblity and authority.

2. Quality System: Its major elements are:

(*a*) To provide for quality engineering and quality planning and prepare a quality manual

(*b*) To identify and prepare quality records.

3. Contract View: This clause enables to build and maintain procedures to review all the supply orders and order amendments before they are accepted to make sure that the orders will be satisfactorily fulfilled.

4. Design Control: The objective of this element is to establish and maintain procedures to control and verify the design of the product in order to ensure that the design outputs meets the design input requirements and conforms to the appropriate regulatory controls.

5. Document Control: It consists of:

(*a*) Document approval and issue to the concerned departments and sections. It must be ensured that documents are approved by a competent authority before their issue is permitted.

(*b*) Review of documents for any changes or modifications.

(*c*) Re-issue of amended documents and recalling the older ones to avoid any confusion or misuse.

6. Purchasing: This clause provides for the following provisions:

(*a*) To monitor continuously the performance of suppliers or those providing sub-contracted services.

(*b*) To verify that the purchace orders are clear and complete and are free from any ambiguity which may lead to controversy or mistrust.

(*c*) To establish and maintain procedures for verification and subsequent storage and maintenance of purchase supplied items or products and to ensure that these items are incorporated into the supplies.

(*d*) To provide for disposal of damaged items, obsolete items or surplus stocks not expected to be of use.

(*e*) Product identification traceability.

7. Process Control: Process control refers to the functions of identifying, planning and controlling. The process of production or service by laying down docuemented procedures, instruction or codes of practice to define how. The desired control are to be achieved and maintained.

8. Inspection and Testing: This clause provides to ensure proper and adequate inspection and testing at receving, in-process and final stages and to maintain complete records of the same, which can serve as an objective evidence of conformance of the product quality to the specified requirements. In case of service, it would involve monitoring the quality of service offered.

9. Inspection of Testing and Measuring Equipment: This clause is provided to ensure maintenance, calibration and control of all instruments, gauges, apparatus and equipments used for inspection and testing work so as to demonstrate of the product to its specifications. It gives an assurance of the accuracy achievable and to select right equipment capable of giving. The desired accuracy for the inspection, measurement and testing.

10. Control of Non-conforming Products: Proper procedures are to be maintained by the supplier for ensuring. That the items not conforming to the specifications or specified requirements and which are cleared for despatch or use are prevented fromindvertent use or despatch. The items concerned have to be properly segregated, and clearly marked as non-conforming.

11. Corrective Action: This clause requires the supplier to provide for arrangement to investigate and rectify. The cause(s) of rejections or defective items so as to prevent. Their re-occurrence in subsequent batches. It involve maintaining records of the actions taken and results achieved.

Proper procedures have to be devised, maintained and documented for the following purposes:

(*a*) Handing of the product of goods to be supplied to customer.

(*b*) Storage at a safe place in such a manner, that they are not damaged and can be easily identified at the time of despatch.

(*c*) Packaging and delivery to destination such that the items are not deterierated, damaged, multilated, broken or piffered during transportation and handling and their quality, accuracy, sensitivity etc. as required in the specifications.

12. Quality Records: A proper system is established for identifiying, collecting, indexing, filling, storing, safely preserving and maintaining and finally disposing of all. The records concerned with quality. The records of quality must be kept clean, legible with no over-writting or erasing and must be easily retrievable and shared at a safe place to minimise their deterioration, damage, tempering or loss.

13. Internal Quality Audits: It is desired that regular quality checks are planned and executed on the quality assurancce system in order to verify whether the quality activities comply with the planned arrangement and further to determine, control and report effectiveness of the quality system.

14. Training and Human Development: It is essential for any organisation to lookafter the development of its employees and in that direction, training and retraining of its staff and workers is essential. Procedures should be built to identify training needs at various levels and the training should be imparted as planned so that all personnal perform their duties satisfactorily.

15. Servicing: It is necessary to establish and maintain procedures for performing and verifying that the service provided after sales meets the specified requirement and the customer is satisfied with the service provided.

16. Statistical Techniques : It is provided in the standards that wherever appropriate, statistical methods shall be adopted and procedures be laid down for identification and adoption of these techniques, particularly for the purpose of:

(*a*) Verifying acceptability of the process capability.

(*b*) Control of material, in-process and product characteristics and

(*c*) Lay down inspection plans and sampling schemes.

9.9. ELEMENTS OF A QUALITY SYSTEM

The applicability of various clauses in the three standards are indicated below for general guidance. These clauses refer to the numbering shown in ISO : 9004 "guidelines on elements of quality management system".

Table 9.2. Quality System Elements Applicable to Different Standards

Clause in ISO: 9004	Title	ISO 9001	ISO 9002	ISO 9003
4	Management responsibiilty	***	**	*
5	Quality system principles	***	***	**
5.4	Quality audit	***	**	—
6	Economics quality costs	—	—	—
7	Contract review	***	***	—
8	Quality in design and speci.	***	—	—
9	Quality in procurement	***	***	—
10	Quality in production	***	***	—
11	Control of production	***	***	—
11.2	Material control	***	***	**
11.7	Control of verification status	***	***	**
12	Product verification	***	***	**
13	Control of meas, test equpt.	***	***	**
14	Control of non-conformities	***	***	**
15.	Corrective action	***	***	—
16.	Handling storage etc.	***	***	**
16.2	After sales service	***	—	—
17.	Quality documentation	***	***	**
17.3	Quality records	***	***	**
18.	Training of personnel	***	***	*
19.	Product safety and liability	—	—	—
20	Statistical techniques used purchase-supplied product.	***	***	—

Keys : *** Full requirement

** Less stringent requirement than in ISO: 9001

* Still less requirement

— Element not specified.

9.10. ESSENTIAL STEPS IN IMPLEMENTING ISO: 9000 QUALITY SYSTEM

(i) Initiating Total Quality Management: It would involve making a decision about going ahead with ISO certification, and intiating the work by defining the scope for the certification, *e.g.*, entire company, single location, project, etc. and setting up project team to take care of the work, defining the responsibilities, assigning authorities, and carrying out a detailed analysis of the company's activities and performance. The analysis would include a delailed and environment analysis using SWOT analysis techniques and establishing distinctive competence of the company in its business policy.

(ii) Planning to Achieve Objectives in a Coordinated Manner: This is done by introducing quality engineering and quality planning, planning/reviewing inspection and quality control on the shop/floor and standardisation for technical specifications of raw materials, in-process materials, test requirements and test criteria, product specifications, instruments, gauges and equipment for testing and test-procedures, tooling, manufacturing procedures and technology of production.

(iii) Orientation of Staff: Staff has to be oriented about total quality management and requirements of ISO:9000, followed by training of persors concerned with problems solving, brain storming, SWOT analysis, statistical techniques statistical process conrol just in time technique etc. specifically, training has to be organised for lead assessors and internal auditors but-other officials and staff who are connected with quality management and implementation, or also various line managers and union officials have to be given suitable training inputs.

(iv) Implementation: This step involves all those functions which have been planned are to be carried out. In general, it may consist of:

(*a*) Creation of quality mannual
(*b*) Assessment of current position by internal audit.
(*c*) Managements review of internal audit reports and plan action.
(*d*) Pre-audit review by the certification body.
(*e*) Audit by the certification body.
(*f*) Certification body reports for review and action.
(*g*) Corrective action
(*h*) Clearing of non-confirmities by certification body and recommendation for certification
(*i*) Award of ISO:9000 certificate

(v) Monitoring: It is necessary to monitor continuously the working of the quality system and the procedures and practices introduced for it. If ignored, there is every likelihood of people reverting to old practices. A proper system of monitoring has to be built and maintained.

(vi) Consolidation: The process of quality improvement is never ending one. As new ideas are received, customer is feedbcak is known new technologies are available and experience is gained from current practies, the quality system should be strengthened continously and consolidated.

ISO 9000: Quality management system has been explained here by taking example an implementation of the same in a company. All the discussions are as applicable to this company.

9.10.1. Management Responsibility (ISO 9001 Clause 4.1)

Quality Policy

The company's Quality Policy, objectives and commitment to quality are defined by the president and chief executive officer as policy statement. The policy and the means of implementation are advised to all personnel by the publication of the Documented Quality system, communication on quality matters by the president of Quality and by the induction routines to employes and as introduction to new employees.

Organisation

Responsibility and Authority

The management organisation chart, defines the lines of responsibility of all personnel. The schedule of key personnel responsibilities and authorities details job functions and level of authority. Additionally procedures clearly indicate responsibilities for their implementation.

Resources

Resources: Resources are demonstrated through the Documented Quality system and include Internal quality Audit and verification covering administration, customer service; and all terms of design and process control.

Management Representation

The company's management Representative, for all quality matters, both internal and external. In this case he is vice president of quality. The responsibilities and authorities are defined in detail in the schedule of Responsibilities.

Management Review

The effectiveness of the quality system's formally reviewed by management chaired by the President and Chief Executive officer. The Management Review meeting is scheduled annually, 2 months after the end of the company's financial year. The objective of the review is to examine irregularities in the operation of the quality system itself and its contents. To this end, the vice president of quality maintains an on going analysis of corrective and preventive actions taken in the year and non-compliances identified by audit, appraisal by outside bodies, customers complaints and non-conformity.

Master: The total as presented for the year and as compared by the meeting with data for earlier periods. As appropriate, the master determines any changes of policy course or action.

A formal of management Review maintained, together with details of decisions taken.

Level 2 Procedure References:

MP-01 Review of the Quality system
MP-02 Corrective Action

MP-20 Analysis of Review I/P data

MP-23 Preventive Action

Schedule of Key Personnel Responsibilities and Authorities

The company's management lies in the hands of the Directors, who are shareholders. Inview of the fact that the company is relatively small responsibility and authority are shared.

The President and Chief Executive Offices

He is also the **Accounts Director** having total responsibility for office administration matters, invoicing, credit control and budgeting. He is authorised to purchase material and services as required for the company's operations. He is also actual point for customer contact and as such he effects contractor view.

The Vice-President of Technical Services

The Vice-President of technical services has responsibility and authority for the control of customer contact. Contract review, technical matters, control of day to day work. He is authorised to purchase materials and services as may be necessary for the implementation of the work. He is the Deputy Management Representative.

The Vice-President of Quality

The Vice-President of quality is responsible for and authorised to set up and implement a quality system in accordaule with the requirements of ISO 9001.

He is the management Representation for all quality matters. In addition, he works in parallel with the vice president covering all the responsibilities and authorities assigned to him.

The Technical Staff

Carry out work at customers premises and in the company's work shop. These activities include a sales element with a vice president carrying out the contract view.

Management Organisation Chart

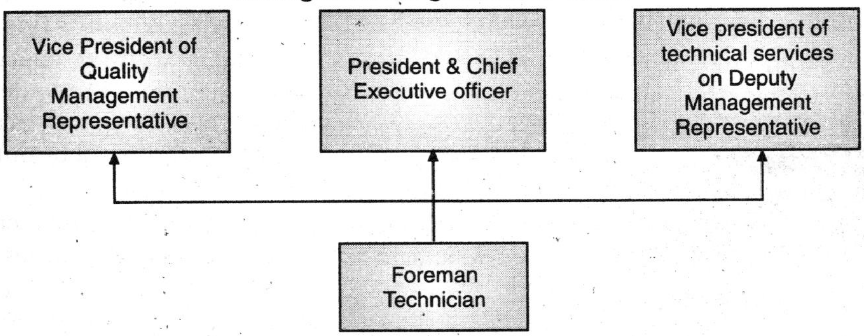

9.10.2. Quality System (ISO 9001 Clause 4.2)

General

The company's documented Quality system is structured in two levels.

Level I: The Quality Assurance manual defines policy, demonstrates scope and commitment; and governs the implementation of the level 2 Procedures manual.

All quality related documentation is formally issue controlled updated and reissued by means of an issue and retrival system designed to ensure withdrawal of out-dated copy.

Quality System Procedures

The content of the level 2 procedures manual is the company's guide for implementation of all activities. In the event that a customer stipulates requirements beyond the scope of or differing from the company's standards, a project quality plan is prepared listing the requirements, standards, and specifications to be observed, acceptance criteria; and tests and inspections to be carried out. This Quality plan shall take precedence over the company's documented quality system, if any conflict arises.

Definitions used throughout the documented quality system are defined in the latest edition ISO 8402.

Quality Planning

The documented quality system is comprehensive and covers all aspects of both the company's activities and the requirements of ISO 9001 as applicable to those activities. Level 2 procedures supporting this manual cover all requirements for quality planning particularly the need for total review of the Quality System.

Level 2 Procedure References

MP-01 Review of the quality system
MP-02 Corrective Action
MP-03 Control of company documentation
MP-29 Quality planning.

9.10.3. Contract Review (ISO 9001 Clause 4.3)

General

The company maintains procedural instructions to govern contract Review.

Review

Each inquiry, invitation to bid to supply or perceived market opportunity is reviewed by the vice president of quality with others as appropriate in order to establish the acceptability of the inquiry content ability of the company to supply based on complete data furnished on receipt of an order. The response to the inquiry and the order are compared. If any discrepancy is identified. The matter is referred to the customer to a point of agreement. There after, the order is accepted, acknowledge and executed.

Amendment to Contract

In the event that the company or the customer should seek to vary the contract condition or requirements , reversions shall be made to the details above to status of agreement with the customer.

Records:

Formal Records of all contract Review activities are maintained.

Level 2 Procedure References

MP-04 Contract Review

9.10.4. Design Control (ISO 9001 Clause 4.4)

General

The company maintains procedural documentation to control the design and development functions.

Design and Development Planning

Every contract accepted by the company which involves design and development activity results in a plan being created outlining the activities involved. Specific staff assigned to each activity and the estimated resources identified for both personnel and equipment reflect progress.

Organisation and Technical Interfaces

The organizational and technical interfaces having input to the process are identified within the plan. Regular meetings of this group ensure that all necessary information is regularly reviewed. Minutes of these meeting are maintained including all meetings with the client.

Design Input

All input information is documented within the activity, including relevant statutory and regulatory requirements. Review process ensures that such input are adequate for the requirements of the design process. Any incomplete or ambiguous requirements will be identified in the the appropriate parts before the design activity proceeds. Contract review is considered to be an essential put of this activity.

Design Review

The design plan identifies appropriate review stages for the project. Meetings involving all relevant personnel are held as required to ensure that the design and development activity is proceeding in an appropriate manner. Minutes of all such meeting are maintained by the design team leader.

Design Output

The output from each stage of the design plants is documented as part of the design review activity and is stated in terms of measurable requirements which are evaluated for:

(*a*) Campliance with the design input requirements.

(*b*) The appropriateness of the acceptance criteria either contained or referenced within the documentation.

(*c*) The proper and completes identification of any or performance critical as peets of the design.

No design documentation is permitted to be released prior to the successful completion of the necessary design view activity.

Design Verification

Each stage of the development activity is subject to verification activities which form parts of the design review. These can take several forms, including.

(*a*) The used of alternative methods to validate design calculation.

(*b*) Comparison of the new design with a existing proven design sufficiently similar in nature to provide a degree of assurance that the new design is adequates for the purpose for which it is intended.

(*c*) The performance of alpha and beta testing of prototypes.

Design Validation

At the conclusion of the design activity and prior to the final design view and releases to production design, validation is performed on design with condition which replicate actual usage is as closely as feasible for product involved. The results of this validation process are compounded to the original design I/P requirements to ensure that an O/P from the design process meets the original intent of the customers.

Design Change

Design change control is exercised from that stage of any project when the original design of any stage of the productre leases the design to other team members for evaluation, testing or integration.

A documented procedure controls the review, approval and release of changes. Prior to such release, individual design engineers maintain designed note books in which all change to the design concept or realisation are recorded.

Level 2 Procedure References:

MP-05 Design control

MP-06 Design Requirements documents

MP-07 Design verification and Review

MP-08 Design Acceptance procedure

MP-09 Design change.

9.10.5. Document and Data Control (ISO 9001 clause 4.5)

General

Quality Manuals. Procedures manuals, reference, standards, formats, services manuals and specifications, are all subject to document control procedure. The president and chief executive office is the only person authorised to issue or reissue controlled manuals. Insertive or hand written changes are not permitted.

Document Approval and Issue

All controlled Documentation is revieved by the vice president of quality prior to issues. A master Document, control list is pubslished to demonstrate issue convey. Obsolete documents age removed from circulation on a basis of one-for-one retrieval under issue. All personnel have access to pertinent data and outdated documents are retained for reference purpose.

Document Changes

Only the vice president of quality make change to documents. The nature of changes is identified by means of vertical bar in the right-hand margin opposite the change.

Level 2 Procedure Reference

MP-03 Control of company Documentation

MP-10 Reference Standards control

MP-20 Document Transmittals.

9.10.6. Purchasing (ISO 9001 clause 4.6)

General

The Company maintain procedures to ensure that purchased product conforms to specified requirements.

Evaluation of Sub-contractors

All sub-contractor and suppliers are assessed in accordance with documented criteria and if acceptable are placed on the Approved suppliers list. The company transactions monitoring and controlling is done as per laid down procedure. Subcontractor records are maintained.

Purchasing Data

Purchasing Documents detail the full data describing the produce or service required including reference to standards and code when applicable. All such documents are reviewed for completeness of specified requirements prior to issue.

Verification of Purchased Product

Supplier Verification at Sub-contractors

Instrument that the company shall opt to verify purchased product at the subcontractor is specified. Details of verification arrangements and the means of product release are stated in the purchasing documents.

Customer Verification of Sub-contracted Product

When specified at control Review Stage the customer, or agent is afforded the right to verify product at the sub-contractor's premises.

Level 2 Procedure References

MP-12 Goods Receipt Routine

MP-13 Purchasing

MP-14 Maintenance of the Approved Supplies List.

9.10.7. Control of Customer Supplied Product (ISO 9001 Clause: 4.7)

In this context, the receipt and handling of customers owned equipment together with the free issue of any component, shall be viewed as customer supplied product.

All product not purchased by the company and at the property of the company shall be subject to Goods Receipt Inspection. Company purchases shall also be identified by label or tag identifying ownership. Any discrepancy or damage shall be reported to the customer.

Level 2 Procedure References

MP-11 Inspection Policy
MP-12 Goods Receipt Routine
MP-28 Control of Customer Supplied Product

9.10.8. Product Identification and Trace Ability (ISO 9001 Clause: 4.8)

Identification and Trace ability of product in progress and stored product is maintained by goods Receipt Routines accompanying documentation and inspection records respectively.

Level 2 Procedure Reference

MP-11 Inspection Policy
MP-12 Goods Receipt Routine
MP-18 Handly, Storage, Packaging Preservation and Delivery.

9.10.9. Process Control (ISO 9001 Clause 4.9)

All process activities experienced by documented procedures for the work is controlled by means of documented instruction and records of work carried out are maintained. No Processes, the results of which cannot be fully verified by subsequent inspection or testing are carried out.

Level 2 Procedure References

MP-24 Servicing
MP-25 Calibrations of Customer is Equipment
MP-27 Overhaul and Repair of Customers Equipment

9.10.10. Inspection and Testing (ISO 9001 Clause: 4.10)

General

The company maintains documented Procedures, all forms of inspection and testing and future cards required to be maintained.

Receiving Inspection and Testing

Receiving inspection and testing is carried out on all goods received which are purchased or customer supplied. The product may be used after goods receipt formality has been completed.

In-Process Inspection and Testing

When applicable in-process inspection and testing is implemented according to procedures or quality plan.

Final Inspection and Testing

Final Inspection is implemented on all services and product supplied.

Inspection and Test Records

Inspection and Test Records are maintained for all form of inspective activity.

Level 2 Procedure References

MP-11 Inspection Policy

MP-12 Goods Receipt Routine.

9.10.11. Control of Inspection, Measuring and Test Equipment (ISO 9001 Clause 4.11)

General

The company maintain procedures for all aspects of control of calibrated equipment. Control and calibration of instrumentation and calibration standards is in accordance with the requirements of ISO 10012 with tolerances amended to the practical usage of the equipment. All such equipment is uniquely identified and each item is supported by a record documents treating item identity of the calibrations carried out, remedial actions and state of accuracy prior to and after calibration.

Control Procedures

All calibrations are supported by criteria of referring traceability to material standard. Full calibration records are maintained.

In the event that equipment demonstrates that it is out of tolerance prior to calibration remedial restrospective action shall be implemented and the customer shall be advised in writing, if such discrepancy may have affected the accuracy of data supplied to the customer.

Level 2 Procedure References

MP-02 Corrective Action

MP-10 Control and Calibration of Inspection Measuring and Test Equipment.

MP-17 Non-compliances

MP-25 Calibration of Customer's Equipment.

9.10.12. Inspection and Test Status (ISO 9001 Clause 4.12)

Each inspection status stage is recorded with endorsement by the inspector and all equipment is labelled accordingly.

Level 2 Procedure References:

MP-11 Inspection Policy

MP-12 Goods Receipt Routine

9.10.13. Control of Non-conforming Product (ISO 9001 Clause 4.13)

General

The company maintains procedures for the control of nonconforming product. This include those instances where any failure to provide service in accordance with agreed condition is observed.

Non-conforming Product Review and Disposition

Product which is identified as nonconforming shall be reported to the vice president of quality for determination of disposition. Customer complaints are dealt with according to procedure and observed non-compliances are similarly regularised by procedures.

Level 2 Procedures Refere**nce**

MP-02 Corrective Action
MP-15 Customer Complaints
MP-17 Non-compliances.

9.10.14. Corrective and Preventive Action (ISO 9001 Clause 4.14)

General

The company maintains procedures of implementing for both corrective and preventive action changes in the documented quality system resulting from these actions are implemented and recorded.

Corrective Action

Corrective action shall be instituted in all cases of failure factors.

These are:

- Non-conforming items and/or work in progress.
- Failures of personnel to observe the mandates of the quality system.
- Recorded audit discrepancies.
- Recorded accrditation authority discrepancies.
- Customer complaints.

All the above are recorded in the corrective Action Register and are subject to verification and periodic review in order to identify trends and re-currences. An analysis of corrective actions is submitted to the management.

Preventive Actions

Preventive Actions are implemented as a result of the identification of trends and re-currence identified by corrective action. Additionally preventive action is applied to the insurances of usability of equipment by periodic maintenances:

Level 2 Procedure References

MP-02 Corrective Action
MP-17 Non-compliances
MP-21 Internal Quality Audits
MP-23 Preventive Action.

9.10.15. Handling Storage, Packaging, Preservation and Delivery (ISO 9001 Clause 4.15)

General

The company maintains procedures for this subject.

Handling

Product is handled inorder to preserve product integrity.

Storage

Storage of all product is secure and identified.

Packaging

Packaging is according to the nature of the product to ensure absence of damage in transist.

Preservation

Preservation is applied as required by the nature of the pro[illegible]

Delivery

Delivery is as specified by or collection by the customers or by company transport.

Level 2 Procedures Reference

MP-18 Handling Storage Packaging and Delivery.

9.10.16. Quality Records (ISO 9001 Clause 4.16)

All job related records are maintained in dedicated files by customer name or contract reference as applicable.

Company records pertaining to quality matters held by the vice president of quality, the responsibilities for preparation, maintenance and destruction, methods of destruction, filing and indexing system and periods of retention are laid down by procedure.

Level 2 Procedure References

MP-19 Preparation and Maintenance of Quality Records

9.10.17. Internal Quality Audits (ISO 9001 Clause 4.17)

All areas of the company other than financial, health and safety and fiscal are audited as scheduled calender basis such that all are covered at least once annually.

The audit system uses serialised reports and corrective action requests which are verified as pointed out by the Internal Auditor. When satisfied objective evidence of all items documents, activities and personnel views is recorded.

Level 2 Procedures Reference

MP-02 Corrective Action

MP-21 Internal Quality Audit

9.10.18. Training (ISO 9001 clauses 4.18)

Training, requirements are planned annually in Decembers of each year using a formal Training plan which highlights requirements for the coming year. Additionally, the training plan acts as record of training and qualification, achievements and is subject to periodic review and update as necessary.

All new employees receive Induction Training and Quality Awareness Addresses given initially and at two-year intervals.

The vice president of Quality maintains individual personnel training records.

Level 2 Procedure References

MP-22 Training

9.10.19. Servicing (ISO 9001 Clause 4.19)

When deliver or services of the company engages in servicing activities, these are viewed as the company's named process.

Level 2 Procedure References

MP-24 Servicing
MP-25 Calibration of Customers Equipment
MP-2 Overhaul and Repair of Customers Equipment

9.10.20. Statistical Techniques (ISO 9001 Clauses 4.20)

As a supplier of service to the customer the, company does not employ statistical techniques.

Level 2 Procedure References

None

9.11. CONTENT OF LEVEL 2 PROCEDURE MANUAL

Reference	Contents	Clause Reference
MP-01	Review of the Quality System	4.1/4.2
MP-02	Corrective Action	4.14
MP-03	Control of company Documentation	4.5
MP-04	Contract Review	4.3
MP-05	Design Control	4.4
MP-06	Design Requirements Documents	4.4
MP-07	Design Verification and Review	4.4
MP-08	Design Acceptance Procedure	4.4
MP-09	Design Change	4.4
MP-10	Reference Standard Control	4.5
MP-11	Inspection Policy	4.10
MP-12	Goods Receipt Routine	4.6/4.7/4.10
MP-13	Purchasing	4.6
MP-14	Maintenance of the Approved Supplier lost	4.6
MP-15	Customer Complaints	4.13/4.14
MP-16	Calibration and Control of Inspection Measuring and Test Equipment	4.11
MP-17	Non-compliances	4.13
MP-18	Handing, Storage, Packaging, Preservation and Delivery	4.15
MP-19	Preparation and Maintenance of Quality Records	4.16
MP-20	Analysis of Review I/P data.	4.1
MP-21	Internal Quality Audits	4.17
MP-22	Training	4.18
MP-23	Preventive Action	4.14
MP-24	Servicing	4.19
MP-25	Calibration of Customer's Equipment	4.11
MP-26	Document Transmittals	4.5
MP-27	Overhead and Preparing Customer's Equipment	4.9
MP-28	Control of Customer Supplies Product	4.7
MP-29	Quality Planing	4.2

SUMMARY

1. **Quality and Quality Control:** Simplest way to define quality would be "The degree to which a product meets the requirements of a customer". All efforts to manage quality and maintain assurance of continued high quality of product or service comes under quality control.
2. **Quality Costs:** Quality costs broadly consists of failure costs, the first step for the top.
3. **Quality Policy:** At the time of introducing the quality system, the first step for the top management should be be announce its quality policy. The management should then ensure this policy is correctly understood and sincerely adopted by all concerned.
4. **Quality Assurance:** This is defined as "All those planned and systematic actions necessary to provide adequate confidence that a product or service will satisfy given requirements for quality".
5. **Quality System and its Activities:** The major activities in a quality system are Marketing, Design and Development, Material Procurement, Production and Process Control, Inspection, Post-production operations.
6. **Quality Audit:** It is an information gathering activity so that the need for improvement or corrective action may be followed. Types of Audit: (*a*) First Party, (*b*) Second Party, (*c*) Third Party.
7. **Quality System Standards:** Quality system standards specify the minimum requirements of quality systems for application to specify products or services.
8. **ISO 9000:** ISO 9000 is a series of International Quality Standards which tell suppliers and purchases, what is required of a quality management system. ISO 9000 companies:

   (*a*) ISO 9000 (*b*) ISO 9001 (*c*) ISO 9002
   (*d*) ISO 9003 (*e*) ISO 9004

   The standards identify the basic disciplines and specify the procedures and criteria to ensure that products or services meet the customer's requirements.
9. **Audit Planning:** The objective of the audit is primarily to ascertain that the documented procedures are being adhered. The ISO 9000 series make it very clear that the task of assuring compliance to the standard is primarily a management function.
10. **Audit Formaility:** This training program focuses on the activity of the auditor acting as an internal auditor working within his or her own corporate environment with auditing system being operated by people who are often friends as well as collegues. Formal audit activity consists of: planning, notification, an opening meeting, the audit itself, the closing meeting, the audit report, followup and closed out and management review.
11. **Audit Notification:** The final part of the planning activity is to make the arrangement with the superviser or manager of the area to be audited for the audit to proceed as planned when the audit schedule is prepared and finished well in advance of the activity, the formal notification to the area concerned is little more than a reminder.
12. **Closing Meeting:** At the end of audit, team will hold a closing meeting with the managers responsible for the functions audited. This takes place immediately after the conclusion of the audit activity and prior to the preparation of the audit report. The objective of the closing meeting is to present the audit observations in such manner that the managers clearly understand the outcome of the audit.

13. **Audit Reporting:** The team leader is responsible for the preparation of the audit report ... accuracy and completes. The audit report should reflect both the ... and content of the audit and be signed and dated by the team leader.

14. **Corrective Action:** This is the part of operation with which the auditor should not get involved. Ownership of the corrective action must by all times remains with the auditee.

15. **Management Review:** The suppplier's management with executive responsibility for quality shall review the quality system at defined intervals sufficient to ensure its continuing suitability and effectiveness in satisfying the requirements of this internationl standard and the suppliers stated quality policy and objectives.

EXERCISES

1. What is a "quality policy"? Explain.
2. What do you understand by TQM? Describe in detail.
3. Explain management of reliability and quality.
4. What are elements of quality systems?
5. Give essential steps in implementing.
6. Explain Quality system ISO: 9000.
7. How is quality linked to the cost of the product? What are the components of quality costs? Explain their effect on the total cost of production?
8. What are the external and internal failure costs? How can these be controlled? Explain with examples?
9. What is the significance of prevention costs? What are their main constituents?
10. What is the role of a management representative in a ISO certified company? What are its responsibilities?
11. How is quality linked to the cost of a product? What are the components of quality cost? Explain their behaviour and their effect on the total cost of production.
12. What factors affect the selection of an ISO model? What elements are specified in ISO 9001?
13. Discuss the need for quality improvement and control for modern industrial requirements.
14. Explain management responsibility as per ISO9001 clause 4.1.
15. Give brief description of quality system of ISO 9001 clause 4.2
16. What is contract review as per ISO 9001 clause 4.3.
17. Describe design control as per ISO 9001 clause 4.4.
18. Explain document and data control as per ISO 9001 clause 4.5.
19. How is purchasing done as per ISO 9001 clause 4.6.
20. What is the method of controlling quality of customer supplied product as per ISO 9001 clauses 4.7.
21. How is process control done as per ISO 9001 clause 4.9.
22. Describe inspection and testing as per ISO 9001 clause 4.10.

10 CHAPTER Ten

TAGUCHI QUALITY METHOD AND JIT

CHAPTER OUTLINE

- Taguchi Quality
- Toguchi Quality Loss Function
- Object and steps of Taguchi methods
- Just-in-Time (JIT)
- Push System
- Kanban and Pull System
- JIT Implementation
- JIT Production Process

10.1. TAGUCHI QUALITY

Genichi Taguchi developed the concept of the Quality Loss Function in the 1970's. The traditional mentality of good quality is that a product is either good or it isn't. It depends on whether or not it is within the specification range. In this approach, the specification range is more important than the nominal or target value. The product is as good or it can be or should be just because it is within specification range. The quality loss $L(y)$ is a step function as shown in Fig. 10.1.

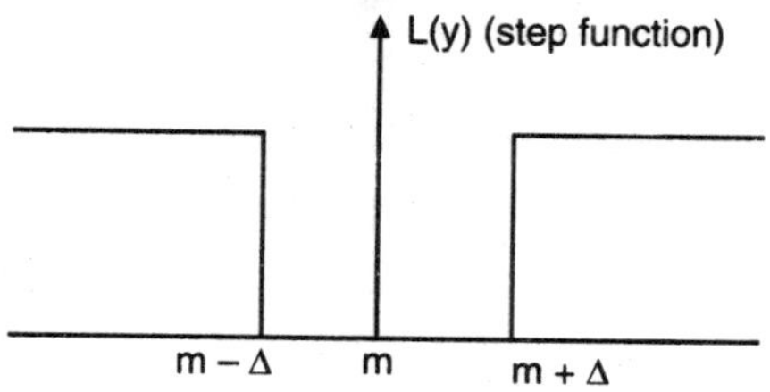

Fig. 10.1. Tradition quality loss function

Taguchi defines the quality as deviation on-target performance. He says that the quality of manufactured produt is total loss generated by that product to society from the time it is shipped.

Financial loss or quality loss $L(y) = k(y - \text{cm})^2$

where Y is objective characteristics

M is target value

$$k = \text{contant} = \frac{\text{cost of defective product}}{(\text{tolerance})^2} = \frac{A}{\Delta^2}$$

10.2. TAGUCHI QUALITY LOSS FUNCTION

Figure 10.2 shows the Taguchi's quality loss function.

Taguchi's developed the concept of the quality loss function in the 1970's. Genichi Taguchi's impact on the word quality scene has been tremendous. His quality engineering philophy has been used successfully by many companies in Japan and elsewhere.

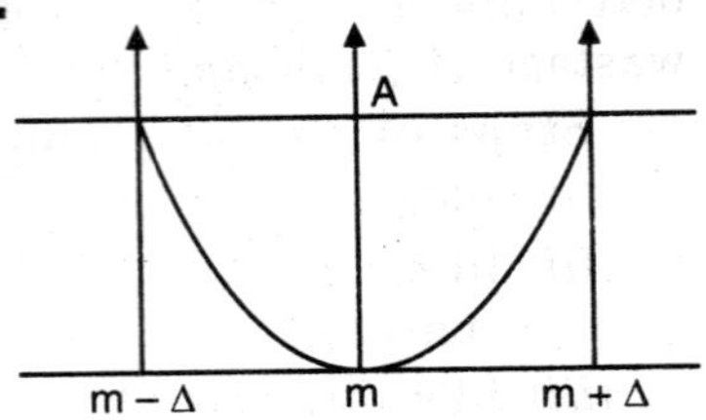

Fig. 10.2. Taguchi's quality loss function

The stress is on the importance of designing quality into product and into processes. There is not dependence on the more traditional tools of on-line quality control. Taguchi focuses more on the engineering apects of quality rather than on management philosphy of statistics. He uses experimental design primarily as a tool to make products more robusts to make them less sensitive to noise factors. He considers experimental design as a tool for reducing the effects of variation on product and process quality characteristics. Taguchi's quality philosophy has seven elements which are as follow:

(*i*) An important dimension of the quality of a manufactured product is the total loss generated by product to society.

(*ii*) In a competitive economy, continuous quality improvement and cost reduction are necessary for staying in business. He identifies three types quality costs – R & D costs, manufacturing costs, and operating costs. All three costs can be reduced through use of suitable experimental designs.

(*iii*) A continues quality improvement program includes continuous reduction in the variation of product performance characteristics about their target values.

(*iv*) The customers loss due to a products performance variation is often approximately proportional to the square of the deviation of the performance characteristics from its target value.

(*v*) The final quality and cost of manufactured product are determined to a large extent by the engineering designs of the product and its manufacturing process.

(*vi*) A product or process performance variation can be reduced by exploiting the nonlinear effects of the product or process parameters on the performance characteristics.

10.3. OBJECTIVE AND STEPS OF TAGUCHI METHODS

The objective of Taguchi's effort is improvement in process and product design. This is by using the identifications of easily controllable factors and their settings. This minimizes the variation in product response simultaneously keeping the mean response on target. If all these factors are set at their optimal levels, then the product becomes robust to changes in operating and environmental conditions. All this is achieved during Taguchis parameter-design stage by removing the bad effect of the cause instead of the cause of the bad effect. The method is applied in a systematic way at a pre-production stage, therefore, it can reduce the number of time measuring tests for determining cost-effective process conditions. This gives savings in costs and wastage of products.

Steps of Taguchi method are as follows:

(*i*) Identification of the main function, side effects and failure mode.

(*ii*) Identification of the noise factor, testing condition and quality characteristics.

(*iii*) Identification of the objective function to be optimized.

(*iv*) Identification the control factor and their levels.

(*v*) Selection of the orthogonal Array and Matrix experiment.

(*vi*) Conducting the matrix experiments.

(*vii*) Analyzing the data, prediction of the optimum levels and performance.

(*viii*) Performing the verification experiment and planning the failure action.

10.4. JUST-IN-TIME (JIT)

Japanese production management technique just-in-time (JIT) is the most successful strategies of the production system. JIT is a manufacturing philosophy, which aims at having the right part at precise right time, and the right quantity to go into assembly.

JIT system was developed at the Toyota Motor Company in Japan in mid 1970s due to Japanese environment where in lack of space and lack of natural resource were considered as waste of resources. This is a system reduces waste of any type, scrap and rework, thus, it strived to perfect quality. Anything not contributing any value to the product is a waste. The full capability of the workforce is utilized. The workforce, itself is made responsible for producing quality products/parts just-in-time to support the next production process.

It is a management philosophy based on elimination of all unnecessary works. It saves time, inventory, scrap etc. There are no problems or bottlenecks. A streamlined production is achieved. It requires employee participation, small lot size, total quality control, and basics of industrial engineering. It considers stable environment, commitment to quality and reduction of waste, and involvement at all levels of organisation.

10.4.1. Characteristics

Ideal lot size in JIT is one piece aiming at driving all queues towards zero to:

(*i*) Minimize inventory

(*ii*) React faster to changes

(*iii*) Shorter production lead times and setup times

(*iv*) Focussed factory

(*v*) GT manufacturing system

(*vi*) Quality problem and its control at source

(*vii*) Total preventive maintenance

(*viii*) Cross-trained worker and worker flexibility

(*ix*) Uniform work load

(*x*) Decentralization

It encourages workforce to participate in the programmes for work improvement, profit increase through cost reduction, inventory reduction and quality improvements leading to waste elimination. This approach is based upon two criteria.

1. It is for production and supply of required number of parts as per need. It is called JIT production system.
2. Another criteria is self-actuation *i.e.*, Whenevery any unusual situations arises in a production line, the incharge stops the line and takes the help in fixing the problem.

The JIT production system uses the pull system rather than traditional push system.

This system concentrates on the end of the line and pulls work-through from preceding operations. Workers perform the required operation on the material or workpiece, drawn from preceding workstations at the necessary time with the instruction card (a Kanban). Thus, work-in-process inventory is minimized, and over-production can be eliminated completely.

10.4.2. Preliminaries to JIT Production

For implementing JIT production, the following should be done first:

(*i*) The individual operations need to standardize through the help of work-study methodology by reducing the non-value adding work-elements to minimum.

(*ii*) Layout provide U-shape so that each operator can handle more than one machine work-centre. This reduces the labour cost.

(*iii*) Multiple-job work by each operator. Every operator is required to master multiple jobs. However, this may lead the lack of enrichment and satisfaction.

(*iv*) Some degree of automation in the work-centre or using machining centre NC system will be reducing set-up times. The reduction of batch sizes and work-in-progress inventory happens as consequence.

(*v*) Stabilization of a process for multiple product in small batch production for production smoothing.

(*vi*) When unusual events happen in production-line, the work-in-charge should stop the work and take help to remove the cause of trouble such self-actuation will help.

10.4.3. Push System

Push system has master schedules and output of Material Requirement Planning (MRP) for determining product schedules and the movement of material in the factory.

This approach determines demand forecast for each item of production. There is difference in sales forecast and demand forecast due to adjustments on the basis of past history of company's sales and company's financial outlays for next few months. This demand forecast considers the status of finished goods inventory. After proper adjustment for the already available finished inventory, production plan for manufacturing is decided. Each unit of the system produces output irrespective of queue before the succeeding units. The overall system thus has many queues of in-process/finished inventory. Material flow and building of schedule in a push system is shown in Fig. 10.3.

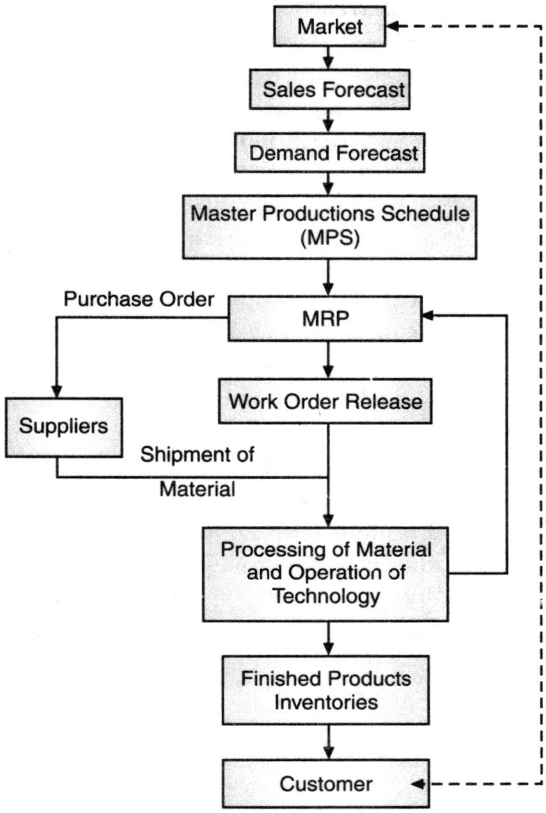

Fig. 10.3. Material Flow and Building of Schedule in a Push System

10.4.4. Kanban and Pull System

In JIT, Kanban cards are used to move material through the production. Two common types of kanban are **withdrawl kanban** and **production kanban**. (*i*) Withdrawl kanbon which is as follows:

<table>
<tr><td colspan="3">Part Number ....</td><td>Preceeding process</td></tr>
<tr><td colspan="3">Part Name ....</td><td></td></tr>
<tr><td>Box capacity</td><td>Box type number</td><td>Issue</td><td rowspan="2">Subsequent process</td></tr>
<tr><td></td><td></td><td></td></tr>
</table>

(*ii*) Production kanban is as follows:

Part Number	Process
Part Name	
Stock location at which to store:	
Container Capacity	

The withdrawl kanban is for releasing material from previous output area or vendor. The production card is for instruction to start production from a previous work centre, material processing or material supplier.

The process of kanban system is as follows:

(*i*) The withdrawl kanban reaches the output area of previous stage. The worker brings the card to the store location of the previous level.

(*ii*) There could be a situation when enough material is available. After comparing and matching the part required in production kanban with the withdrawn kanban, the worker removes the production kanban from the container places the withdrawl card. The removed production kanban is placed on the production kanban post.

(*iii*) The production at the previous work-centre starts when the post for production kanban contains specified number production kanban.

(*iv*) The material at the previous work-centre, which has just been tagged with withdrawl kanban, is transported to next work-centre. It is placed in this work-centre till the required production stands here. In case of material requisitioned from vendors, the work centre relaxes withdrawl kanban at the time when material is just needed from vendors.

(*v*) Once the material reaches at a work-centre, the worker checks the withdrawl kanban, removes it and places the withdrawl kanban-post. The material, however, is used for production or shipment.

The complete kanban system is shown in Fig. 10.4.

The material flow and building of schedule in a pull system is as given in Fig. 10.4.

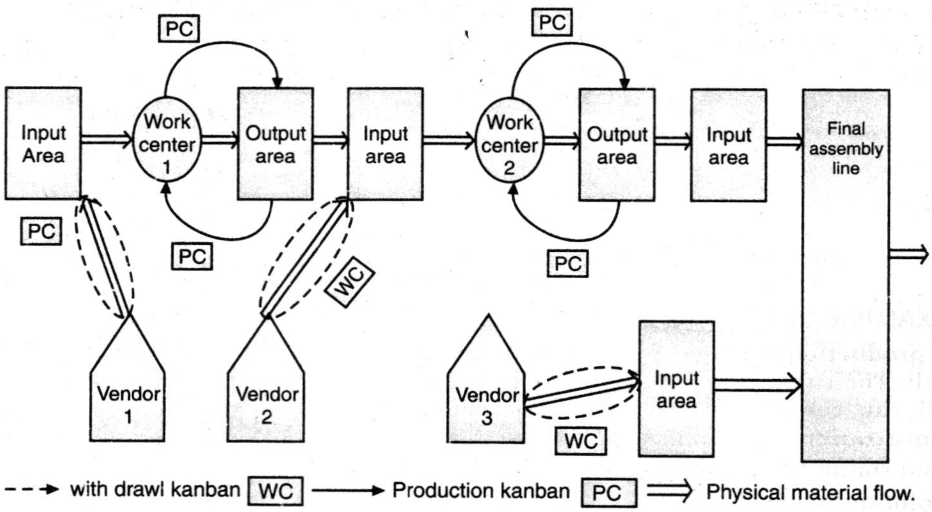

Fig. 10.4. Complete Kanban System

Kanban system attempts at continual reduction of inventory. The number of Kanbans are calculated as follows:

$$N = D \times L + \frac{S}{C}$$

where, *N:* Number of Kanban or containers

D: Demand units *i.e.*, average number over a given period of time

L: Lead time, *i.e.*, time to replenish an order, expressed in the same time unit as expressed in demand

S: Safety stock as a percentage of demand during lead time, based on service level and variance of demand.

C: Container size

Container size should be kept much smaller say 10 to 16% than the average demand during the lead time.

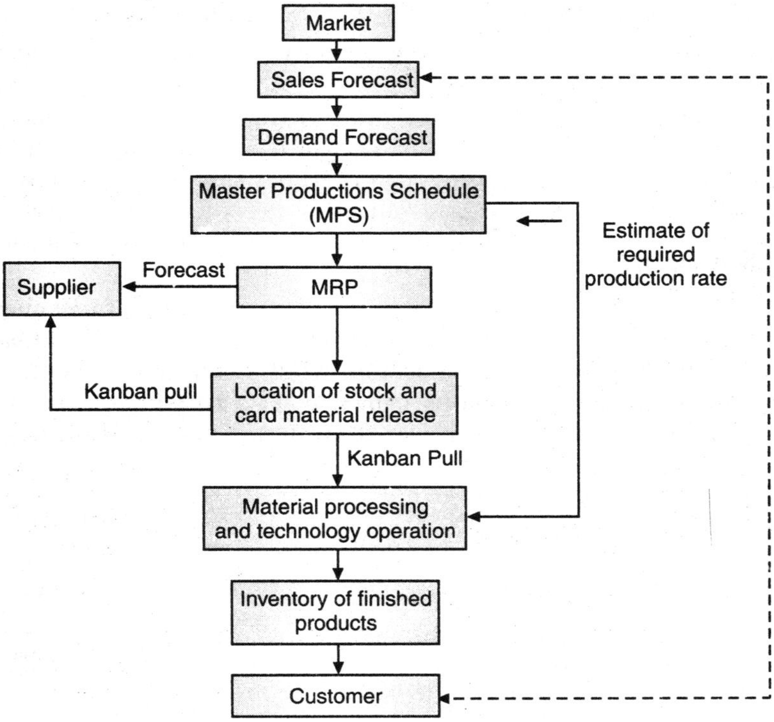

Fig. 10.5. Material Flow and Building of Schedule in a Pull System

EXAMPLE 10.1.

A production manager has to process an average of 250 parts per hour in the cell. The capacity of each container is 30 parts and one Kanban attached to all the containers. The time to receive new parts from the previous workstations is 25 minutes. Factory maintains a safety stock factor of 15%. Determine the Kanban needed for the plant.

Solution: Given,

$$D = 250 \text{ parts per hour}$$
$$L = 25 \text{ minutes} = 0.4167 \text{ hour}$$
$$S = 0.15\,(250 \times 0.4167) = 15.6$$
$$C = 30 \text{ parts}$$

Now, since number of Kanban is

$$N = D \times L + \frac{S}{C} = (250 \times 0.4167) + \frac{15.6}{30} = 3.99 \simeq 4 \text{ Kanbans}$$

10.4.5. Demerits

JIT production is useful for inefficient multi-production, small-batch production. It is a Toyota production system and is the most significant development in production system. This system production is very efficient, but has some demerits, such as:

(*i*) It is effective only when the daily demands are quite stable.

(*ii*) Stocks, are maintained to assure JIT delivery to customers, therefore, the minimum zero inventory can be theoretically achieved.

(*iii*) Only principle of first come first served is used by manufacturing items in order of releasing cards.

(*iv*) Lot needs to be done for employer-employee cooperation, daily workstation rotation, training operators for different kinds of jobs, and systems adaptability to market fluctuation. A labour environment conducive to variety of serious issues are needed.

10.4.6. MRP vs JIT system

MRP stands for material resource planning. JIT system and MRP system are opposite in nature. JIT is for minimization of the inventory while MRP is for planned inventory. MRP is a push-system while JIT is a pull-system. Push-system produces inventory for future use. Pull system demands material from a previous work-centre or vendor at a time when material is just needed. Comparison JIT versus MRP are as follows:

S. No.	Factors	JIT System or Pull System	MRP System or Push System
1.	Information processing	Kanban cards: simple visual control	Computationally extensive process of records
2.	Inventory	Treated as liability. Efforts to minimize. Reduced inventory, exposes the production problem	Treated as asset. Maintained to cover the market uncertainty as safety stock. Planned inventory keeps production problem under inventory cover.
3.	Lot sizes	Requisitioned for immediate use	Requisitioned for a schedule, calculation for each time-bucket. It should balance set-up cost and holding cost
4.	Demands variations	Works well	Works poorly.
5.	Vendors	Treated as co-worker Effort on developing one reliable vendor at a location nearest to plant.	Treated as adversaries. Effort on having multi-sources to advantage of competitors
6.	Quality	Efforts on zero-defect	Tolerant on scrap by building formula of predicting it.
7.	Machine maintenance	Critical activity as no machine failure is desirable	Not a critical issue as queues are admissible.

Reduced inventory or zero inventory benefits are as follows:

(*i*) Warehouse and work space saving.

(*ii*) Inventory cost reduction leads to saving in working capital.

(*iii*) Low inventory level feasible only when the major productivity related problems are eliminated. Subsequently, quick response to problems during operation is must.

Soon after, quick response to problems and prevention from disrution are necessary for smooth operation.

In short, lower asset level leads to better utilization of resources as shown in Fig. 10.3.

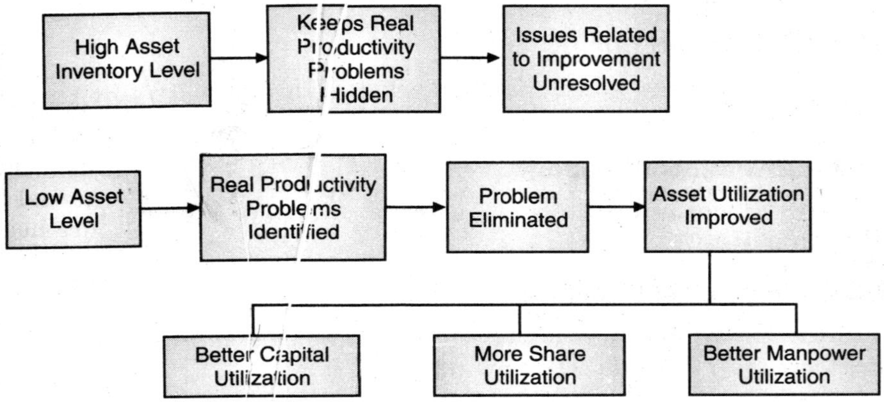

Fig. 10.6. Lower Asset Level Leads to Better Utilization of Resources

10.5. JIT IMPLEMENTATION

JIT requires long time to pursue the basic ingredients of JIT. Some requirement is controllable by the management, while rest is uncontrollable. JIT implementation is shown in Fig. 10.7.

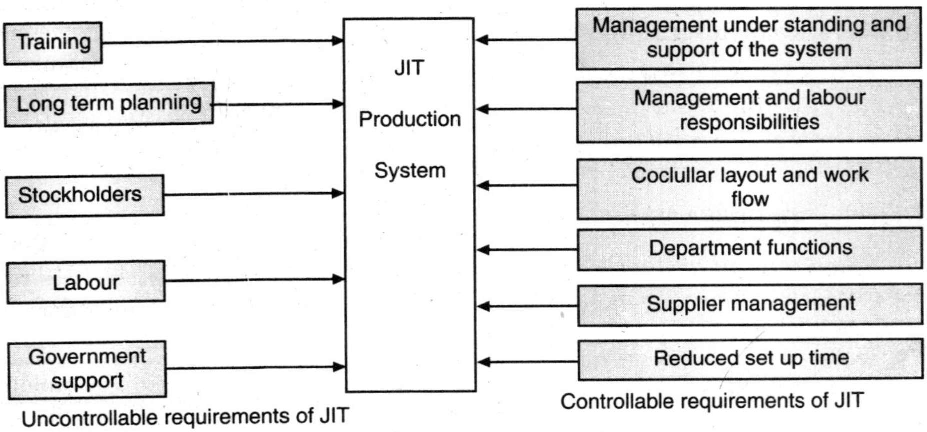

Fig. 10.7. JIT Implementation Requirements

The details are given below:

10.5.1. Training

With investment in training and development firm upgrades skill of workers. Better quality production is achieved through training. In addition, the productivity also improves.

10.5.2. Long Term Planning

This requires continuous improvements and trial and error approach.

10.5.3. Stock Holders

The earned profit is reinvested in the firm stockholders, who are "management worker" among stockholders as dividend.

10.5.4. Labour and Union

JIT demands for multifunction worker with flexibility to switch over. Union must be enterprise-oriented.

10.5.5. Government Support

Better "labour-union-manage government" relation is a must for conducive JIT environment.

10.5.6. Management Support

For elimination of productivity-related problems and reduction of asset/inventory level, the support of management is a must.

10.5.7. Management and Labour Responsibilities

Healthy criticism should be encouraged and hidden problems should be eliminated. Team decision-making and cooperative problem solving facilitates.

10.5.8. Cellular Layout and Work Flow

Group technology based cellular layout is needed. The flow of materials and arrangement of machines in the cell may preferably be of U-type.

10.5.9. Department Function

Workers are better trained to solve quality and engineering problems. In case of any problem, at any worker level, it is attended by entire department even if total production remains disrupted for few minutes. Traditional functions of departments get modified.

10.5.10. Supplier Management

Suppliers are treated as partner rather than competitors. They trained in quality and production so that no defective should reach at the shop floor or assembly line.

Some of the ways to reduce set-up time are:

(*i*) Standardization the external set-up actions need adjustable gauges to be replaced with permanent ones.

(*ii*) Colour-code all hole connections; air, hydraulic, water etc.

(*iii*) Design a quick locating for positioning pins and hols, with quick fasteners.

(*iv*) Add longer monitor on press to detect two piecess in die before damaging dies.

(*v*) Involve tool-and-die designers in set-up reduction programs or all new design incorporate quick change-over conceptor.

(*vi*) Photograph completed operation or a guide for set-up man such as location of tables, branches, baskets, etc.

(*vii*) Review material flow charts with reduced movements in mind.

(*viii*) Standardize all bolt sizes.

(*ix*) Code parts on dispatch list for major or minor set-up to aid scheduling.

10.6. JIT PRODUCTION PROCESS

Master production plan is for final stage of the whole production process. The worker perform the required work on the work-material preceding by the providing workstation at the necessary time. Hence, it is called pull-through system. We need to use two kinds of Kanban for the purchase of providing production information need be used which are:

(*i*) withdrawl information kanban.

(*ii*) production ordering information kanban.

The withdrawl Kanban indicates the parts or items and their quantities to be brought from the previous stage to the current workplace. Once this Kanban is issued the parts and the quantities indicated on Kanban are withdrawn and brought to the current workplace together with the Kanban.

Kanban allows to withdraw parts from the store of previous stage so as to move them to the next stage. The production starts and continues uninterrupted as quantities of the withdrawn parts have been completely replenished.

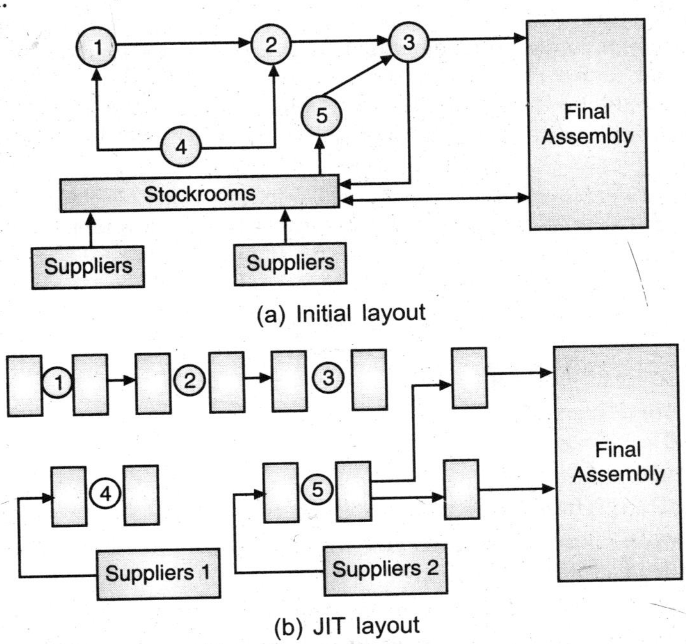

Fig. 10.8. Structured Production

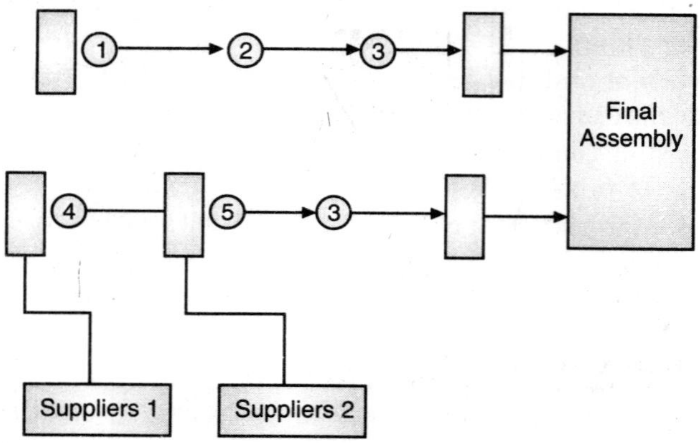

(c) JIT layout with GT

Fig. 10.8. Structured Production

The Fig. 10.8. indicates the pull-through type production system receives the master production schedule at in the beginning. Required parts are brought from the previous stage with the circulation of Kanban. This plays a role of information flow. It shows different layouts configurations.

SUMMARY

1. Japanese production management technique just-in-time (JIT) is the most successful strategies of the production system. JIT is a manufacturing philosophy, which aims at having the right part at precise right time, and the right quantity to go into assembly.
2. JIT system was developed at the Toyota Motor Company in Japan in mid 1970s due to Japanese environment where in lack of space and lack of natural resource were considered as waste of resources.
3. MRP stands for material resource planning. JIT system and MRP system are opposite in nature. JIT is for minimization of the inventory while MRP is for planned inventory. MRP is a push-system while JIT is a pull-system.
4. Push system has master schedules and output of Material Requirement Planning (MRP) for determining product schedules and the movement of material in the factory.

EXERCISES

1. What is JIT? Explain.
2. Give comparison of MRP vs JIT system.
3. Describe Push system.
4. Explain Kanban.
5. Describe JIT Implementation.
6. Explain JIT production process.
7. Describe Taguchi Quality.
8. What is Tagushi's Quality loss function?
9. Give objectives and steps of Taguchi method.

❏❏❏

11 CHAPTER Eleven

ISO-14000 QUALITY MANAGEMENT SYSTEM

CHAPTER OUTLINE

- Introduction
- ISO 14000
- ISO 14001
- ISO 14004
- ISO 14064
- Other ISO 14000 Series Standards
- Similarities between ISO 9001 and ISO 14001
- Differences Between ISO 9000 and ISO 14000

11.1. INTRODUCTION

The International Organisation for standardisation widely known as ISO, is an international standard-setting body composed of representatives from various national standard organisations. Founded on 23 February 1947, the organisation promulgates worldwide proprietary industrial and commercial standards. It has its headquarters in Geneva, Switzarland. While ISO defines itself as a non-governmental organisation, its ability to set standards than often become law, either through treaties or national standards, makes it more powerful than most non-governmental organisation. In practice, ISO acts as a consortium with strong links to governments.

ISO is a voluntary organisation whose members are recognized standard authorities, each one representing one country. The bulk of the work of ISO is done by the 2700 technical committees, sub-committees and working groups. Each committee and sub-committee is headed by a secretariat from one of the member organisations. The American National Standards Institute (ANSI) is the united states representative to ISO. ISO's main products are the International Standards. ISO also publishes Technical Reports, Technical Specifications, Publicly, Available Specifications, Technical Corrigenda, and Guides.

11.2. ISO 14000

ISO 14000 standards are developed by the International Organisation for Standardization (ISO), which has representation from committees all over the world. ISO 14000 is a series of international standards on environmental management. It provides a framework for the development of an environmental management system and the supporting audit programme. These standards exist to help organisations minimise how their operations negatively affect the environments and promote more effective and efficient environmental management in organisations and to provide useful and usable tools ones that are cost effective, system-based, flexible and reflect the best organisations and the best organisational practices available for gathering, interpreting and communicating environmentally relevant information.

It offers a source of guidance for introducing and adopting environmental management systems based on the best universal practices, in the same way that the ISO 9000 series on quality management system, which is now widely applied, represents a tool for technology transfer of the best available quality management practices. In structure, the ISO 14000 series is similar to ISO 9000 quality management and both can be implemented side-by-side.

ISO 14001 standard represents the core set of standards used by organisations for designing and implementing an effective environmental management system. ISO 14004, gives additional guidelines for good environmental management system and more specialised standards dealing with specific aspects of environmental management.

11.3. ISO 14001

ISO 14001 specifies a framework of control for an Environmental Management System against which an organisation can be certified by a third party. It sets out requirements for an Environmental Management System (EMS) which can be employed by an organisation to measure and document their environmental impact.

In 2004, the ISO 14001 standard was updated to ISO 14001: 2004 to reflect changes in environmental consciousness. It can lead to benefits such as the reduced cost of waste management, savings in consumption of energy and materials, lower distribution costs, improved corporate image among regulators, customers and the public, and framework for continual improvement of environmental performance. It applies to those environmental aspects that the organisation identifies as those which it can control and those which it can influence. It does not itself state specific environmental performance criteria.

It is applicable to any organisation that wishes to establish, implement, maintain and improve an environmental management system to assure itself of conformity with its stated environmental policy.

ISO 14001:1996 has been introduced which provides framework for pollution prevention and waste reduction and compliance with legislative and regulatory requirements with:

(*i*) Reduction in liability/risks.

(*ii*) Competitive advantage for 'Green' products Globally.

(*iii*) Creation of an improved community goodwill.

(*iv*) Improved environmental performance.

(*v*) Improved Industry Government relations.

(*vi*) Helps in conservation of raw materials and energy.

(*vii*) Reduces incidents that results in liability.

11.4. ISO 14004

ISO 14004 provides guidance on the development and implementation of environmental management system. The guidelines in ISO 14004: 2004 are applicable to any organisation, regardless of its size, type, location or level of maturity. While the guidelines are consistent with the ISO 14001: 2004 environmental management system model, they are not intended to provide interpretations of the requirements of ISO 14001:2004.

ISO 14004:1996 contains guidelines on the development and implementation of environmental management systems and principle, and their coordination with other management systems. The guidelines are intended for use as a voluntary internal management tool.

11.5. ISO 14064

ISO 14064 standards provide tools for accessing and supporting greenhouse gas (GHG) reduction and emissions trading. Implementing ISO 14064 is intended to achieve the following benefits:

(*i*) Promote consistency, transparency and credibility in GHG quantification, monitoring, reporting and verification.

(*ii*) Enable organisations to identify and manage GHG-related liabilities.

(*iii*) Facilitate the trade of GHG allowances or credits, and

(*iv*) Support the design, development and implementation of comparable and consistent GHG schemes or programmes.

ISO 14064 comprises three standards: They can be used independently, or as an integrated set of tools do meet the varied needs of GHG accounting and verification. They are:

(*i*) ISO 14064-1:2006, Greenhouse gases - Part-I: Specification with guidance at the organization level for the quantification and reporting of greenhouse gas emissions and removals.

(*ii*) ISO 14064-2:2006, Greenhouse, gases-Part 2: Specification with guidances at the project level for the quantification, monitoring and reporting of greenhouse gas emission reductions and removal enhancements.

(*iii*) ISO 14064-3:2006, Greenhouse gases-Part 3: Specification with guidance for the validation and verification of greenhouse gas assertions.

11.6. OTHER ISO 14000 SERIES STANDARDS

Other standards in the series are actually guidelines, many to help to achieve registration to ISO 14001. These include the following:

(*i*) ISO 14010 provides general principles of environmental auditing.

(ii) ISO 14011 provides specific guidance on audit an environmental management system.

(iii) ISO 14012 provides guidance on qualification criteria for environmental auditors and lead auditors.

(iv) ISO 14013 provides audit program review and assessment material.

(v) ISO 14015 provides guidelines on environmental assessment of sites and organisations.

(vi) ISO 14031 provides guidelines on environmental performance evaluation.

(vii) ISO 14062 discusses making improvements to environmental impact goals.

(viii) ISO 14063 provides guidelines and examples on environmental communication.

11.7. SIMILARITIES BETWEEN ISO 9001 AND ISO 14001

ISO 9000 is an international standard for the development of quality management system that can potentially be adopted by any business. ISO 14000 is an international standard for environmental management systems than can also be potentially adopted by any organisation.

There are many similarities between ISO 9000 standards and ISO 14001. All management systems, whether focused on quality, safety, or the environment, have certain core elements. These includes policy, defined organization and responsibilities, training, records system, internal audits, corrective action system, management review for continual improvement.

In case of ISO 9001 and ISO 14001, there are similar requirements for all of these core elements.

11.8. DIFFERENCES BETWEEN ISO 9000 AND ISO 14000

ISO 9000 purpose is to enable companies to demonstrate that they are in a position to provide products or services that meet customer expectations and are focuses on total customer satisfaction. ISO 14000 serves as tool for defining and implementing activities that meet environmental business concerns.

A quality system will include evaluation of suppliers and review of customer contracts. An environmental system will include methods of evaluating environmental impacts and systems for responding to emergencies.

In a quality system, the major system requirements are defined by customers. This means that the goals of a quality system are generally quite clear to meet or exceed customer requirements. Differing requirements from various customers are the only complexities which may complicate the system. The situation for environmental systems is not so straight forward. Customers may still be interested in a company's environmental impacts, but there are many additional stakeholders to consider as well. Regulatory bodies, local communities, and environmental activist groups all have an interest in a business environmental performance. Additional system requirements may also come from international treaties, industry-specific programs, and codes of practice. This all adds upto very complex and constantly changing set of requirements from a wide variety of stakeholder.

SUMMARY

1. The International Organisation for standardisation widely known as ISO, is an international standard-setting body composed of representatives from various national standard organisations.
2. ISO is a voluntary organisation whose members are recognized standard authorities, each one representing one country. The bulk of the work of ISO is done by the 2700 technical committees, sub-committees and working groups. Each committee and sub-committee is headed by a secretariat from one of the member organisations.
3. ISO 14000 is a series of international standards on environmental management.
4. In structure, the ISO 14000 series is similar to ISO 9000 quality management and both can be implemented side-by-side.
5. ISO 14001 specifies a framework of control for an Environmental Management System against which an organisation can be certified by a third party. It sets out requirements for an Environmental Management System (EMS) which can be employed by an organisation to measure and document their environmental impact.

EXERCISES

1. Explain ISO-14000 Quality Management System.
2. Discuss ISO-14004 Quality Management System.
3. Explain ISO-14064 Quality Management System.
4. What are the similarities between ISO-9001 and ISO-14001?
5. Differentiate between ISO-9000 and ISO-14000.

12 Chapter Twelve

Quality Audit

CHAPTER OUTLINE

12.1. INTRODUCTION

A quality system assessment comprises of a number of key elements which are:

(*i*) Audit Planning,

(*ii*) Adequacy audit,

(*iii*) Compliance audit,

(*iv*) Audit reporting,

(*v*) Obtaining corrective action;

(*vi*) Follow-up and closeout of non-conformances.

12.2. AUDIT PLANNING

When planning the audit it must always be remembered that the best of audits are only a sample of the system. In an attempt to make that sample as representative as possible, it is necessary to identify the critical activities, processes and departments and ensure that these are audited first.

A matrix for planning purposes is used for creating horizontal and vertical audit trails which can then be used to determine the best use of the audit team's resources. The primary focus of this training program is on the conduct of internal audits, and it is extremely common sense for those to take place with only one auditior in the team. In this situation, when time and resources are at maximum premium, careful planning is of even greater importance than when there is a complete team of staff available to conduct the audit.

Horizontal audits trails are process related, while vertical audit trails are department important. Wherever possible the same auditor should be allocated to individual horizontal or vertical audit trails thereby giving the auditor the opportunity to follow them all the way through and obtain a total picture of the process or department under view. The allocation of vertical and horizontal trails presents the audit planner with certain difficulties. Part of the planning task is to ensure that no one in the company or department being audited has to be interviewed by more than one auditor, and this can militate against an individual auditor following a complete horizontal or vertical. If the internal audits are undertaken, individuals trails will often be followed during different parts of the audit cycle.

Table 12.1. Audit Matrix

	Sales	Purchasing	Goods in	Rec. Imp.	Stores	Design	Manufact.	Indust. Eng.	Inspection	Test	QA	Pack/Desp.
Quality system	X	X	X	X	X	X	X	X	X	X	X	X
Documentation	X	X	X	X	X	X	X	X	X	X	X	X
Software	—	—	—	X	—	X	X	X	0	X	X	—
Material	—	X	X	X	X	0	X	0	X	X	X	X
Returns	0	—	0	X	0	0	X	0	X	X	X	0
Calibration	—	—	0	X	X	X	X	X	X	X	X	—
Training	X	X	X	X	X	X	X	X	X	X	X	X

Relevance:

X = High

0 = Low

..... None

12.3. PERFORMING AUDIT

The objective of the audit is primarily to ascertain that the documented procedures are being adhered. The ISO 9000 series make it very clear that the task of assuring compliance to the standard is primarily a management function. ISO 9001: 1994 paragraph 4.1.2.3 states that the management representative shall have defined authority for:

(*a*) "ensuring that the quality system requirements are established, implemented and maintained in accordance with this international standard".

Arguably the internal auditor has no concern with issue of system adequacy, and is solely concerned with issues of system compliance. In reality, particularly in the earlier days of system implementation, the internal auditor becomes the individual who is expected to flag any problems with system adequacy to the management.

Arguably the internal auditor has no concern with issues of system adequacy, and is solely concerned with issue of system compliance. In reality, particularly in the earlier days of system implementation, the internal auditor becomes the individual who is expected to flag any problems with system adequacy to the management.

Therefore, if becomes necessary to split the audit into two parts:

First: The auditor performs an adequacy audit of the documentation against the standard;

Second : The auditor performs a compliance audit of the actual working practices against the procedures.

The auditors stock in traid is objective evidence of compliances, and of non-compliances. The customer who is the management of the company, will not be interested in hearing about how the auditor "felt" about things, now what the auditor "thinks" about something. What they want to hear is that there is "Objective evidence that ..."

Objective evidence was identified in section one as being "Observations and statements of facts and the information contained in records which can be verified". A some what more worthy definition is provided in ISO: 10011 – 1 The internation standard which lays down a formal basis for the management of audits. The definition contained within ISO 10011 – 1 is as follows.

"Qualitative or quantitative information, records or statements of facts pertaining to the quality of an item or service or to the existence and implementation of a quality system element are based on observation, measurement or test and which can be verified".

Objective evidence is be collected through interviews, examination of documents, and observations of activities and conditions in the relevant areas. No auditor should even accept as gospel any information gathered from an interview unless it can be verified by cross checking the same information from an independent source, such as records or observation. The internal auditor, even more than the second and third party auditor, needs to be aware of the risk that he or she might get used as a pawn in managerial "turf wars" and other manifestations of company politics. The next time that an employee tried to "score one over" a colleagues by trying to plant evidence of non conformance through an audit interview will not be first time that situation has been encountered.

A typical list of the type of problems which will be found most regularly is shown here:

- procedure not being followed;
- unauthorised handwritten changes to documents and drawings;
- operation performed with no work instruction;
- obsolete documents and drawings at place of work;
- measuring and test equipment out of calibration;
- measuring equipment used by support departments, *e.g.*, maintenance, industrial engineering not included in calibration system;
- measuring and test equipment in the design area not included in the calibration system;

– purchasing from unaproved vendors;
– poor FIFO and shelf life control in stores;
– incorrect use of statistical methods;
– no formal contract review;
– no follow up of corrective action;
– audit findings not closed out;
– no records of management review.

12.4. AUDIT FORMALITY

This training program focuses on the activity of the auditor acting as an internal auditor working within his or her own corporate environment auditing system being operated by people who are often friends as well as colleagues. This personal relationship which exists between the internal auditor and the auditee makes the task one of the most difficult of all audit activities. Despite the fact that a good auditor and a committed management will always emphasise that it is the system not the person who is being audited, it is often difficult for staff to accept this in the early days of operating an audit program. A great deal of tact and diplomacy is therefore required by the individual who act as an internal auditor.

One way in which the impact of the personal side of the audit process can be minimised, and the auditor gain recognition for the activity from managers and colleagues.

Who will often be senior to themselves in the company hierarchy, is to introduce a degree of formality into the process, and behave more like an external auditor. How formal a company wishes to make the audit will depend vary much on company size and its culture but experience has shown that when the full formal structure of an audit is followed, the entire process is taken much more seriously by all connected.

Formal audit activity consists of:

1. Planning
2. Notification,
3. An Opening meeting
4. The audit itself
5. The closing meeting
6. The audit report
7. Follow up and closed out
8. Management review.

12.5. AUDIT NOTIFICATION

Audit planning activity is to make the arrangement with the supervisor or manager of the area to be audited for the audit to proceed as planned. When the audit schedule is prepared and published well in advance of the activity. The formal notification to the area concerned is little more than a reminder. None the less, it is a step worth taking. The internal audit process must contain no surprises for the auditee.

Most frequently the contact will be either personal or by internal telephone. The use of e-mail is becoming increasingly common, and when this media is chosen, a copy of the e-mail note should be taken and placed in the audit file. When the contact has been personal or by telephone, follow up with a short written note to the individual concerned identifying the key facts about the audit:

(a) date and time
(*b*) number of auditors
(*c*) scope of the audit
(*d*) number of guides needed, if any
(*e*) duration of audit
(*f*) invite staff to attend opening meeting
(*g*) indicate likely time of closing meeting.

The larger the organisation, the more important this level of formality is. In very large companies, guides can be almost essential while in small companies they will nearly always be unnecessary. Ask the addresses to confirm the arrangements.

Both this notification letter and the reply should be placed in the audit file.

12.6. OPENING MEETING

When the audit is a team effort, it is important that auditors arrive in the area or department as a united team; there is nothing worse than the team ariving in ones twos . As soon as all the auditors and all the management personal are assembled, the lead auditor should take control of the situation and get straight into opening meeting. There should be a written agenda in front of him or her. The key points of which are as follows:

(*a*) When necessary introduce the members of the audit team to the auditee and have the auditee introduce his or her staff;
(*b*) Pass round attendance record;
(*c*) Review the scope and objectives of the audit;
(*d*) provide a short summary of the methods and procedures to be used to conduct the audit;
(*e*) clarify any ambiguities make sure that every one is clear about events;
(*f*) detail how non-conformance will be recorded. Remind the auditee of the use of the report form and of witnessing findings;
(*g*) if necesary arrange for escorts;
(*h*) confirm the approximate time for the closing meeting;
(*i*) if necessary ask for a quick four of the area.

The team leader acts as the chairman during this meeting which should be relatively short and last no longer than 10–15 minutes establish if there are any questions then close the meeting. Each auditor will then leave with their respective guides or escorts and starts the audit.

12.7. AUDIT

The audit itself will be conducted through the use of the interviewing techniques which have already been discussed and the collection of the all important objective evidence. The ultimate auditors question when pursuing an audit trail is to close out the engineering with the request "show me". For the system implementation, the evidence. It is either there or it isn't there which are called conformance or non-conformance.

The audit findings will be a series of notes about what has been observed during the audit possibly with photocopies of documents attached. Once they have been collated, some of these notes will be translated into non-conformance reports and some of them will remain as observations. There will also be actual non-conformance reports which have been raised during the audit.

Each non-conformance report must either be reference back to the procedure, the standard, or a particular contract as appropriate. While generating the non-conformance report, the auditors must remember the golden rule given in section one-if you cannot express the problem in the words of the procedure, standard or contract against which the audit is being conducted, then this is no non-conformance.

12.8. NON-CONFORMANCE REPORT

A typical non-conformance report is Fig. 12.1. The report is split into five parts:

Part 1: Identifies the auditor, the area or procedure audited, the date and the report number.

This part of the report is completed by the auditor.

Part 2: Details the non-conformance found and the objective evidence. It also carries signatures from the auditor and the auditee.

This part is completed by the auditor either during the audit. The finding should be witnessed by the auditee or the guide or when other observation have been collated, the auditors should seek a signature on the form of the auditee at the closing meeting.

Part 3: Sets out the corrective action agreed by the responsible manager. It defines the completion date and is signed and dated by the responsible manager.

This part should be completed by the manager or supervisor responsible for the area being audited. Sometimes this will happen at the closing meeting. More often, it will be completed and returned to the audit manager within an agreed time after completion of the audit and submission of the audit report. The action recorded in this part of the form should address two aspects of corrective action-primary action to connect the immediate problem identified, and secondary action to ensure that the problem does not recur. Often times, this will be one and the same action.

Part 4: Records that the corrective actions have been implemented are effective and are signed off.

This part of the form will be signed off by the audit manager once the corrective and preventive action identified in part 3 has been verified.

Verification may in the form of a review of modified documents, or it may take the form of a further audit.

Part 5: Records any comments or action from the quality system management review.

	Quality System Non-conformance Report		Issue-I
Part-1	Report No.	Auditor	
	Department or Procedure audited	Date	
Part-2	Detail of Non-conformance		
Part-3	Proposed corrective action to be completed by : Date .... Responsible manager .... Signed .... Date		
Part-4	Corrective action completed Signed .... Date		
	Management Review Commnets		
Part-5	Management Approval Signed .... Date		

Fig. 12.1 Non-conformance report

Certain audit finding may be offered to the management review meeting in their entire, rather than as elements of a roll-up report on the overall audit process. This part of the form makes provision for such further review and for the management review team to annotate the finding when so required.

Every non-conformance report should be fully completed by the various individuals responsible for each part of the form and carefully filled in an easily retrievable manner. All corrective actions should be time bound. Failure to identify a time frame for corrective action will always result in it being taken tomorrow.

The way in which an audit finding is written up can have a significant impact on the reception when the non-conformance is received. The report should report only the facts which have been second not the assumptions which may or may not follow from the evidence collected. Always remember that the absence of evidence is not perse the evidence of absence. For example, if a record or other document cannot be produced during an audit there is a temptation to assume that the record or document does not exist and to write up the non-conformance in that way. The auditor then gets to look very foolish when the response to the presentation of the non-conformance is to have the missing item in question could not be produced during the audit. In this way the requirement for availabilities is still recorded as a non-conformance, but if the missing paperwork is around, no one ends up with egg on their face.

The corrective action may be to train the staff to know that the missing item exists or to make it accessible. There can be no argument that corrective action is needed, where as a statement that something didn't exists when it did can potentially result in an argument in the first place, and a total lack of corrective action in the second, as production of the goods must result in the nonconformance being withdrawn. So next time, the auditor will face the same problem.

One of the responsibility of the lead auditor is to classify severity of the finding. There are three options which were authined in section one. The likelihood of an internal audit revealing critical non-conformance in a well structed system is remote, but it could happen in which case the auditor would immediately stop the audit and report to the department manager.

A major non-conformance might well be found even through a complete procedure is not being followed at all. In terms of reporting. It is straight forward and will either lead to a change in working practice or a rewrite of the procedure to bring it in line with current activities.

The area within which one is an auditor, one will have to exercise careful judgement in minor non-conformance. It is of vital importance than the auditor does not degenerated into a nitpicking exercise. The auditor must try and engaging the real intent of those being audited and treat minor finding accordingly.

A sensible auditor might even decide that a couple of minor non-conformances noticed during the audit of an area obviously totally dedicated to implementing their system effectively, will not merit a nonconformance report at all, but merely an audit observation.

The audit has already been described as a means of identifying opportunities for improvement. This positive attitude, and the way in which the auditor conducts will have a great deal to do with the success of the message of improvement can be put across. Under no circumstances should audit become a mere paper machine devalued by careless use, and regarded as a necessary evil attached to ISO: 9000.

The skilled auditors use all their senses to the full to observe, listen, speak, and above all use common sense.

12.9. CLOSING MEETING

At the end of audit the audit team will hold a closing meeting with the managers responsible for the functions audited. This takes place immediately after the conclusion of the audit activity and prior to the preparation of the audit report.

The objective of the closing meeting is to present the audit observations in such manner that the managers clearly understand the outcome of the audit. Like the opening meeting this is a formal activity chaired by the lead auditor and should have a formal agenda. This agenda is as follows:

(*a*) say thank you to everyone for co-orperating

(*b*) pass around the attendance sheet again and get everyone to sign in.

(*c*) re-confirm the scope of the audit

(*d*) present an objective overview of the audit results.

(*e*) make sure that the managers understand the findings and help with any clarification which may be needed.

(*f*) identify response dates for corrective action.

(*g*) advise the issue date of the audit report.

The manager or supervisor should be given copies of the non-conformance reports at the closing meeting. The lead auditor may ask each individual assessor to state their findings or he or she may choose to present all the findings personally. In order to keep the meeting under control, all question should be placed and answered through the team leader.

The written audit report usually with NCR's attached should be presented within ten working days. The managers can then respond to the report and detail the corrective action they intend to implement.

12.10. AUDIT REPORTING

The team leader is responsible for the preparation of the audit report, its accuracy and completeness. The audit report should reflect both the tone and content of the audit and be signed and dated by the team leader.

A typical format for the audit report would be:

Summary sheet:

- Summarises the audit and lists the audit findings.

Report Sheet:

- Opening meeting, attendees and outcome.
- Overviews of the audit, with details of findings and observations.
- Closing meeting attendees and outcome.
- Follow up of previous audit findings.

Audit findings:

- Originals of any non-conformance reports.

Covering memo:

- Include the date by which a response to the audit finding is required.

If for any reason the audit report cannot be issued within the ten working days described at the closing meeting, the reasons for the delay must be given to the auditee together with a revised issue date.

12.11. CORRECTIVE ACTION

This is the part of operation with which the auditor should not get involved. Ownership of the corrective action must by all times remains with the audite. The minute that the auditor becomes part of the solution to the finding, the audit program is under threat. There is no way that the auditors can become the owners of all the corrective actions which will be required in any organisation over any period of time. Ownership of system corrective actions must stay with the owner of the system and they are the functional managers.

With internal audits the auditee will usually be able to identify quite easily the actions to be taken by taking advantage of any opportunities for improvement which the audit has revealed. If this is so, the use the remaining sections of the form can be speedily completed, the corrective action verified and signed off and the audit closed.

There are times, however, when this speedy resolution is not possible. It may be that the designated auditee jacks the necessary authority within his or her function to take wide ranging actions which impact on many other department or may be the recommendation for corrective and preventive action requires sweeping changes to many procedure.

It may be that the problems incurred are too wide spread to enable any form of 'quick fix' and re-audit. Under these circumstances the diplomatic skills of the auditor are hard pressed as he or she seeks to find a way of telling the auditee, or more probably his or her senior manager, that the opportunities for improvement are greater in number than the area of compliance.

Time bound corrective action must be agreed by the responsible manager and signed off.

Once corrective action has been implemented, the audit manager should allow several weeks to pass before initiating a follow-up audit. This will allow the action to stabilise, and give some visibility of the effectiveness of the corrective action.

12.12. ANALYSIS OF FINDINGS

Analysis take two forms. The first is the collation and reporting of all deficiencies identified in anyone audit. That is the activity addressed by the audit report. The other form of analysis is the global review of all audit results for the benefit of management review meetings. The latter is only an extension of the former, but of course carries policy implication.

The most important aspect of any analysis is to provide data which will enable the management to drive the continual process of system review and improvement. This being the case, like any other such activity it is important to identify the key areas to address. The application of the Pareto approach is often as good a method of analysis as any. It has the advantage of being simple and easy to understand it to force simply classification within data files. It provides clearly identifiable improvement targets, and best of all the results which feed back into the system are usually quite speedily seen.

12.13. MANAGEMENT REVIEW

"The supplier's management with executive responsibility for quality shall review the quality system at defined intervals sufficient to ensure its continuing suitability and effectiveness in satisfying the requirements of this international standard and the suppliers stated quality policy and objectives".

ISO 9001: 1994 Paragraph 8.1.3.

Note: The results of internal quality audits form an integral part of management review activities.

ISO 9001: 1994 Paragraph 8.17.

The purpose of management review of audit result is to ensure that management objectives, assignments, delegations and methods are:

(*a*) achieving the desired results;

(*b*) revealing defects or irregularities in any of the systems audited;

(*c*) indicating possible improvements;

(*d*) effectively monitoring management at all levels;

(*e*) uncovering potential weak sports and eliminating waste or loss;

(*f*) verifying that corrective and preventive action procedures are effective.

Obviously management review consist of far more than just reviewing system audit findings and corrective action. However, this analysis should form a substantive part of the discussion, comments or actions resulting from the management review as well as being minuted may be detailed in part 5 of the non-conformance report and signed off by the appropriate manager.

By following simple procedures, the system audit activity can be turned into the single most powerful driver of continuous improvement which any management can implement within a company.

SUMMARY

1. A quality system assessment comprises of a number of key elements which are:
   (*i*) Audit Planning,
   (*ii*) Adequacy audit,
   (*iii*) Compliance audit,
   (*iv*) Audit reporting,
   (*v*) Obtaining corrective action;
   (*vi*) Follow-up and closeout of non-conformances.
2. When planning the audit it must always be remembered that the best of audits are only a sample of the system. In an attempt to make that sample as representative as possible, it is necessary to identify the critical activities, processes and departments and ensure that these are audited first.
3. The objective of the audit is primarily to ascertain that the documented procedures are being adhered. The ISO 9000 series make it very clear that the task of assuring compliance to the standard is primarily a management function.
4. Formal audit activity consists of:
   (*i*) Planning

(*ii*) Notification,
(*iii*) An Opening meeting
(*iv*) The audit itself
(*v*) The closing meeting
(*vi*) The audit report
(*vii*) Follow up and closed out
(*viii*) Management review.

5. The audit itself will be conducted through the use of the interviewing techniques which have already been discussed and the collection of the all important objective evidence. The ultimate auditors question when pursuing an audit trail is to close out the engineering with the request "show me". For the system implementation, the evidence. It is either there or it isn't there which are called conformance or non-conformance.
6. The team leader is responsible for the preparation of the audit report, its accuracy and completeness. The audit report should reflect both the tone and content of the audit and be signed and dated by the team leader.
7. With internal audits the auditee will usually be able to identify quite easily the actions to be taken by taking advantage of any opportunities for improvement which the audit has revealed. If this is so, the use the remaining sections of the form can be speedily completed, the corrective action verified and signed off and the audit closed.
8. The most important aspect of any analysis is to provide data which will enable the management to drive the continual process of system review and improvement. This being the case, like any other such activity it is important to identify the key areas to address. The application of the Pareto approach is often as good a method of analysis as any.
9. The results of internal quality audits form an integral part of management review activities.
10. By following simple procedures, the system audit activity can be turned into the single most powerful driver of continuous improvement which any management can implement within a company.

EXERCISES

1. What is an quality audit planning? Explain.
2. What are steps of performing the quality audit?
3. What do you understand by audit formality?
4. What is purpose of audit notification? How is it done?
5. What is "opening meeting" in context with auditing?
6. Describe in detail the actual auditing?
7. What is significance of "Closing meeting"?
8. How is quality audit reporting done?
9. Describe "Corrective action" part of audit report.
10. What is analysis of findings after auditing?
11. What is "management review" in context with auditing?

13 Chapter Thirteen

Reliability

CHAPTER OUTLINE

- Introduction
- Redundancy
- Unit Redundancy
- Optimisation
- Introduction
- Testing Methods
- Reliability Improvements
- Element Redundancy
- Stand by Redundancy
- Reliability-Cost trade-Off
- Testing Requirements
- Acceleration

13.1. INTRODUCTION

Many failure can be attributed to bad design and overstressing of components as per a large number of reliability studies. Economist, Pareto, pointed out that in many real life problems, a relatively small proportion of the independent variables accounts for the characteristic behaviour of the dependent variable. This is also true of the reliability of a system, *i.e.*, the factors leading to a majority of the defects may be relatively few. A plot of the failure rates and the several causative factors is shown in a pot of Fig. 13.1. This plot is known as the Pareto chart or curve. It illustrates the principle that a majority of the failures may be due to only a small proportion of the many possible reasons. Some elementry methods of statistical analysis can be used to identify these few causes. The chart shown refers to an automobile.

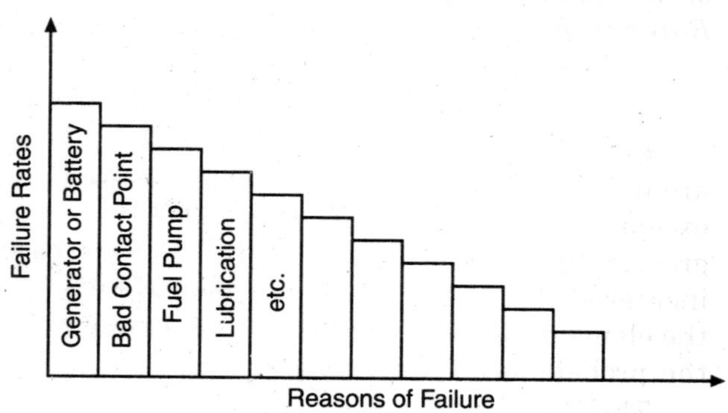

Fig. 13.1. Pareto chart

13.2. RELIABILITY IMPROVEMENTS

One can have superior components and parts with low failure rates, as a first approach to improving reliability. But the components of high reliability needs more time and money for development. Systems may be larger in size and weight. Normally, the aim is not merely to produce a system with the highest reliability, but to evolve a system which reflects an optimum total cost.

Research and Development, production, spares and maintenance are the major items contributing to the total cost. Figure 13.2 shows the cost of these individual items versus reliability.

A highly reliable component or unit, would undoubtedly be required a correspondingly high investment in the research and development activities, and this will be reflected in considerable measure in the total cost.

But, reliability will be of greater significance when the goal is not so much the cost, but rather the requirement to achieve a set mission. Space craft must have components with exceptionally high reliability factor.

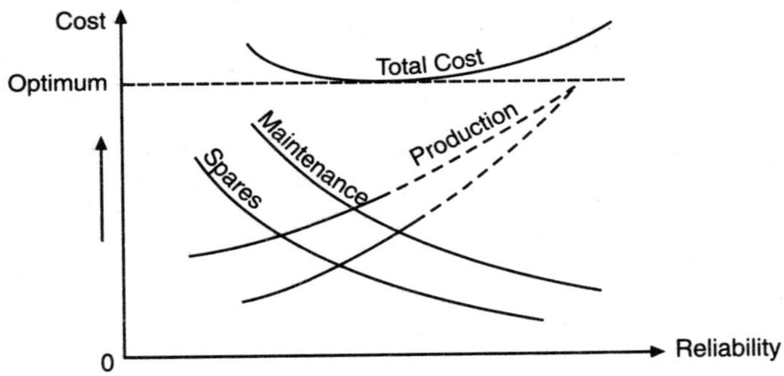

Fig. 13.2.

13.3. REDUNDANCY

It is not possible to produce highly reliable components or the cost of manufacturing such components is very high, one can improve the system reliability by the technique of introducing redundancies *i.e.*, delibrate creation of new parallel path in a system. If the elements with reliabilities of success $R(a)$ and $R(b)$ are connected in parallel, the reliability $R(a \text{ or } b)$ is given by:

$$R(a \text{ or } b) = R(a) + R(b) - R(a \text{ and } b)$$
$$= R(a) + R(b) - R(a) \times R(b)$$

Considering that the elements are independent. As both $R(a)$ and $R(b)$ are individually less than one. Their product is always less than $R(a)$ or $R(b)$ except the condition when $R(a) = R(b) = 1$. Thus, $R(a \text{ or } b)$ is all the time greater than either $R(a)$ or $R(b)$. When the element reliability cannot be increased this is a method of improving the reliability of a system. None of the elements is sufficient for the proper operation of the system, so to increase the probability of success both elements have been used.

Thus, system becomes redundant.

13.4. ELEMENT REDUNDANCY

Take a_1 as an element having reliability of $R_1(t)$. Figure 13.3 shows than another element a_2 having reliability $R_2(t)$, is connected to it in parallel. Suppose the system is operable with either a_1 or a_2 , the reliability of the system is $R_1(t) + R_2(t) - R_1(t) \times R_2(t)$. This is better than the individual reliabilities of elements a_1 and a_2.

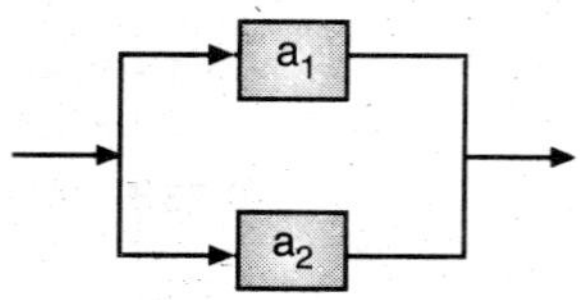

Fig. 13.3.

Constant Failure Rate: In constant failure rate, the reliabilities $R_1(t)$ and $R_2(t)$ are given by exp $(-\lambda_1 t)$ and exp $(-\lambda_2 t)$, respectively. The system reliability will be given as:

$$R(t) = \exp(-\lambda_1 t) + \exp(-\lambda_2 t) - \exp[-(\lambda_1 + \lambda_2)t]$$

In case the elements are identical, *i.e.*, $\lambda_1 = \lambda_2 = \lambda$, then

$$R(t) = 2e^{-\lambda t} - e^{-2\lambda t}.$$

Linearly Increasing Failure Rate

In linearly failure rate the reliabilities are given as:

$$R_1(t) = \exp(-K_1 t^2/2),\ R_2(t) = \exp(-K_2 t^2/2)$$

The system reliability is given as:

$$R(t) = \exp(-K_1 t^2/2) + \exp(-K_2 t^2/2) - \exp[-(K_1 + K_2)t^2/2]$$

For identical elements, $K_1 = K_2 = K$, and

$$R(t) = 2\exp(-Kt^2/2) - \exp(-Kt^2)$$

The element reliabilities and the system reliability for the two cases considered are shown in Fig. 13.4 (*a*) and (*b*), taking the elements to be identical.

The reliabilities are plotted as functions of λt and $t\sqrt{k}$, respectively, for the conditions of constant failure rate and linearly increasing failure rate. By putting $\lambda t = \tau$ in the Fig. 13.4(*a*), the system reliability with two identical elements in parallel is $2e^{-\tau} - e^{-2\tau}$, with $t\sqrt{K} = \tau$ in Fig. 13.4(*b*), the system reliability is 2exp $(-\tau^2/2)$ – exp $(-\tau^2)$. The time scale has been normalised in each case.

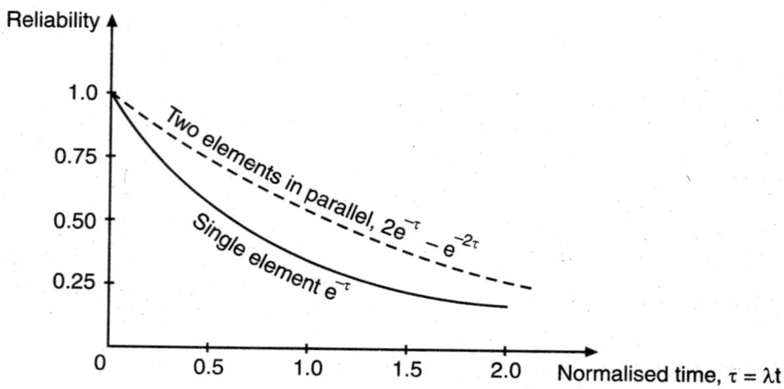

Fig. 13.4. (a) Constant failure rate

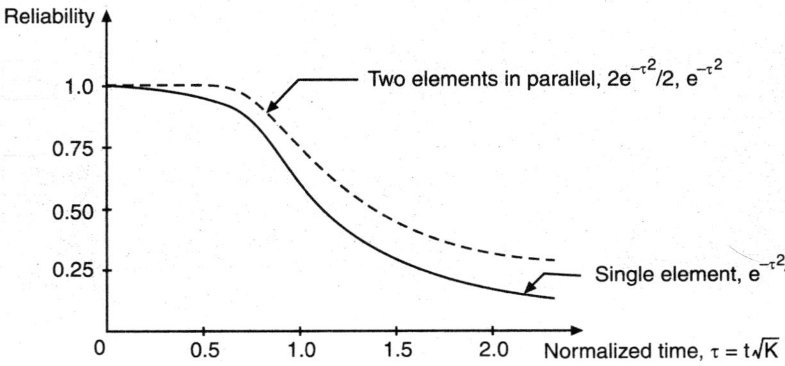

Fig. 13.4. (b) Linearly-increasing failure rate

13.5. UNIT REDUNDANCY

Suppose two elements a_1 and b_1, having reliabilities $R(a_1)$ and $R(a_2)$, respectively, are connected in series as given in Fig. 13.5(a). Successful operation of the system the probability is given:

$$R(s) = R(a_1 \text{ and } b_1) = R(a_1) \times R(b_1) \quad \text{...(13.1)}$$

Consider that the two elements are independent. Fig. 13.5(b) shows redundancies in the form so that the path with elements a_1 and b_1 in series is duplicated by another path with elements a_2 and b_2 in series.

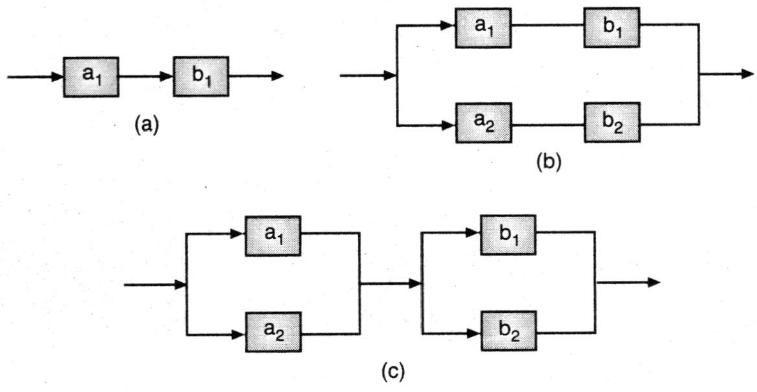

Fig. 13.5.

The system operation success depends on the proper functioning of path $a_1 b_1$ or path $a_2\, b_2$. Thus, the probability of successful operation of the system is given as:

$$\begin{aligned} R(s) &= R[(a_1 \; and \; b_1) \text{ or } (a_2 \text{ and } b_2)] \\ &= R(a_1 \text{ and } b_1 + R(a_2 \text{ and } b_2) - R[(a_1 \text{ and } b_1) \text{ and } (a_2 \text{ and } b_2)] \\ &= R(a_1) \times R(b_1) + R(a_2) \times R(b_2) - R(a_1) \times R(b_1) \times R(a_2) \times R(b_2) \end{aligned} \quad \text{...(13.2)}$$

An alternative is given in Fig. 13.5(c). The reliability is given as:

$$\begin{aligned} R(s) &= R[(a_1 \text{ or } a_2) \text{ and } (b_1 \text{ or } b_2) = R(a_1 \text{ or } a_2) \times R(b_1 \text{ or } b_2) \\ &= [R(a_1) + R(a_2) - R(a_1) \times R(a_2)] \times [R(b_1) + R(b_2) - R(b_1) \times R(b_2)] \end{aligned} \quad \text{...(13.3)}$$

Let the elements be identical, each of reliability p.

Equations (13.1), (13.2),(13.3) give:

$R_2(s) = R^2$ for Fig. 13.5 (a) ...(13.4)

$R_b(s) = 2R^2 - R^4 = R^2(2 - R^2)$ for Fig. 13.5 (b) ...(13.5)

$R_c(s) = (2R - R^2)(2R - R^2) = R^2 (2 - R)^2$ for Fig. 13.5 (c)...(13.6)

As R is less than one, $(2 - R^2)$ is greater than one. Therefore, $R_b(s)$ as given by equation (13.5) is greater then $R_a(s)$. In addition, one can easily show that $R_c(s)$ is greater than $R_b(s)$ as:

$$\frac{R_c(s)}{R_b(s)} = \frac{(2-R)^2}{2-R^2} = \frac{4-4R+R^2}{2-R^2} = \frac{(2-R^2)+2(1-R)^2}{2-R^2}$$

This quantity is greater than one. Thus, the system given in Fig. 13.5(c) has the greater reliability.

EXAMPLE 13.1.

Two elements a and b have reliabilities 0.7 and 0.8 respectively. These are connected in series as shown in the Fig. 13.6(a). If redundancies are introduced in the form shown in Figures b and c. Find the resultant system reliabilities.

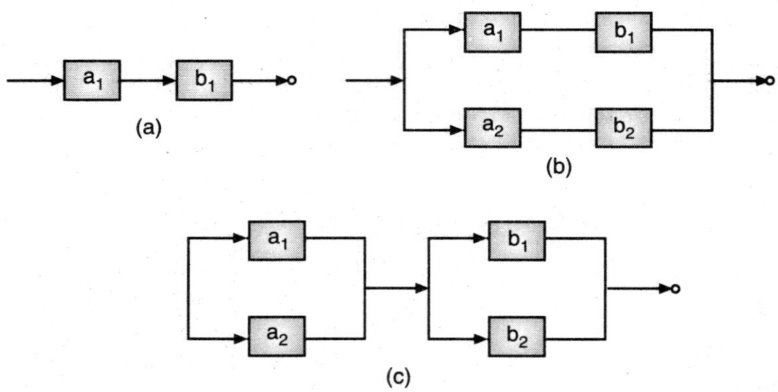

Fig. 13.6.

Solution: Consider the series configuration with no redundancy as shown in Fig. 13.6(a)

$$R(s) = R(a) \times R(b) = (0.7) \times (0.8) = 0.56$$

In the unit redundancy of the type shown in Fig. (b), the system reliability from Equation 13.2 is given by:

$$R(s) = R(a_1) + R(b_1) \times R(a_2) \times R(b_2) - R(a_1) \times R(a_2) \times R(b_1) \times R(b_2)$$

$$= 2(0.7)(0.8) - (0.7)^2 (0.8)^2 = 0.806$$

Now, for element redundancy of the type shown in Fig. 13.6(c), the system reliability from equation (13.3) is given by:

$$R(s) = (0.7 + 0.7 - 0.7^2)(0.8 + 0.8 - 0.8^2)$$

$$= (0.91)(0.96) = 0.874$$

Thus, even if the elements are not identical, better reliability can be achieved by element or component redundancy them by unit redundancy.

13.6. STAND BY REDUNDANCY

Figure 13.7(*a*) shows a two element stand by system and Fig. 13.7(*b*) shows a two element parallel system for comparison. All the channels or paths are active from beginning of the operation of the system till its failure in the parallel system. All the paths are not active at the same time in stand by system. When the system is put on operation initially, the switch (s) connects the input end to element a' while the other element b' is kept in reserve in a turned off condition. The switch senses by some built in mechanism if element a' fails due to some reason, and the connection is made to the standby element b'. The time taken for the switch to shift the connection from a' to b' and also the warm-up time taken by b' to become fully operational is short enough as per assumption to assure that the system operation is uninterrupted.

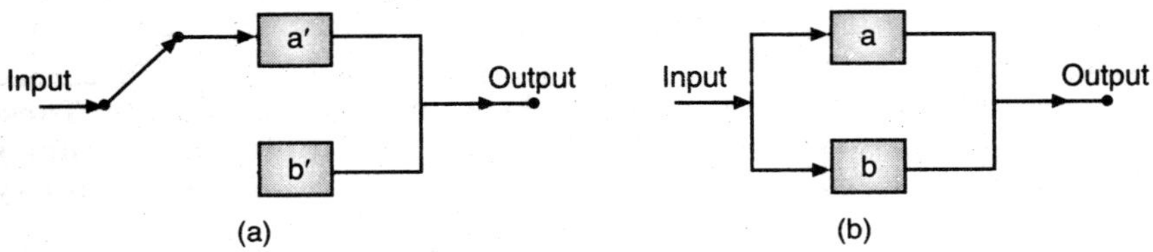

Fig. 13.7.

The system fails when both channels fail for the parallel system as shown in Fig. 13.7(*b*). Thus, the reliability of proper operation is:

$$R(s) = 1 - R(\bar{a} \text{ and } \bar{b}),$$

For $R(\bar{a})$ and $R(\bar{b})$ being the reliabilities of failure of elements a and b respectively. The equation can also be

$$R(s) = 1 - R(\bar{a}) \times R(\bar{b}/\bar{a}), \qquad ...(13.7)$$

For $R(\bar{b}/\bar{a})$ being the reliability of failure of element b, considering that element a has failed. If the reliability of each is P and the elements are independent and identical then we get:

$$R(s) = 1 - (1 - R)(1 - R) = 2R - R^2 \qquad ...(13.8)$$

If first channel a' fails the standby system shown in Fig. 13.7(*a*) fails and then when the decision switch shifts to channel b'. Channel b' also fails. Thus, the reliability probability of proper operation for the stand by system is given as:

Thus,

$$R(s) = 1 - (\bar{a}' \text{ and } \bar{b}') = 1 - R(\bar{a}') \times R(\bar{b}') \qquad ...(13.9)$$

We must be careful in interpreting this expression. It would appear from equations (13.7) and (13.9) that the two reliabilities would be identical if units a and a' are identical and if units b and b' also are identical. But actually it is not the same. $R(\bar{b}/\bar{a})$ may be equal to $R(\bar{b})$ if elements a and b are independent for the parallel system, or $R(\bar{b})$ may be affected a little if the two elements interact. $R(\bar{b}/\bar{a})$ in the stand by system is always a conditional probability as one can see. The two reliabilities are different as b' in Fig. 13.7(*a*) becomes only after element a' has failed which is clear whereas in Fig. 13.7(*b*), elements a and b are simultane active. Similarly, the reliability of a standby system is greater than that of a parallel system.

The reliability of the two identical a' and b' being R, it can be seen that the reliability for a standby system is given by:

$$R(s) = R(1 - \log_e R),$$

Considering that the switch is perfect in its operation. If one makes a comparison of the reliabilities for three configurations shown in Fig. 13.8, observations are the reliability is R for a single element, the reliability is $2R - R^2$ for two active identical elements in parallel, the reliability is $R(1 - \log_e R)$ for a two elements in stand by system with identical elements.

13.7. OPTIMISATION

A large number of stations are set up to send a signal for launching missiles to attack enemy war aircrafts based on the detection of the approach of the planes. Although the reliability of the launching of missiles increases with each increase of a station, simultaneously, the probability of receiving a false message sent by accident also increases. Hence, one should take into account not only the reliability of the system operating properly at a time when it is supposed to, but also the reliability of the system operation when it should not. To do this, let us assume that, for each component, there is a probability that the component will function at a time when it is not supposed to. Assume that every component operates when it is supposed to, with a reliability R, and does not operate when there is no signal with a reliability r. That is, a component will not operate every time when it should, and a component will not operate every time when it is not supposed to (reliability being r, where $0 < r < 1$).

Figure 13.8 shows a simple system having four components connected in parallel. In order for the system to operate properly. It is essential that at least one of the elements operates when the signal is given and, simultaneously none of them operates when no signal is available.

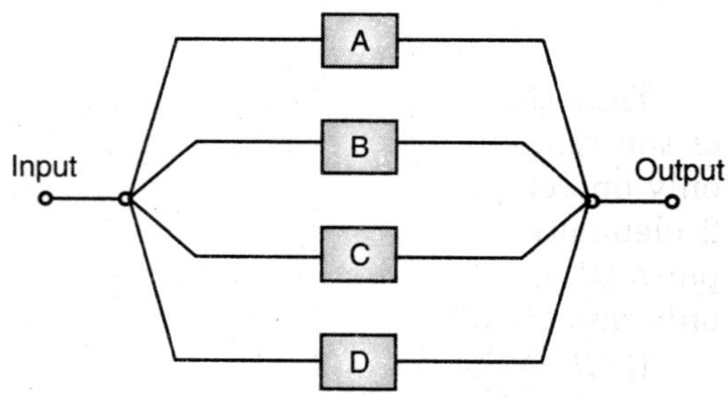

Fig. 13.8.

The reliability that atleast one of lest one of them operating when the signal is available is equal to the reliability that none of them fails when the signal is available, and which is equal to $[1 - (1 - R)^4]$.

The reliability of none of them operating when no signal is available is r^4. Hence, the system reliability is $R = r^4\,[1 - (1 - R)^4]$. Now, if the system has m components connected in parallel, then the system reliability is given by:

$$R_m = r^m\,[1 - (1 - R)^m] \qquad ...(13.10)$$

If for a system $r = 1$ and the system reliability is given by:

$$R'_m = 1 - (1 - R)^m \qquad ...(13.11)$$

By taking m large enough, the reliability of the system, equation 13.10, can be made as small as desired. Take a system with m elements and another system with $(m + 1)$ elements to see the dependence of R on m. If t_m is the ratio of the reliabilities of these two systems, *i.e.*, and $t_m = R_{m+1}/R_m$, then as long as $t_m > 1$, the system reliability increases with an increase in the number of elements. If $q = (1 - p)$, then we get:

$$t_m = \frac{r^{m+1}(1-q^{m+1})}{r^m(1-q^m)} = \frac{r(1-q)(1+q+q^2+...q^m)}{(1-q)(1+q+q^2+...+q^{m-1})}$$

If $m = 1, 2, 3, ...$ then we get:

$$t_1 = r(1 + q)$$

$$t_2 = r\left(\frac{1+q^2}{1+q}\right)$$

$$t_3 = \left(\frac{1+q^3}{1+q+q^2}\right)$$

It can be easily see n that as m increases, the value of t_m decreases. It is due to being less than one and therefore q_m decreases as m increases. But, we are interested in increasing R_m until $t_m > 1$. That is, we are interested in finding the first stage for $t_m < 1$, depend on r and P. The plot of the following equations for $t_m = 1$:

$$t_1 = r(1 + q) = 1$$

$$t_2 = r\left(\frac{1+q^2}{1+q}\right) = 1$$

$$t_3 = \left(\frac{1+q^3}{1+q+q^2}\right) = 1$$

Take the first equation having $t_1 = 1$. It has been seen that t_r is the ratio of the reliabilities of two systems, one with 2 elements and the other with only one element. In case of a given set of values for R and r. One can use 2 elements if t_1 is greater than or else one uses only one element. For all pairs (R, r) which lie below the line $t_1 = 1$ as per Fig. 13.10. One should use only one element.

If (R, r) lies between $t_1 = 1$ and $t_2 = 1$, it is best to use two elements. For Say, if $R = 0.85$, one should use one component when r is less than 0.87, and the two components when r is between 0.87 and 0.98 one can give similar interpretations for other values of P and r.

13.8. RELIABILITY-COST TRADE-OFF

The reliability of the system depends on a large number of parameters which directly affect the behaviour of the system such as electrical, electronic, or mechanical. Generally these parameters, are not deterministic, but are as per probability laws. The design of the system will be better if estimation of these parameters are closer. Consider the problem of designing a slender column to carry on axial load considered, young's modulus for the material of the column is as per a log-normal probability law with a mean value $\mu_E = 205$ KN/ mm^2 and a coefficient of variation equal to 0.0215.

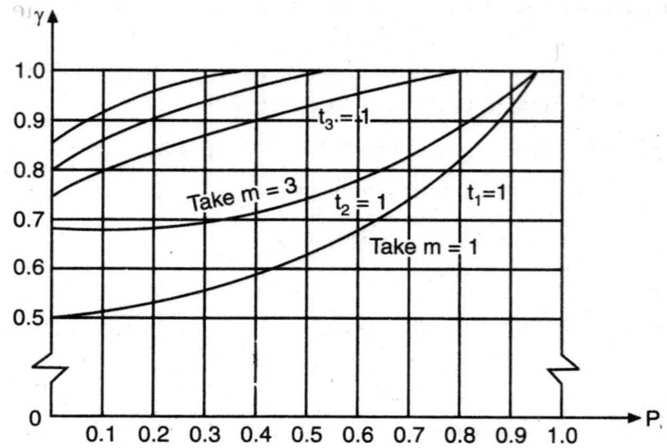

Fig. 13.9.

By utilising a higher quality material this variablity could have been reduced, but one would bave bad to pay a higher cost for it. One could have determined the load carrying capacity of the column with a smaller variance using such a higher quality material, but this improvement at such higher cost may not be desirable *i.e.*, there is a trade-off between reliability and cost. the functioning of a delicate electronic instrument with several protective devices. Can be another example say, voltage stabiliser, vibration isolation system, and thermal cooling system, each device are as per the probability laws corresponding to it and its own variability factors. Reduction of for being the variabilities will be accompanied by increased cost.

13.9. TESTING REQUIREMENTS

A reliability programme requires the largest share of manpower and funds for a test programme of an item. This is because test programme is considered as one of the most important activities in a reliability programme. The designer bases his design and subsequent designs on the vital inputs provided by the test programme. Testing data is the source of almost all meaningful data from inception thought the entire life. This applies to the hardware, corrective actions on design process, and planning for maintenance facility to supper the equipment in actual. Most of the troubles are unexpected hardware failure, take of maintenance and repair facility, accumulation of stress beyond the capabilities. These troubles are precluded by a complete, intergrade, planned and document test programme.

A subjection to conditions which show the real character of the item is known as testing. Testing is a continuing operation to provide information throughout the complete evolution of the system and is no unique operation performed at a specific point in the development of a complex system.

There are several reasons of testing. Testing gives information at various stages of an item. Various stages where "tests" can be performed are given as follows:

(*i*) **Intant Stage:** Testing is done to find out the feasibility of certain configuration or item.

(*ii*) **Development Stage:** Testing is done to determine the optimal configuration out of several configurations available. Optional configuration is with respect to performance, reliability, cost and modes of behaviour under varying conditions.

(*iii*) **Evolution Stage:** To improve economy, maintence, use of standard parts etc. comparisons are made.

(*iv*) **Performance Stage:** Adequacy of an item in a system to meet requirements of performance and reliability is tested.

(*v*) **User Stage:** Testing is done to know whether the inherent design can withstand handling, transportation and other diversified usage of the item.

13.10. TESTING METHODS

A test programme includes many kinds of tests covering all tests on the hardware from inception through the final use and disposition of hardware. An understanding of different kinds of available tests are essential so that a correct choice can be made. The different kinds of test methods are marginal testing, Non-destructive testing, Destructive testing and accelerated testing.

13.10.1. Marginal Testing

It is a method of predicting the probability of an imminent failure due to degradation . Periodic testing on a programme basis is conducted in this test. For example electronic equipment testing may involve imposing voltage limits on certain electronic equipments. Subsequently, resulting symptoms are observed. Isolation of degraded components or parts and replacement of the same is feasible before the occurrance actual failure.

13.10.2. Non-destructive Testing

Radiation ultrasound, magnetic field or liquid penetrants type of probing agents are used in a non-destructive testing. Observation of the relative response of the porting agent helps in detection of a plan whose size is measured. It may be noted that response may not always be directly related to the relative severity of the plan.

Fatique is the cause of a majority of mechanical failures. It is implied that the failure is initiated by an alternating stress. The alternating stores may be part of the load. However, in many eases it arises from within the unit itself. There must be out of balance forces associated with any machine having moving parts as perfect balance can never be achieved. These unbalances can be generated by load fit due to poor design, manufacture, wear or poor installation, Irrespective of the cause, such an alternating stress mill lead to a vibration. The level of vibration is basically a measure of the alternating stress in a unit. This decides the ability of the unit to with stand failure by fatigue.

13.10.3. Destructive Testing

Isolation of potential failures are done by a destructive test. Abnormal stresses are put on the component. If component survives this stress, than the component is very reliable. A beam or truss will definitely withstand a lighter load than that which was used destructive testing. Thus, destractive testing is very effective for static situations. If the destructive tests data are time ordered, the weabest part fails first in time, the second weabest part fails next, and so on. The life of ball bearings the life or an electric bulb are examples of such trends. Progressively consored samples are progressively added samples are the two models which will be discussed here as these have immense practical utility.

13.10.3.1. Progressively Censored Samples

Initial censoring leads to withdrawal of a portion of the survived items. There are some items which remains on continue test and observation until ultimate failure or until a subsequent state of censoring is performed. Censoring may be progressive for sufficiently large samples. When certain specimens must be withdrawn form a life test prior to failure, such samples censoring may be progressive. This is for use in related experimentation or for some other urgent need. A compromise between the need for rapid testing and the desire to include at least some extreme life spans in the sample data give rise to progressively censored samples. The early ceinsuring of a substantial number of sample specimens frees facilities for other tests as such very useful approach if test facilities are limited and prolonged tests are inexpensive.

13.10.3.2. Progressively Added Samples

In case records are available on dates of installation of n items. There will be same samples which fail and remaining samples will survive to remain in use. An estimate of the average can be made using the dates of installation and the lives of the failed ones. Such situation arises if there is no complete control by the experimenter over the life test or all the items are not available for test at one time or there is time limit prescribed in advance by the management for the experiment.

13.10.3.3. Accelerated Testing

There are occasions when a tightly compressed schedule exists for releasing the design for production although research and development process of the product is continuing. It is essential to find out that the product has adequate life and no serious weakness exist. In such case, accelerated testing is a must.

A test technique where item failure occurance is accelerated by subjecting items to stresses in excess of their use or normal conditions, is known as accelerated testing. In the accelerated testing, the item is found to have a shorter life than normal conditions of operation. Accelerated test establishes a relationship between an items failure behaviour, stresses and time. This relationship helps in prediction of behaviour of the item under various tests or operating condition. This saves time and expenditure as compared with testing in normal conditions. However, some products have such a short life that accelerated testing in such a short cases is meaningless.

These are occasions when information on the life of items over a range of conditions that they encounter in use. Testing over a range of conditions desired or over more severe conditions give such informations which are extrapolated to the desired range of conditions. The conditions can be temperature, voltage, pressure, vibration, cyclic rate, load etc. or combination of these. Suppose temperature is the only variable in an accelerated testing some standard model can be used with mathematical equations. An accelerated test can be conducted properly if:

(*i*) Through engineering knowledge should be used in designing tests which are commensurate with actual use.

(*ii*) The factors must be kept constant so that variable of interest are varied in the test to achieve proper results.

(*iii*) Variations due to materials or other conditions in manufacturing and usage, must be kept minimal.

(*iv*) The model chosen for testing should have sufficient flexibility to very various parameters of interest.

(*v*) Failure conditions must be defined properly for example number of operations before failure.

13.10.4. Reliability Tester

Inspite of all testing contributes data for reliability calculations and hence could be considered in a larger sense to be reliability testing, there are specific tests which are performed for no other purpose than to gather these data. These are the tests referred to in this section, and for purposes of this discussion they have been grouped into peripheral testing, life testing, accelerated life testing, service-life evaluation testing and surveillance testing. The data from reliability testing are used to determine mean time for cycles to and between failure, to calculate or verify attained reliability, to establish storage and operating life limits on critically age sensitive parts (and from both of these come the depth requirements for spare parts), and to determine modes of failure. Reliability tests are performed at all stages of the project and all levels of assembly. They are performed both in ambient and enviromental conditions, and they include both destructive and non-destructive tests inspections and examinations. They may also include some actual use tests, although they are usually confined to the laboratory to ensure control of input conditions to the system.

13.10.4.1. Peripheral Testing

In testing the parts are done in environments and input conditions which stimulate as nearly as possible the actual range of use conditions, and a successful test is one in which the part functions properly in these conditions. From such testing, however, it is not possible to determined how much margin of safety has been designed and built into the product, since the part has not been stressed to functional yield. It is useful in predicting reliability of a population from data gathered on a limited sample to test the parts to environments and input conditions which are more rigorous than the expected service conditions by a substantial enough margin that failures can clearly attributed to the peripheral conditions. This margin should be at least 10 to 15 per cent of the spread from ambient to the limit of the service environment of the item.

13.10.4.2. Life Testing

Reliability prediction and reliability assessment are generally concerned with the determination of the mean time for cycles to and between failures as this number is basic in reliability calculations. The number can be calculated directly from the data collected from the life test program, where tests are conducted not only on samples of completed assemblies but on spare parts and piece parts as well. The tests are conducted in the laboratory on test equipment which, for economy of testing cost, is designed to function continuously on cycle the hardware automatically. The function in interrupted at regular intervals and functionally tests or non-destructive inspections are made to know whether there has been any determination of the operability of the part with time or cycles of operation. Normally, the most severe expected service environments are chosen and samples are utilised in a statistical design of experiments help the interpretation of results of the system.

Life testing is very slow and costly and can take six months to a year to complete. In some situations ,where real time is the same as operating time, the test programme may take years. Typical of these are tests of paints, where the actual service conditions are exposure to outdoor weather, or of sub-marine cable and equipment, where the actual service conditions is exposure ocean depths. In these circumstance, it is essential that the life testing program be instituted on the earliest production prototype, so that field failures of serviced equipment delivered at a later time can be predicted prior to occurance or that corrective action on the design or production process can be instituted before production actually begins.

13.10.4.4. Surveillance Testing

The last test program in the reliability test group is surveillance testing. These tests, which are performed on samples drawn at regular intervals, from the actual field service stocks, consist of ambient tests and examinations performed on the samples at progressive levels of disassembly. The object of the testing is to discover evidence of failure or incipient failure in the hardware, including not only shifts in values of components in functional hardware but chemical deterioration of materials fatique cracks, corrosion, whiskers, hardening of rings and seals, and any other unanticipated modes of failure of the part. The two characteristics differentiating surveillance testing from other kinds of reliability testing are the limitation of testing to ambient examinations and the complete disassembly of the specimen of the parts.

13.11. ACCELERATION MODELS

1. Arrhenius Model

A common approach is based on the arrhenius model, when failures are accelerated primarily as a result of an increase in Temperature.

$$\gamma = Ae^{-B/T}$$

Here γ is the reaction or process rate, A and B are constants, and T is temperature measured in Kelvins. Its acceleration factor may be determined from the following equation:

$$AF = \frac{Ae^{-B/T_2}}{Ae^{-B/T_1}} = \exp\left[B\left(\frac{1}{T_1} - \frac{1}{T_2}\right)\right]$$

Value of B can be approximated by testing at two different stress termperature and calculating the acceleration factor on the basis of the fitted distributions. Thus,

$$B = \frac{InAF}{\frac{1}{T_1} - \frac{1}{T_2}}$$

For $AF = \frac{\theta_1}{\theta_2}$; with θ_i representing a scale parameter or a percentile at the stress level corresponding to T_i.

2. Eyering Model

Additional stresses are allowed in the model and can be derived from quantum mechanics. A simple form is given by:

$$\gamma = AT^{a}e^{-B/T}e^{CS}$$

γ is the process rate, A, a, B, and c are constants, T is temperature (in Kelvins); and S is a second stress. The first exponential factor and its coefficient account for the teperature and except for T^{a} , behave as in the arrhanius model. The second exponential factor involves a second, non-thermal stress. If additional stresses are present, additional factor like the second can be included (with constants different from C).

If a is close to zero, than the T^{a} factor will be close to 1 at all temperature, and its effect can be included as part of the constant A. In the absence of a second stress, the similarity with the arrhenius model is apparent and explains why the arrhenius model works as well as it does although it is strictly an empirical model. Theoretical considerations is the Eyring model.

The constants must be estimated from test data to apply this model. Estimating the four constants in this model will requires at least four data points at two different temperature levels and two different stress levels. This model acceleration factor is given by:

$$AF = \left(\frac{T_2}{T_1}\right)^{a} \exp\left[B\left(\frac{1}{T_1} - \frac{1}{T_2}\right)\right] e^{C(S_2 - S_1)}$$

3. Degradation Model

It may be possible to predict the time to failure by extrapolating degradation in performance overtime, if a product has an descrvable performance measure that changes with time. Performance may be measured at either stressed or normal operating conditions, and a critical level of performance that will result in a failure must be specified. Corosion, crack propagation, and the shelf life of pharmaceutical products are examples of degradation process. Regression analysis can be used to develop empirical models that relate degradation in performance to time. Relationship in the simplest form is linear with $y = a - bt$, where y is the performance measure (or frequently the log of the performance measure), t is the amount of tme the product is exposed at a constant stress level, a and b are constant to be determined experimentally.

This time to failure is treated as a 'typical' value; Therefore, it may be interpreted as the mean or median of the failure distribution. Alternatively, n units may be tested, performance measured several times for each unit and separate regression lines fitted to each unit. Then a sample of n predicted failure times is generated.

4. Cumulative Damage Model

The following generalisation of miner's rule may be used, if component damage that will lead to failure accumulates continuously, and if the damage rate depends only on the amount of damage and not on any past history:

$$\sum_{i=1}^{n} \frac{t_1}{L_1} = 1$$

where t_i = The amount of time at stress level i

L_i = The expected lifetime at stress level i

Consider two stress levels-one normal (t_1) and the other high(t_2) to apply this model then

$$\frac{t_1}{L_1} + \frac{t_2}{L_2} = 1 \quad \text{or} \quad t_2 = L_2 - \frac{L_2}{L_1} t_1$$

A comulative damage model is shown by the failure line in Fig. 13.10.

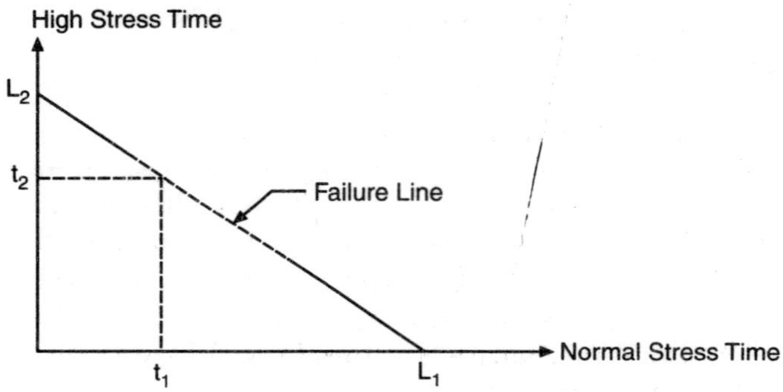

Fig. 13.10.

5. Step Stress Model

Testing begins with normal stress in a step stress accelerated life test. After a period of time, the stress is increaed. Such step-wise increases are then continued until all the test units fail. The increase in stress is equivalent to a linear change in the time scale. This primary assumption in developing the step stress model. These models are more complex than the constant stress models.

SUMMARY

1. **Reliability Improvement:** One can have superior component and parts with low failure rates as a first approach to improving reliability. But the component of high reliability needs more time and money for development.
2. **Redundancy:** It is not possible to produce highly reliable components or the cost of manufacturing such component is very high, one can improve the system reliability by the technique of introducing redundancies, *i.e.*, delibrate creation of new parallel path in a system. If the elements with reliabilities of success $R(a)$ or $R(b)$ are connected in parallel, the reliability $R(a \text{ or } b)$ is given by:

$$R(a \text{ or } b) = R(a) + R(b) - R(a \text{ and } b)$$
$$R(a \text{ or } b) = R(a) + R(b) - R(a) \times R(b)$$

3. **Stand *Y* Redundancy :** All the channels or paths are active from beginning of the operation of the system till its failure in the parallel system. All the paths are not active at the same time in stand by system.
4. **Reliability cost Trade off:** The reliability of the system depends on a large number of parameters which directly affect the behaviour of the system. The design of the system will be better if estimation of these parameters are closer.
5. **Reliability Testing:** A subjection to conditions which show the real character of the item is known as testing. Testing is a continuing operation to provide information throughout. The complete evolution of the system and is no unique operation performed at a specific point in the development of a complex system.
6. **Non-destructive Testing:** Radiation ultrasound, magnetic field or liquid penetrants type of probing agents are used in non-destructive testing. Observation of the relative response of the perting agent help in detection of a plan where size is measured.
7. **Destructive Testing:** Isolation of potential failures are done by a destructive test. Abnormal stresses are put on the component. If component survives this stress, than the component is very reliable. Destructive testing is very effective for static situation.
8. **Accerlated Testing:** A test technique where item failure occurrence is accerlerated by subjecting items to stresses in excess of their use or normal condition is known as accerlated testing. In the accerlated testing the item, *i.e.*, found to have a shorter life than normal conditions of operation. Accerlerated test establishes a relationship between an items failure behaviour, stresses and time.
9. **Life Testing:** Reliability prediction and reliability assessment are generally concerned with the determination of the mean time for cycles to and between failures as their number is basic in reliability calculations. The number can be calculated directly from the data collected from the life test program. Life testing is very slow and costly and take six months to a year to complete.
10. **Surveillance Testing:** The last test program in the reliability test group is surveillance testing. The object of the testing is to discover evidence of failure or incipient failure in the hardware, including not only shifts in values of components in functional hardware but chemical deterioration of material fatigue, cracks, corrosion etc. and any other unanticipiated modes of failure of the part.

EXERCISES

1. Describe reliability improvement.
2. Explain various types of redundancies.
3. What is optimisation? Explain.
4. Explain reliability cost trade-off.
5. What is the need of testing a system?
6. What are various types of testings?
7. Describe non-destructive testing.
8. Explain destructive testing.
9. What are progressively censored and added samples?
10. What do you understand by accelerated testing?
11. What is peripheral testing?
12. Explain life testing.
13. What is surveillance testing?
14. Explain various acceleration models of testing.

14 CHAPTER **Fourteen**

PRODUCTION PLANNING AND CONTROL

CHAPTER OUTLINE

- Introduction
- Production Planning Versus Production Control
- PPC Information Requirement
- PPC Organization
- Meaning of Quality Control
- Process Control
- Production Cycle or Procedure
- Introduction
- Cost of Quality

14.1. INTRODUCTION

Production is the transformation of materials into a desirable output. It is a series of sequential operations to produce a desirable products acceptance to customer. It is a well organized activity to meet specific objectives. An efficient production system is one which produces the products with required quantity and specified quality at predetermined cost pre-established time. Production planning and control is a tool to achieve the objectives. Thus, a production system delivers quantity, quality at optimum cost and time.

Production planning analysis the given data of demand delivery scheduled etc. and the information available. Firms resources like machine, materials and men are processed as per prepared scheme to obtain the target in the most economical manner.

Production control helps, if there is any deviation between the actual and planned scheme. Production planning and control is the direction and coordination of firms resources towards attaining the desired goal.

14.1.1. Need for Production, Planning and Control (PPC)

Industries have to streamline the production activities and attain the maximum utilization of firms resources to enhance the productivity, to be in competition. Production planning and control is a tool to coordinate the activities of the production system by proper planning and control system for being competitive PPC achieves effective utilization of firms resources, production objectives with respect to quality, quantity, cost and time schedules of deliveries and helps the company to supply products on the continuous basis at competitive rates without compromising on quality.

14.1.2. Production Planning and Production Control

Production planning is the determination, acquisition and arrangement of all facilities necessary for future production of products. It is the design of production system. It organizes the production. It establishes the production programme to meet the targets set using the various resources based on the estimated demand for company's products.

There may be several factors which affect the production system, and causes deviation from the actual plan. The factors which affect are non-availability of materials, plant, equipment and machine breakdown, changes in demand and rush orders, absenteeism of workers, and lack of coordination and communication between various functional areas of business.

The control function comes into action if there is a deviation between actual production and planned production. The control activity is through initiating the production, progressing, and corrective action based upon the feedback and reporting back to the production planning.

14.1.3. Objectives of Production Planning and Control

Objectives of PPC are systematic planning of production activities to achieve the highest efficiency in production of goods/services. It organizes the production facilities like machines, men etc., to achieve the desired production with proper quantity, quality, time and cost. There is optimum scheduling of resources. There is coordination with other departments relating to production to achieve regular balanced and uninterrupted production flow. It works to conform to delivery commitments, provides material planning and control and allows adjustments due to change in demand.

14.1.4. Functions Production Planning and Control

Functions of PPC are shown in Fig. 14.1.

A macro level planning is pre-planning. It involves analysis of data and an outline of the planning policies based upon the forecasted demand, market analysis and product design and development. This stage is concerned with process design. It involves decision-making with respect to methods, machines and work flow.

The planning function starts once the task is specified and the analysis of machines, methods, materials and manpower (4 M's) are done. Process planning is the next activity with short-term and long-term planning. Due consideration are given for simplification of products and processes with standardization.

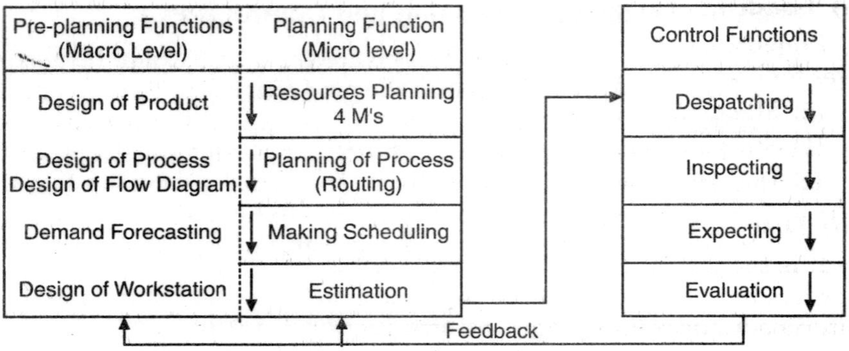

Fig. 14.1. Function of PPC

There is control of despatching, inspection and materials. There is analysis of work-in-process. Evaluation is done and corrective actions are taken through a feedback from analysis. To enhance and ensure effectiveness of PPC, a good communication, and feedback system is essential.

The Main Functions of PPC are Summarised as follows:

(*i*) Material Function

Materials are raw materials, finished parts and bought-out components. These should be available in required quantities and at required time. The specifications of materials quality and quantity with delivery dates and procurement are the part of this functions.

(*ii*) Machines and Equipment

A detailed analysis is done of available production facilities, equipment down time, maintenance policy procedure and schedules. Economy of jigs and fixtures and equipment availability is considered. Minimum down time due to breakdowns are attempted.

(*iii*) Methods

Analysis is done to find alternatives and selection of the best method. Developing specifications for processes and determination of sequence of operations are also done.

(*iv*) Process Planning

Selection of path or route is established for the raw material. The following functions are included.

(*a*) Path of travel with due consideration to layout.

(*b*) Breaking down of operations.

(*c*) Details of the set-up time and process time for each operation.

(*v*) Estimating

The operations times are estimated. Extensive analysis of operations, methods and routing is done and a standard time for operation are established. Work measurement techniques are used.

(*vi*) Loading and Scheduling

Preparation of machine loads and fixation of starting and completion dates for each of the operations are done. The following functions are included:

(*a*) Machines loading is done as per their capability.
(*b*) Start and completion times for each operation is determined.
(*c*) Delivery schedules are coordinated with sales department.

(*vii*) Dispatching

It is the process of setting production activities in motion through release of orders and instructions. The production activities are started by releasing materials, compensator, tools, fixtures and instruction sheets to the operator.

This function involves:

(*a*) Definite work to definite machines, work centres and men are assigned.
(*b*) Required materials from stores are issued.
(*c*) Jigs and fixtures at correct point of use are made available.
(*d*) Necessary work orders are released to authorise timely start of operations.
(*e*) Start and finish time of each job on each machine or by each man are recorded.

(*viii*) Expediting

Follow-up is done which coordinates to execute the production plan. Follow up of materials, follow up of work-in-process and follow up of assembly are done. This function includes:

(*a*) Identification of bottlenecks and delays and interruptions.
(*b*) Correction of the errors and action plans are devised.
(*c*) Ensuring production rate in line with schedule is ensured.

(*ix*) Inspection

The limitations with respect to methods, processes, etc., are known through inspection.

(*x*) Evaluation

The factors influency the production planning and control are analyzed and evaluated to identify the weak spots for the corrective action. Replanning and planning will be effected by a feedback. The communication data and information gathering and analysis are the key to success.

14.2. PRODUCTION PLANNING VERSUS PRODUCTION CONTROL

S.No.	Production Planning	Production Control
1.	A pre-production activity.	In action only after when production activity begins.
2.	Involves the collection and analysis of data, standards, materials, machines and maintenance.	Deals with communication, information and reports like productivity etc.

3.	Useful to anticipate the problems and devising remedies.	Taking corrective steps in case of error to match actual performance against the planned.
4.	A centralised activity dealing with control of materials, tool and process planning.	Includes functions of dispatching, inspection etc.
5.	Ensure availability of all the necessary resources to make quality production at right time.	Tracks the activities and fees schedule.

14.3. PPC INFORMATION REQUIREMENT

Production planning and control effectiveness depends upon the accuracy of the information. Information is vital in the success of PPC function. The required information, sources and the department are tabulated as follows:

S.No.	Required Information	Sources	Department
1.	Production programme consisting of quantity to be produced, delivery date, special features, etc.	The order accepted by the marketing deptt or the sales order	Marketing
2.	Specifications and tolerances as per quality standards.	Design and Engineering translated into specification.	Purchase, stores and engineering.
3.	Materials of production such as types of materials, quality and quantity procurement lead time and stock position.	Bill of materials (BOM), material shock cards and drawing.	PPC
4.	Toolings	Special and standard toolings.	PPC
5.	Details of operations, sequence, process, machines and equipment's capability, jigs and fixtures and process parameters.	Process sheets, load charts and studies of process capability.	Industrial Engineering.
6.	Operation and set up standard times.	Time and motion study data.	PPC
7.	Dates of starting and finishing	Machine loading and scheduling charts.	Manufacturing.
8.	Work status.	Report of production.	Manufacturing.

14.4. PRODUCTION CYCLE OR PROCEDURE

Production procedure is dependent on the technology, manufacturing processes, types of layouts, organisation structure and management policies. Thus differences are due to different practices of production, planning and control systems. The PPC functions varies for manufacturing system like jobbing, batch, mass and flow production systems. All the functions of PPC are needed in the job shop.

Production cycle starts with the customer and ends up with satisfying the needs of the customer. Production procedure cycle is comprised of:

(i) **Sales Forecast:** Marketing or sales department does it after a through analysis and market research.

(ii) **Preparation of Production Budget:** It is prepared by the finance department in consultation with production department.

(iii) **Engineering Details:** The engineering department prepares drawing, B.O.M. (Bill of materials) specifications.

(iv) **Production Planning:** It begins as soon as the technical information is received. It results in a schedule or time table of production. The inventory levels are for procurement of materials. The production planning provides the complete data on methods, process sheets, machine loading and production schedule.

(v) **Dispatching:** Dispatching of detailed production orders are done to the shop giving information on what, how and when and where the operations are to be done.

(vi) **Progressing:** Suitable corrective steps are taken in case of difference between planned and actual production. These are done through the control action exercise during the manufacturing period and progress is continuously compared with planned schedules.

(vii) **Inspection:** Inspections are done at various stages. The quality control ensures the desired specifications.

(viii) **Evaluation:** After and before production, evaluation is carried out. This helps in taking corrective actions to improve methods and down times. Finally reports from both production and financial departments are provided to the management.

(ix) **Stacking:** The finished product is transferred to stock.

(x) **Product Delivery:** Finally, the product is delivered to the customer.

Fig. 14.2., shows the complete production procedure.

14.5. PPC ORGANIZATION

PPC organization depends upon factors such as degree of centralization desirable, internal structure composition of PPC, system and management policy. A brief summary is as follows:

(i) In case of repetitive type of production with higher degree of automation, the PPC work is done by line staff.

(ii) If the sequence of operations is prefixed for the product type of layout, then PPC may be the part of production department itself.

(iii) If each product requires varying capacity, then an independent PPC department is to be set up.

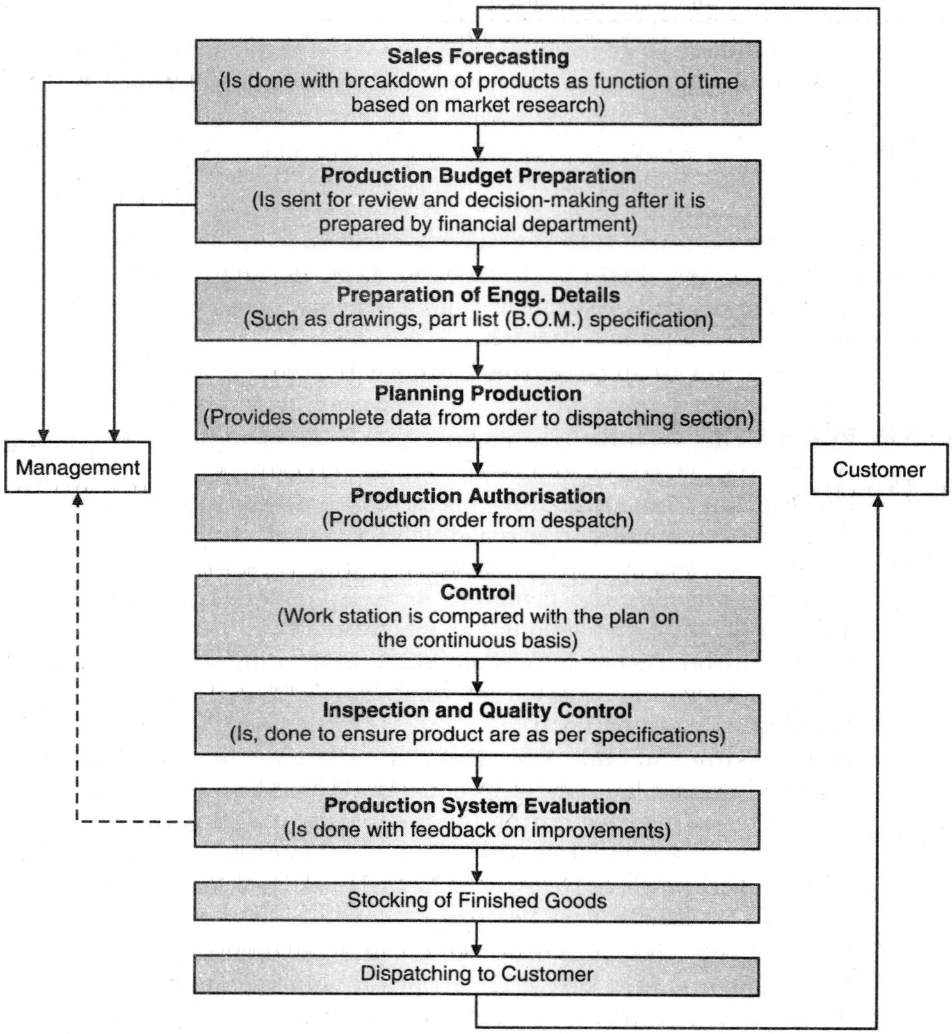

Fig. 14.2. Production Cycle or Procedure

14.5.1. Centralized and Decentralized PPC

A staff specialists controls production planning in centralized planning. In case of multi-product and multi-plant organizations centralized planning is more effective.

The decentralized planning and production involves the line staff in planning production. Multi-product, multiple plants, large workforce requires higher degree of centralization. The organizations having the prefixed sequence of operation and having small workforce require low degree of centralization.

PPC organization is shown in Fig. 14.3.

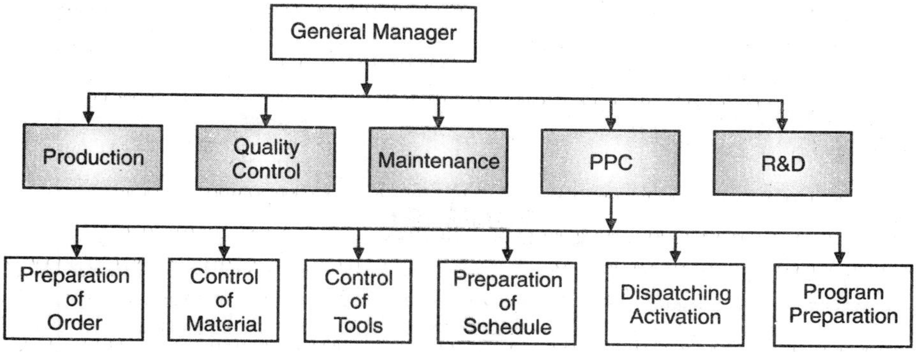

Fig. 14.3. PPC Organization

14.5.2. PPC and Manufacturing Methods

(*i*) Functions of PPC in Job Production

(*a*) On receipt of the order the materials are purchased.

(*b*) Special tools are either house made or purchased from outside.

(*c*) The work methods, the machines and time estimates for completion of the operation are made on the basis of drawings and specification and supervisor decision. Process planning activity normally does not exist.

(*d*) Preparation of schedule is done to mark the beginning and finish of each activity. The supervisor decides the day-to-day scheduling.

(*e*) Progress is made through the meeting with supervisors.

(*ii*) Functions of PPC in Batch Production

(*a*) There is more importance to material control and tools control and systematic stock replenishment.

(*b*) Route sheets or Process sheets are prepared in detailed form.

(*c*) Greater details are to be worked out for loading and scheduling.

(*d*) The detailed data is to be collected on the progress of the work.

(*iii*) PPC Function in Continuous Production

(*a*) Critically in materials function.

(*b*) Nature of the plant is such that no tools control is feasible.

(*c*) Process planning activity does not exist.

(*d*) Final quantity determines scheduling.

14.5.3. PPC Problems

Macro level and long term design of system is a problem. Short term at micro level is another problem. Problems are summarized below:

(*i*) Macro Level Decisions

(*a*) Process design.

(*b*) Equipment selection and work station design.

(*c*) Facilities planning and layout in the form of flow of work.

(*d*) Policies of replacement.

(ii) Micro Level Decisions

(a) Material and tool control

(b) Production control

(c) Quality and cost control

(d) Labour control

SUMMARY

1. Production is the transformation of materials into a desirable output. It is a series of sequential operations to produce a desirable products acceptance to customer. It is a well organized activity to meet specific objectives. An efficient production system is one which produces the products with required quantity and specified quality at predetermined cost pre-established time.
2. Production planning is the determination, acquisition and arrangement of all facilities necessary for future production of products. It is the design of production system. It organizes the production.
3. Objectives of PPC are systematic planning of production activities to achieve the highest efficiency in production of goods/services. It organizes the production facilities like machines, men etc., to achieve the desired production with proper quantity, quality, time and cost.
4. PPC organization depends upon factors such as degree of centralization desirable, internal structure composition of PPC, system and management policy.
5. A staff specialists controls production planning in centralized planning. In case of multi-product and multi-plant organizations centralized planning is more effective.

   The decentralized planning and production involves the line staff in planning production. Multi-product, multiple plants, large workforce requires higher degree of centralization.

EXERCISES

1. Explain the need for production planning and control in the organization.
2. What are the functions of PPC?
3. Justify that planning the production is the pre-operation activity.
4. Explain PPC in relation with the manufacturing organizations.
5. Explain objectives of PPC and how they are achieved.
6. Give comparison between production planning and production control.
7. Explain essential information for efficient working of PPC.
8. What are the relationships of PPC with other departments.
9. Describe production or procedure cycle.
10. What are centralized and decentralized PPC?
11. The nerve centre of the organization is PPC. Explain in detail.
12. Describe the production problem area.
13. How does PPC contributes to effective utilization of resources.
14. Explain PPC functions for various types of manufacturing methods.

15 CHAPTER Fifteen

MAINTENANCE FUNCTIONS

CHAPTER OUTLINE

- Maintenance History
- Maintenance Function
- Maintenance Staff
- Need of Planned Maintenance
- Maintenance Objective
- Maintenance Organization
- Administration
- Maintenance Policy

15.1. MAINTENANCE HISTORY

Before World War II, maintenance was considered as an added cost to the plant which did not increase the value of finished product. The maintenance was restricted to make the unit operational when it breaks down. It was taken as the cheapest alternative. When the advances of engineering and scientific technology developed during and after the World War II, other types of maintenance were developed. These were much cheaper for example preventative maintenance.

Maintenance was classified as a function of the production system. Presently, issues of environment, safety, quality of product and services have made maintenance as one of the most important for success of the industry. There is continuous need of a very well organised maintenance programme to complete world-wide in the world class companies.

15.2. MAINTENANCE OBJECTIVE

Maintenance objective needs to be consistent with the production goals. Maintenance objectives and production goals relationship can be seen in the action of keeping production machines and facilities in the best possible condition in an industry.

Maintenance objectives are to maximize production or increase facilities availability at the lowest cost. The quality should be highest and safety standards should be met. Other objectives of maintenance are:

(*i*) Downtime reduction
(*ii*) Reduction of breakdowns and emergency shutdowns.
(*iii*) Equipment efficiency improvement and reduction of scrap rate.
(*iv*) Equipment useful life optimisation.
(*v*) Energy usage minimisation.
(*vi*) Resources utilisation optimisation.
(*vii*) Spare stock control improvement.
(*viii*) Identification and implementation of cost reductions.
(*ix*) Provide reliable cost and budgetary control.

15.3. MAINTENANCE FUNCTION

Maintenance function is either repair or replacement with a minimum possible cost. With this an item should reach its acceptable productivity condition. Maintenance is an organised activity which is carried out at a minimum cost acquired to keep an item in its best operational condition.

British standard glossary of terms defines maintenance as: "The combination of all technical and administrative actions, including supervision actions, intended to retain an item in, or restore it to, a state in which it can perform a required function".

A maintenance function is shown in Fig. 18.1 which depicts various maintenance types.

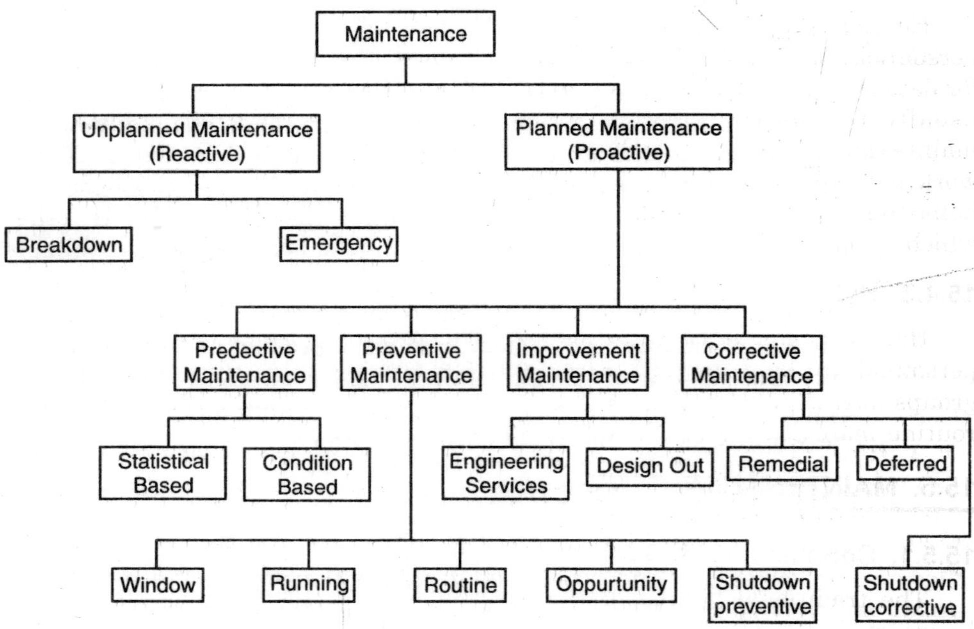

Fig. 15.1. Types of maintenance

15.4. MAINTENANCE ORGANIZATION

Maintenance organization can be based on geographical organizational structure, maintenance staffing or reporting structure.

Geographical organizational structure can be as centralized organization, organization by areas, hybrid organization. Maintenance staffing structure can have complete inhouse staff, combined in house/contract staff, contract maintenance staff and complete contract maintenance. Report structure type organization can have maintenance centric structure, production centric structure and engineering centric structure.

The maintenance organization and administration are described here.

15.4.1. Centralized Organization

All personal report to one central location from which they are directed to work locations. It provides the benefit of more extensive use of the personnel.

Technician can always be directed to the highest priority work. A central maintenance organisation reduces, the amount of nonproductive time for maintenance. However, it has slower response time caused by increased travel time due to re-allocation of personnel. If there is a problem in one area of the plant and the workers are in other areas of the plant, it takes time to find them, redeploy them, correct the problem and then return them to their original assignment.

15.4.2. Organization by Areas

Here specific areas within a plant or facility are assigned to maintenance personnel. Few maintenance personnel are always kept in a control location for data collection, analysis, crew scheduling, work planning etc. Organizations usually respond in a timely manner in the area configuration as the maintenance personnel are close to the equipment. However, finding enough work to keep all the maintenance personnel in an area busy is difficult. It helps to instill in maintenance workers a sense of ownership of the equipment which is one of the biggest advantages of area organizations.

15.4.3. Hybrid Organization

Here some maintenance personnel are assigned to areas and the remaining personnel are kept in central location. The central group provides the area groups during shutdowns, major maintenance etc. The area personnel do the routine maintenance build relationships with the operations personnel.

15.5. MAINTENANCE STAFF

15.5.1. Complete In-house Staff

The traditional approach in most companies is to have a complete in-house staff. The craft technicians who perform maintenance are direct employees of the company under this approach.

The company is responsible for all administration functions for each employee, as well as salary and benefits.

15.5.2. Combined In-house/Contract Staff

In the 1980s combined in-house/contract staff became a more common approach to maintenance. Most of the maintenance is performed by the in-house staff. Certain maintenance tasks such as service on air conditioners, equipment rebuilds, and insulation are performed by contractors. The amount of staff required for specific skill functions can be reduced by this method. This approach can contribute even further savings, if the contract personnel are not required full time.

15.5.3. Contract Maintenance Staff

This method is gaining popularity in the United States and it is already common in Japan. Contract maintenance staffs combine the company's supervisors with contract employees. Properly-skilled individuals are provided by the contractor. This removes the burden of training and personnel administration from the company. There is no need of same employees all the time in this approach. The interaction between the in-house supervision and the contract personnel helps to compensate in case contract employees have less familiarity with the equipment.

15.5.4. Complete Contract Maintenance

All craftsmen, planners and supervisors are included in the complete contracting maintenance staff. Supervisors generally report to a plant engineer or plant manager. The need for any in-house maintenance personnel is eliminated in this approach. Although this is not yet popular, this program can prove to be cost effective and a good alternative to conventional maintenance organizations. It is difficult to manage a contract work force in most companies.

15.6. ADMINISTRATION

15.6.1. Reporting Structure

The reporting structures may be divided into three models which are Engineering centric structure/model, Production centric structure/model and Maintenance centric structure/model, see Fig. 18.2, 18.3 and 18.4.

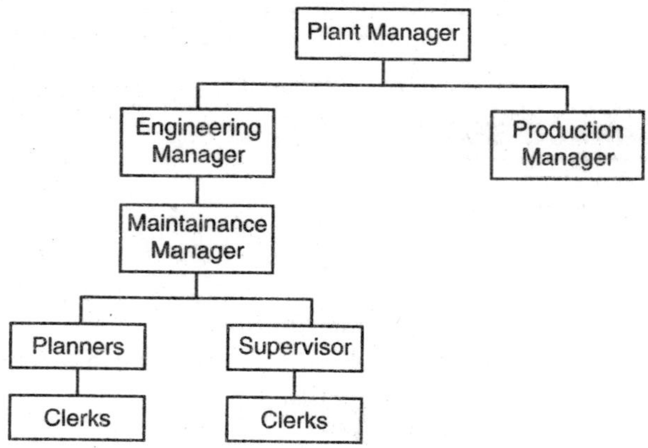

Fig. 15.2. Engineering centric structure/model

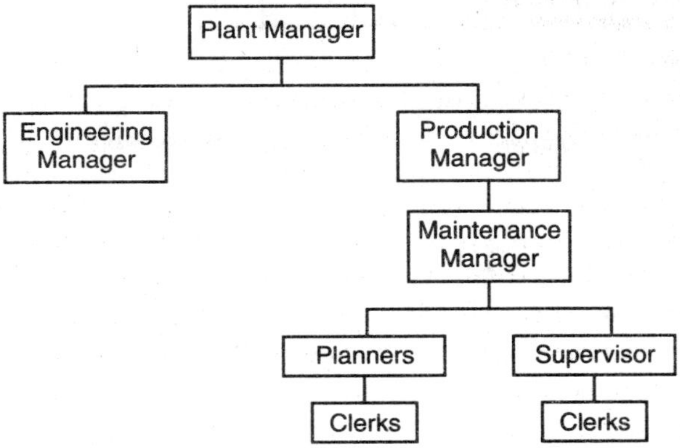

Fig. 15.3. Production centric structure/model

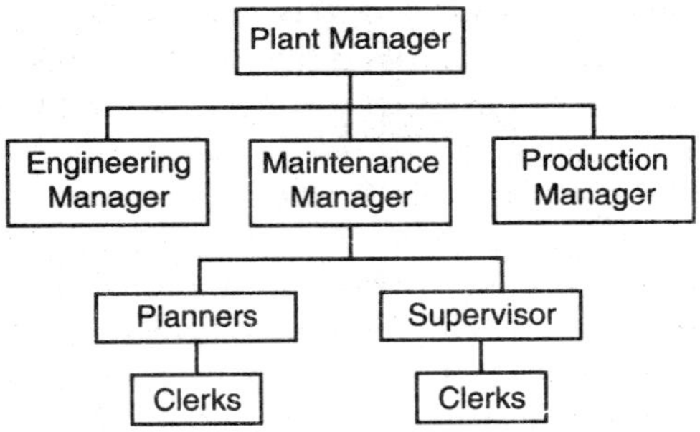

Fig. 15.4. Maintenance centric structure/model

15.6.2. Functions of Key Personnel

The maintenance manager is responsible for the entire maintenance function including the planning, supervising and engineering staff. He has responsibility for delegeting assignments to the appropriate personnel. He coordinates closely with counterparts in other in-house organisations. He promotes proper understanding of the maintenance function to other organisation. He ensures that all supervisor, planners, technicians and maintenance engineers are properly educated and trained. He takes responsibility for planning, cost control, union activities, vacation planning, etc.

Supervisors ensures quality of work. He ensures that work is accomplished in a safe and efficient manner. He reviews work planning and scheduling with the planner. He works with plant or production supervisior to ensure first-line maintenance is being performed by operators and verifies the qualifications of hourly personnel and recommends training as needed.

A maintenance planner develops a weekly schedule and assists the maintenance first-line maintenance foreman or supervisor in determining job priorities, plans, schedules and coordinates corrective and preventive maintenance activities.

15.7. NEED OF PLANNED MAINTENANCE

The modernisation and urgent need for higher productivity have brought with it sophisticated equipment and machines. These incorporate newer, and therefore not adequately proven, technologies and designs. These consist of a larger number of assemblies/subassemblies/components. Hence, have higher probabilities of failure. These failures cause a loss of availability of the system and also result in an increase in the direct cost of maintenance. The cost of maintaining modern industrial systems is very high. There is need of fulfilment of two seemingly contrary objectives, namely the maximisation of availability and the minimization of the life cycle cost of maintenance. In view of all these, there is a need of planned maintenance.

Maintenance control in a mine, fleet or plant can increase profits in two ways.

(*i*) **Increased production:** Reduction of wasteful or unnecessary down time increases production, thereby increasing profits.

(*ii*) **Reduced cost:** Higher productivity, method improvements or material changes can reduce maintenance costs, thereby increasing profits.

Maintenance management continually helps to increase production or reduce costs. Maintenance can help operations achieve higher production by increasing on-line-time or up-time. The maintenance department has the greatest opportunity to increase production. Planned maintenance helps keep downtime at a minimum. Advantages of planned maintenance provides a daily plan for front-line supervisors and a tool for operations to assign priorities. It helps field repairs coordinate work with shop and construction forces. It reduces the time required for critical shutdowns or overhauls, maintenance cost and emergency breakdowns.

15.8. MAINTENANCE POLICY

Maintenance policy has to address the following aspects and prepare procedure manual accordingly:

(*i*) How to set up and run your maintenance department in todays fast-paced and highly competitive business environment.

(*ii*) How to manage all the function of a maintenance department.

(*iii*) How to start and run a complete maintenance work order and call order system.

(*iv*) How to make various maintenance plans work.

The manual is for maintenance manager, maintenance supervisor, plant engineer, facilities manager, plant manager, or the facilities engineer. The maintenance policy has to specify which maintenance types are to be used for various equipments, machineries and plants.

SUMMARY

1. There is continuous need of a very well organised maintenance programme to complete world-wide in the world class companies.
2. Maintenance objectives are to maximize production or increase facilities

availability at the lowest cost. The quality should be highest and safety standards should be met.

3. Maintenance function is either repair or replacement with a minimum possible cost. With this an item should reach its acceptable productivity condition. Maintenance is an organised activity which is carried out at a minimum cost acquired to keep an item in its best operational condition.
4. Maintenance organization can be based on geographical organizational structure, maintenance staffing or reporting structure.
5. Maintenance management continually helps to increase production or reduce costs.
6. Maintenance management continually helps to increase production or reduce costs.
7. Planned maintenance helps keep downtime at a minimum.
8. It reduces the time required for critical shutdowns or overhauls, maintenance cost and emergency breakdowns.

EXERCISES

1. What are maintenance objectives and functions? Explain in detail.
2. Describe one maintenance organization and administration with the help of diagram.
3. What is the need of plant maintenance? Explain with diagrams.
4. Briefly explain historical background of maintenance activity.
5. Give the guidelines followed for maintenance policies.
6. What is the need of maintenance and does it benefit? Relate with some examples.

16 CHAPTER Sixteen

GENERAL MAINTENANCE

CHAPTER OUTLINE

- Breakdown Maintenance
- Age Replacement Policy
- Corrective Maintenance
- Maintenance Planning
- Costs and Delays
- Replacement Policy
- Periodic Replacement
- Preventive Maintenance (PM)
- Scheduled Maintenance
- Inspection

16.1. BREAKDOWN MAINTENANCE

The required repair, replacement, or restore action is performed on equipments/ machines or a sytem after the occurrence of a failure in order to bring to at least its minimum acceptable condition. This type of maintenance is the oldest. Breakdown maintenance is also known as Run to Failure Maintenance (RTF). It can be of two types namely emergency maintenance and breakdown maintenance.

Emergency maintenance is carried out as fast as possible in order to bring a failed machine or facility to a safe and operationally efficient condition. Breakdown maintenance is performed after the occurrence of an advanced considered failure for which advanced provision has been made in the form of repair method, spares, materials, labour and equipment. Whenever, some equipment and machineries aer very critical, they are replaced are block or based on aging.

Breakdown maintenance is useful in the situation where the failure of a component in a system is unpredictable. The cost of performing run to failure maintenance activities is lower than performing other activities of other types of maintenance; and the equipment failure priority is too low in order to include the activities of preventing it within the planned maintenance budget.

Breakdown maintenance activities are expensive in terms of both direct and indirect cost. Using this type of maintenance the occurrence of a failure in a component can cause failures in other components in the same equipment, which leads to low production availability. Breakdown activities are very difficult to plan and schedule in advance.

Time based maintenance is done to avoid breakdown failures. The time based maintenance can be replacement of complete block of equipments and machinery, replacement on the basis of aging or just replacement after fixed periods. Other maintenance methods are corrective maintenance, preventive maintenance, maintenance planning, maintenance scheduling and predictive maintenance.

16.2. REPLACEMENT POLICY

When failure occur, it has to be replaced. As failure is unexpected, one resonably assume that failure replacement is more costly than preventive replacement. This is generally because a preventive replacement can be adequately planned and arranged to reduce both time and cost. However, a failure may cause damage to other components of an equipment, or other equipment and the total replacement cost much higher than failure. For reducing the number of failures and subsequently by reducing the total expected cost of replacement preventive replacement can be carried out at specific intervals. If the period of time over which the equipment or machinery has to be operated is much higher than the interval between preventive replacements, then one may consider only one cycle of operation as shown in Fig. 16.1.

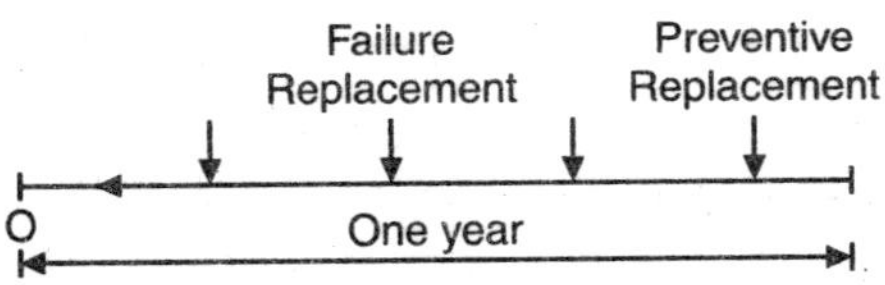

Fig. 16.1. One cycle of operation

Accordingly model for the total expected cost of replacement for this cycle is developed.

16.3. AGE REPLACEMENT POLICY

Making preventive replacement at fixed intervals leads to the possibility of carrying out a preventive replacement shortly after a failure replacement. This preventive replacement depends upon the age of the equipment/machinery which is the elapsed time from the previous replacement, see Fig. 16.2.

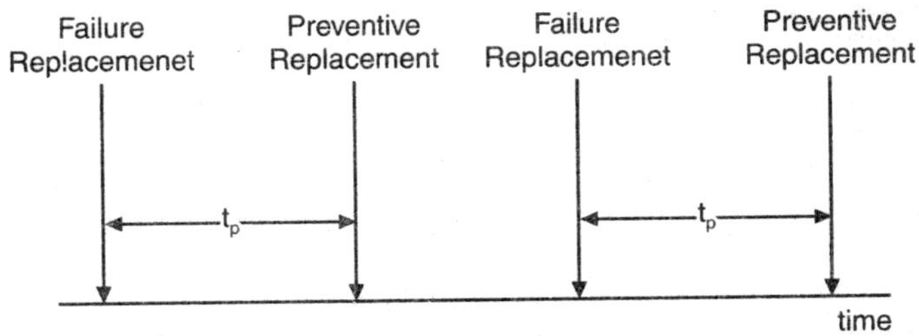

Fig. 16.2. Age replacement policy

Thus, the age replacement policy is to perform a preventive replacement after the equipment has reached a specific age (t_p). Also failure replacement is performed whenever necessary. The optimal replacement age of the item is determined to minimise the total expected cost of replacement per unit time.

A group replacement policy is used in situation where the item is so inexpensive that it is cheaper to replace it in a group rather than individually, as the costs of replacing one at a time is much greater than the per unit cost of replacing the item in a group. In such case, all the items in the group are replaced irrespective of age.

16.4. PERIODIC REPLACEMENT

Consider an equipment which is not subject to sudden failures but whose deterioration through use result in increasing operating and maintenance cost as shown in Fig. 16.3.

To reduce the operating and maintenance cost a replacement can be performed. The operating cost trends following each replacement are identical and the interval between replacements is constant. The operating cost of the equipment increases due to the deterioration of certain parts of the equipment. Some of these deteriorating parts can be replaced to bring down the operating cost of the equipment. A balame has to be struck between the money spent on replacements and the savings obtained by reducing the operating cost, see Fig. 16.4.

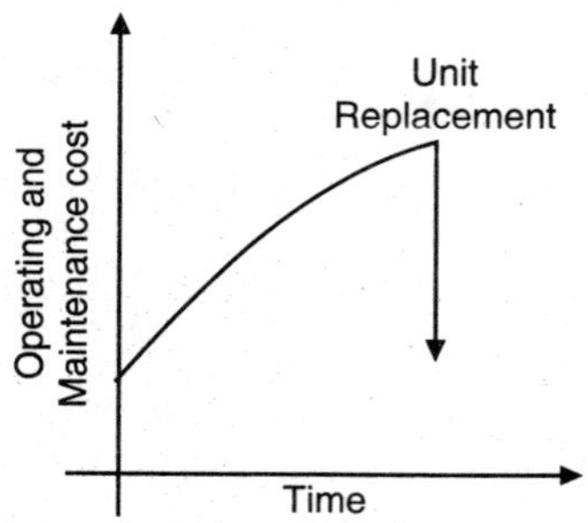

Fig. 16.3. Increasing operating and maintenance cost

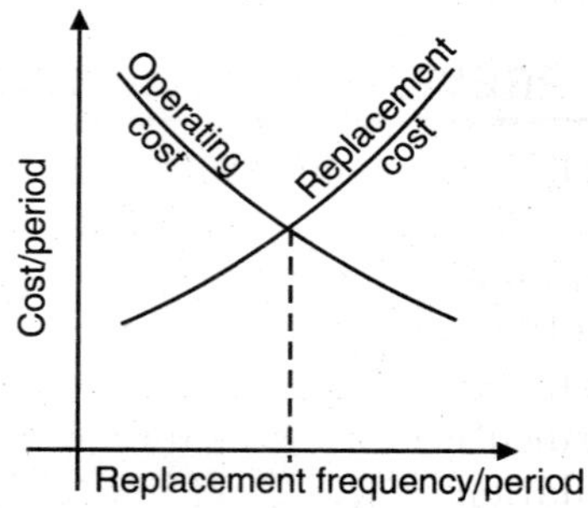

Fig. 16.4. Cost versus replacement

16.5. CORRECTIVE MAINTENANCE (CM)

CM is composed of repair, replacement, or restore activities after the occurrence of a failure to eliminate the source of this failure or reduce the frequency of occurrence. British Standard 3811: 1993 defines the corrective maintenance as "The maintenance carried out after recognition and intended to put an item into a state in which it can perform a required function". Corrective maintenance are of three types:

(*i*) **Remedial maintenance:** These activities eliminate the source of failure without interruping the continuity of the production process. The item to be corrected is taken out of the production line and replaced with reconditioned item or its workload is transferred to its redundancy.

(*ii*) **Deferred Maintenance:** These are concutive maintenance activities which are initiated after the occurrence of a failure such that they will not affect the production process.

(*iii*) **Shutdown corrective maintenance:** When the production line is in total stoppage situation, the corrective maintenance activities are performed. Corrective maintenance maximises the effectiveness of all critical plant systems. It eliminates breakdown and unnecessary repair. It also reduces the deviation from optimum operating conditions.

For the corrective maintenance, the failure should occur before any corrective action is taken. Corrective maintenance activities are planned and regularly taken out to keep plants machines and equipment in optimum operating condition.

By conducting four important steps corrective maintenance can be performed. These are: fault detection, fault isolation, fault elimination and verification of fault elimination.

In the fault elimination, the steps taken can be adjusting, aligning, calibrating, reworking, removing, replacing or renovation. For corrective maintenance, accurate identification of incipent problems is done. Effective planning need to be done requiring the skills of the planner, the availability of well developed maintenance database about standard time to repair, a complete repair procedures, and the required labour skills, specific tools, parts and equipment. It also requires proper repair procedures, adequate time to repair, and verification of repair.

16.6. PREVENTIVE MAINTENANCE (PM)

PM activities are performed on equipment, machinery and systems before the failure. PM presents or eliminates any degradation in operating conditions of the equipment, machinery and system. British standard 3811: 1993 defines preventive maintenance as:

"The maintenance carried out at predetermined intervals or according to prescribed criteria and intended to reduce the probability of failure or the degradation of the functioning and the effects limited."

The efficiency of preventive maintenance are affected by the following factors:

(*i*) In order to perform this type of maintenance, there is the need for an adequate number of staff in the maintenance department.

(*ii*) The production equipment and machinery should be chosen which are suitable for the working environment.

(*iii*) Staff qualification and skills requirements are gained through training.

(*iv*) Executive management support and commitment to the PM programme is essential.

(*v*) PM programme should be properly planned and scheduled.

(*vi*) PM programme should be ability based.

Preventive maintenance activities satisfy most of the maintenance objectives. PM is good for the equipment and machines whose breakdown would cause serious production losses. Breakdowns and emergency repairs are minimised due to PM. Replacements, adjustments, major overhauls, inspections and lubrications are the activities of PM. The preventive maintenance are of different kinds according to the nature of its activities. These are:

(*i*) **Routine maintenance:** The maintenance activities that are repetitive and periodic in nature such as lubrication, cleaning and small adjutment are routine maintenance.

(*ii*) **Running maintenance:** It include the maintenance activities that are carried out while the equipment is running. The activities are performed before the actual preventive maintenance.

(*iii*) **Opportunity maintenance:** It is a set of maintenace activities when an unplanned opportunity exists on a machine during the planned maintenance to other machines.

(*iv*) **Window maintenance:** It is a set of maintenance activities when a machine is not required for a definite period of time.

(*v*) **Shutdown preventive maintenance:** It is a preventive maintenance carried out when the production line is in total stoppage situation.

16.7. MAINTENANCE PLANNING

The continuous process indutries such as thermal power plants, chemical and fertilizer plants, oil refineries and integrated iron and steel-works, maintenance are a high-technology based. They have very demanding functions, therefore, all maintenance activities must be thoroughly planned. It requires identification of corrective maintenance actions from an analysis of past and recurrent failures, preventive maintenance inspection etc. Planning and scheduling of all maintenance activities are for an improvement of efficiencys of the maintenance function. It may be noted that:

(*i*) preventive maintenance is used to cover only periodic preventive maintenance activities and preventive maintenance inpection of the stop-job type.

(*ii*) all the systems are formalised job orders to detail the work to be performed by the maintenance personnel.

(*iii*) all the scheduling procedures, requires a procedure just for preventive maintenance activities and the need for analysis of recurrent failures and the corrective maintenance activities.

An integrated system includes the reporting of failures, analysis of failures, scheduling of corrective maintenance jobs, scheduling of periodic preventive maintenance jobs and scheduling of preventive maintenance inpections.

16.8. PLANNED MAINTENANCE

Maintenance or scheduled planned maintenance is a service carried out by a competient and suitable personnel to ensures that an item of equipment is operating correctly to avoid any unscheduled breakdown and downtime.

Planned maintenance comprises condition based maintenance and preventive maintenane. The maintenance event is preplanned and all future maintenance is preprogrammed. Planned maintenance is created for every item separately according manufactures recommendation. Plan can be based on equipment running hours, date based or for vehicles distance travelled. For example, maintenance planning is done for car maintenance. After a fixed kilometers oil should be changed, parts renewed etc.

Planned maintenance has some advantages over condition based maintenance. Planned maintenance has easier planning of maintenance and ordering spares. The costs are distributed more evenly and there is no initial costs for instruments for supervision of equipment.

However, planned maintenance is less reliable than equipment with condition based monitorily (CBN). Planned maintenance is more expensive due to more frequent parts change. Parts which required scheduled maintenance at fixed intervals are generally due to wearout or a fixed shelf life.

16.9. COSTS AND DELAYS

A replacement decision is done if the cost of an in-service failure exceeds the cost of preventive replacement action. In other words, it is more costly to replace an equipment or machinery after failure than before. There are three fundamental assumptions made in all replacement decisions. First, the state of the equipment whether good or failed, is always known with certainty. Second the replacement action makes the equipment to the new condition. Third, the operating/running and maintenance costs of the equipment subject to replacement are increasing, or the hazard rate of the equipments must be increasing.

The replacement of an operating plant or equipment is made to reduce the maintenance and running costs since most equipment require increased maintenance and running costs with age. The replacement of capital equipment is considered from the engineering economics viewpoint. The capital, running and maintenance costs and the salvage values are considered. There are two kinds of decisions, which need to be taken. First, decision with regard to the replacement interval of a capital equipment. This requires to determine the economic life of a capital equipment so as to minimise the total cost. This must include capital cost, maintenance cost and the salvage value. Second, decision with regard to replacement alternatives which involves the comparison of the asset which the organisation presently owns with the equipment being considered as its replacement.

Replacement of an expensive capital equipment includes the investment of large sums of money. The capital equipment has a useful life of many financial years such as the running and maintenance costs and the benefits accuring from the investment in the form of revenues earned from the product and services produced continue for a number of years. As money can earn interest at a certain rate through its investment for a period of time, it is important to recognize that a sum of money received at some future date is not worth as much as money in hand at the present.

Under the patterns of maintenance costs, there are three possible patterns of maintenance cost over the useful life of an equipment namely sporadic, constant and increasing.

The annual maintenance cost, the equivalent annual cost of maintenance, the equivalent annual cost of capitary recovery and return and the total equivalent annual cost figures are shown in Fig. 16.5.

From the figure it is clear that the minimum total equivalent annual cost is recovered at the end of 6 years. It is also noticed that the total equivalent annual cost curve drops steeply and becomes almost flat in the region of minimum. Hence, for this capital equipment with the given salvage values and with a linearly increasing maintenance.

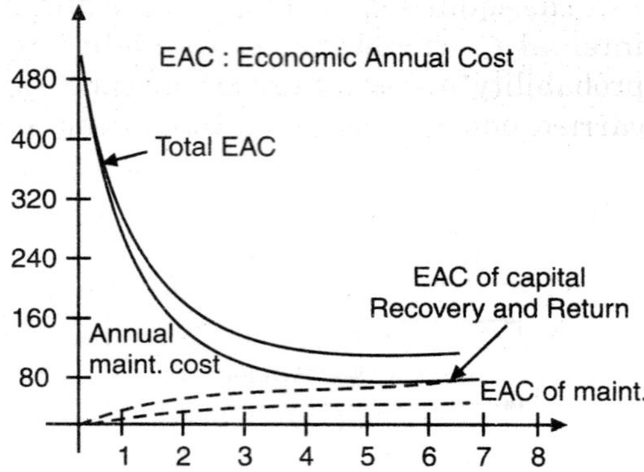

Fig. 16.5. Cost of maintenance versus cost of a capital equipment

Then, we can conclude that the optimal replacement interval for the given capital equipment is 6 years. This illustrates the method used for the evaluation of the minimum cost life, of the optimal replacement interval, of capital equipment.

16.10. INSPECTION

Inspection is carried out to determine the state of equipment for detecting certain conditions which may subsequently develop into major faults and result in equipment breakdown. Measurement of certain indicators of the state of the equipment is done. This inspection is carried out to minimise the equipment downtime due to equipment failures and the consequent corrective maintenance action to maximise equipment availability. Once the pre-specified indicator values are determined through equipment inspection, they are compared with standards, or other acceptance criteria. Corrective actions in the form of adjustment, preventive replacement of components, and minor repairs are carried out if necessary. The inspection frequency should be on the basis of the cost of inspection which is a function of the level of inspection and the expected benefits from such inspection.

16.10.1. Inspection Intervals

The state or condition of an equipment in continuous use is always known. The inspection is needed for equipments which are stored or for emergency usages, for example, military equipments or weapons. Other examples are emergency light, and fire extinguishers etc. Some kinds of equipments may be used intermittently in a process plant or manufacturing organization. In such cases, the equipment must be in an operating condition whenever it is required to be used. Such equipments are inspected after certain intervals of time. Consider that the inspection interval t_i and the time required to carry out inspection is T_i as shown in Fig. 16.6, which is for a cycle when equipment is found to be working.

The second cycle shown is for the equipment which has failed during the inspection.

The equipment availability per unit time will be a function of the inspection interval t_i. Further, the availability of the failed cycle is derived from the probability density function of time $f(t)$, see Fig. 16.7. As the inspection is carried out at time t_i the mean time failure is the shaded area.

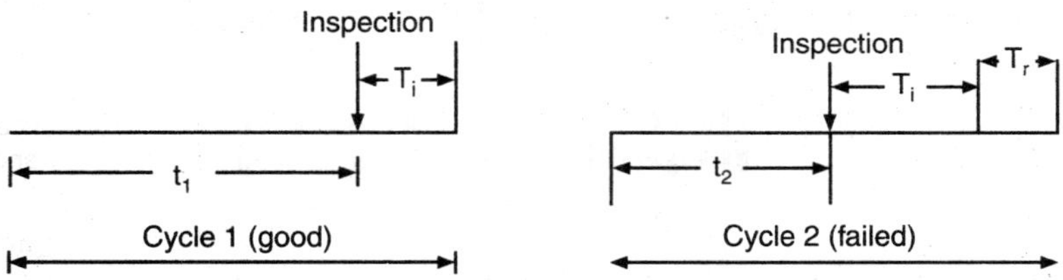

Fig. 16.6. Good cycle and failed cycle

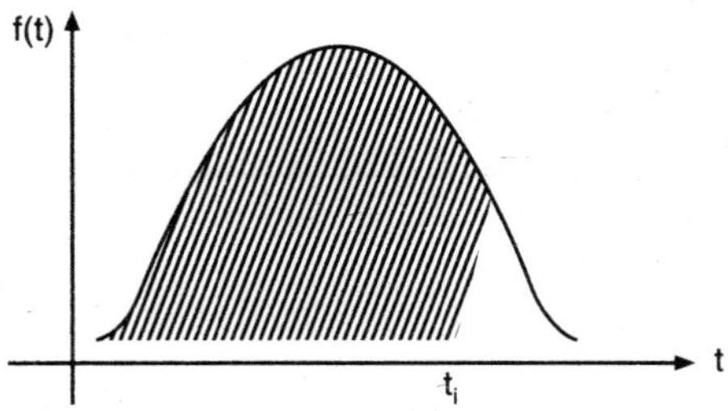

Fig. 16.7. Failed cycle density function

16.10.2. Inspection Reports

A sample form of inspection report is shown in Fig. 16.8. There are different forms for different items, which depend upon the criticality of the failure rate. The form is filled up by the inspector and given to preventive maintenance personnel who works as per the details of instruction sheet filled by the inspector.

16.10.3. Card History System

An equipment history card is shown in Fig. 16.9. The details include emergency repair, corrective maintenance, work actually done, item replaced, cost of spares and labour etc.

<table>
<tr><th colspan="16">Inspection Report</th></tr>
<tr><td>Eqpt. Nomenelture
Descrp:</td><td colspan="2">Equpt No.:</td><td colspan="3">Department
(where installed)</td><td colspan="2">Location</td><td colspan="2">Date
Acquired</td><td colspan="3">Date
commissioned</td><td colspan="3">Descrp. of
Imp. access</td></tr>
<tr><td rowspan="2">Items to be
inspected</td><td colspan="3">Jan.</td><td colspan="3">Feb.</td><td colspan="3">Mar.</td><td colspan="3">Apr.</td><td colspan="3">May</td></tr>
<tr><td>1</td><td>2</td><td>3</td><td>1</td><td>2</td><td>3</td><td>1</td><td>2</td><td>3</td><td>1</td><td>2</td><td>3</td><td>1</td><td>2</td><td>3</td></tr>
<tr><td></td><td colspan="3"></td><td colspan="3"></td><td colspan="3"></td><td colspan="3"></td><td colspan="3"></td></tr>
<tr><td>Notes:</td><td colspan="12">G : good, F : Fair, P : Poor
Inspection initials & date :</td><td colspan="3"></td></tr>
</table>

Fig. 16.8. Form for inspection report

<table>
<tr><th colspan="8">Card History Systems</th></tr>
<tr><td colspan="2">Eqpt. Descrp.</td><td>Eqpt. No.</td><td>Dept./Shop (where installed)</td><td>Location</td><td>Date acquired</td><td colspan="2">Date commisioned</td></tr>
<tr><td colspan="3">Manufacturer</td><td colspan="2">Descript. of Imp. Access</td><td colspan="3">Dates of review of total maint. cost.</td></tr>
<tr><td colspan="8"></td></tr>
<tr><td rowspan="2">Date</td><td rowspan="2">E/C</td><td rowspan="2">Work Actually Date</td><td rowspan="2">Items resplaced</td><td colspan="2">Costs</td><td rowspan="2">Comulative total</td><td rowspan="2">Remarks</td></tr>
<tr><td>Spares materials</td><td>Labour</td></tr>
<tr><td></td><td></td><td></td><td></td><td></td><td></td><td></td><td></td></tr>
<tr><td colspan="5">E→ Emergency Repair
C → Corrective Maintenance</td><td colspan="3">Form No. . . . .</td></tr>
</table>

Fig. 16.9. Card history system

16.11. TOTAL PRODUCTIVE MAINTENANCE (TPM)

Total productive maintenance is a maintenance which involves a newly defined concept for maintaining plants and equipment. The goal of TPM is to increase production, increase employee morale and job satisfaction. It brings maintenance into focus. Down time for maintenance is scheduled as a part of the manufacturing.

TPM helps to avoid wastage in a quickly changing economic environment, producing goods without reducing product quality, reduce cost, produce a low batch quantity at the earliest possible time and goods sent to the customer must be non-defective.

The attributes of TPM are detailed as follows:

(*a*) TPM is unique as operators are also made to involve in the maintenance process.

(*b*) TPM matrices are:

  (*i*) Adopting life cycle approach for improving the overall performance of equipment.

  (*ii*) Productivity improvement by highly motivated workers.

  (*iii*) Use of voluntary small groups activities for identifying the cause of problem and equipment modification.

(*c*) TPM objectives are:

  (*i*) To achieve zero defects, zero breakdowns and zero accidents.

  (*ii*) Involve people in all levels of organization.

  (*iii*) To form different teams to reduce defects and do self-maintenance.

(*d*) TPM direct benefits are:

(*i*) Increase productivity.

(*ii*) Manufacturing cost reduction.

(*iii*) Reduce accidents.

(*iv*) Customer satisfaction.

(*v*) Follow pollution control measures.

(*e*) TPM indirect benefits are:

(*i*) Clean, neat and alternative workplace.

(*ii*) Achieve goals by working as team.

(*iii*) Higher confidence level.

(*iv*) Machine ownership feeling of the workers.

Steps in introduction of TPM in a organization are as follows:

Step A: Prepatory Stage

Step 1: Proper understanding, commitment and active involvement of the top management is needed.

Step 2: Training is to be done based on the need.

Step 3: Setting up TPM and department committees for improvement, autonomous maintenance, quality maintenance etc.

Step 4: Establishing the TPM working system and target by benchmarking each area and fixing up a target for achievement.

Step 5: A master plan for institutionalizing wherein TPM becomes an organizational culture.

Step B: Introduction Stage

With all including suppliers as they should know that there is need of quality supply and affiliated companies who can be customers, sister concerns etc.

Step C: Implementation

In this stage eight activities are carried which are called eight pillars.

Step D: Institutional Stage

Now, TPM is applied to all challenging level to which one can take this movement.

Organization Structure

Organization structure for TPM implementation is shown in Fig. 16.10.

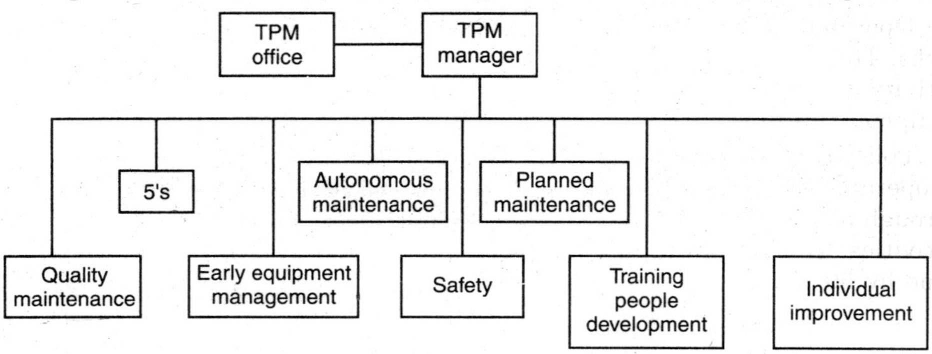

Fig. 16.10. TPM Organization

The eight pillars of TPM are shown in Fig. 16.11, and discussion of the same follows:

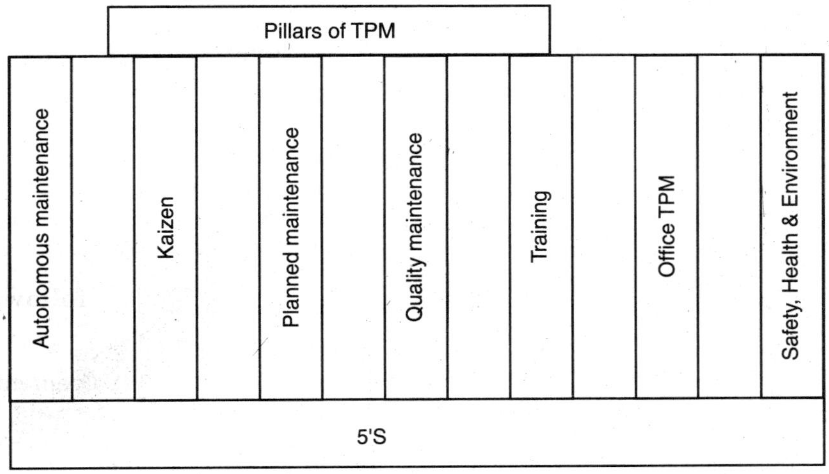

Fig. 16.11. Eight pillars of TPM

(*i*) 5'S

5'S stand for sort out, systematic, sweep, standardize and self-discipline. In brief, all these means sorting and organizing the items as critical, important and frequently, used items. They should be kept as per the concept that "Each items has a place, and only one place." The items should be placed back at the same place. Useless items can be salvaged.

Name plates and colored tags has to be used to identify items easily. The workplace is made free of burns, grease, oil, waste, scrap, etc. There are no loosely hanging wires from machines. Decision by employees are taken on standards for keeping the workplace/machine/pathways clean. Such standards are used in the whole organization which are tested and inspected at random.

Considering 5'S as a way of life indicates self-descriptive among the employees of the organization. This leads to punctuality and dedication to the organization etc.

(*ii*) Autonomous Maintenance

Operators are developed to be able to take care of small maintenance tasks. This makes skilled maintenance personnel work on more value added activity and technical repair. The operator is made responsible for upkeep of equipment to present it from deteriorating.

Policy is for uninterrupted operation of equipments, flexibility for operators to operate and maintain other equipments, eliminating the defects at source through active employee participation and stepwise implementation of these activities. Targets are to reduce oil consumption by 50% and reduce process time by 50%.

(*iii*) Kaizen

Basically Kaizen is for small improvements on a continual basis, involving all people in the organization. A very large number of small improvements in an organizational environment are more than a few improvements of large value. Unnecessary things are reduced at workplace that affect efficiencies. Kaizen policy is to practice concepts of zero losses in every sphere of activity, relentless pursuit in all resources, relentless pursuit to improve overall plant equipment effectiveness and focus of easy handling of operators.

The objectives of TPM is to maximize equipment effectiveness. It aims to maximize machine utilization and also machine availability maximization for other uses. Kaizen pursuses for efficiency of operator and material and energy utilization. Kaizen activities try to thoroughly eliminate sixteen major losses. The sixteen major losses are: breakdown loss, adjustment losses, cutting blade loss, start-up loss, minor stoppage, speed loss, rework loss, scheduled downtime loss, management loss, operating motion loss, line organization loss, logistic loss, measurement and adjustment loss, energy loss, tool breakage loss and yield loss.

(*iv*) Planned Maintenance

It achieves trouble free machines and equipments producing defect free product. The maintenance is broken down into four groups namely preventive maintenance, breakdown maintenance, corrective maintenance, and maintained prevention.

Planned maintenance helps to evolve efforts from a reactive to a proactive method and one trained men power to help train the operators to better maintain their equipment. Policy is to achieve and sustain availability of machines, reduce spares inventory and improve reliability and maintainability of machines.

(*v*) Quality Maintenance

It is for customer satisfaction through highest quality through defect free manufacturing. First step is to gain understanding of equipment affecting product quality and eliminate current quality concerns leading to quality control to quality assurance. Quality defects are classified as customer end defects and in house defect. For customer end, defects are customer end line rejection and field complaints. In-house defects are related to products and process. Target is to achieve and sustain customer complaints at zero, reduce in-process defects by 50%, and reduce cost of quality by 50%.

(*vi*) Training

It creates multi-skilled employees whose morale is high and who has eager to perform all required functions effectively and independently. Operators are upgraded by sufficient "Know-How" and also learn "Know-Why". By experience they solve the problems. The employees are trained to achieve skills. A factory full of experts is the goal.

Steps of training activities are setting policies and priorities and checking present status of education and training, training the employees for upgrading the operation and maintenance skills, preparation of training calender, and evaluation of activities and study of future approach.

(*vii*) Office TPM

Office TPM must improve productivity, efficiency in the administrative function and identify and eliminate losses. The processes and procedures must move towards increased office automation. Office TPM should be a senior person from one of the support function such as head of finance, purchase etc. Members representing all support functions and people from production and quality should be in committee. Office TPM has involvement of all people in support functions for focusing on better plant performance, better utilized work area, reduced inventory carrying cost, reduction in number of files, reduced man power, clean and pleasant work environment etc.

Office TPM are for eliminating production output loss due to various resources, mistakes in preparation of cheques, delays in delivery due to any of the support functions, delays in payment to suppliers, and delays in information.

(*viii*) Safety, Health and Environment

Focus is to create a safe workplace and a surrounding area without affecting production procedures. A committee is constituted which comprises representative of officers as well as workers. Utmost importance is given to safety in the plant. Target is to achieve zero accident, zero health damage and no fire.

SUMMARY

1. Breakdown maintenance is useful in the situation where the failure of a component in a system is unpredictable. The cost of performing run to failure maintenance activities is lower than performing other activities of other types of maintenance; and the equipment failure priority is too low in order to include the activities of preventing it within the planned maintenance budget.
2. Time based maintenance is done to avoid breakdown failures. The time based maintenance can be replacement of complete block of equipments and machinery, replacement on the basis of aging or just replacement after fixed periods. Other maintenance methods are corrective maintenance, preventive maintenance, maintenance planning, maintenance scheduling and predictive maintenance.
3. All maintenance activities must be thoroughly planned. It requires identification of corrective maintenance actions from an analysis of past and recurrent failures, preventive maintenance inspection etc. Planning and scheduling of all maintenance activities are for an improvement of efficiencys of the maintenance function.
4. Scheduled maintenance or planned maintenance is a service carried out by a competient and suitable personnel to ensures that an item of equipment is operating correctly to avoid any unscheduled breakdown and downtime.
5. Under the patterns of maintenance costs, there are three possible patterns of maintenance cost over the useful life of an equipment namely sporadic, constant and increasing.
6. Inspection is carried out to determine the state of equipment for detecting certain conditions which may subsequently develop into major faults and result in equipment breakdown.
7. Corrective actions in the form of adjustment, preventive replacement of components, and minor repairs are carried out if necessary. The inspection

frequency should be on the basis of the cost of inspection which is a function of the level of inspection and the expected benefits from such inspection.

EXERCISES

1. Explain various types of maintenance activities with the help of a diagram.
2. What do you understand by replacement policy in a maintenance? Explain in detail.
3. Describe block replacement, age replacement and periodic replacement policies.
4. What are corrective and preventive maintenance? How do they differ?
5. Describe maintenance planning and scheduled maintenance.
6. Explain cost of maintenance, cost of equipment and production delays with suitable diagrams if necessary.
7. How does inspection of equipment and machinery help in keeping it functional.
8. What do you understand by inspection intervals, inspection reports and card history system?

17 CHAPTER Seventeen

PREDICTIVE MAINTENANCE (PDM)

CHAPTER OUTLINE

- Introduction
- Methods of Condition Monitoring
- Liquid Penetrate
- Ultrasonic Testing
- Radiographic Testing (RT)
- Computerized Maintenance
- Non-Destructing Testing
- Magnetic Particles
- Oil Analysis

17.1. INTRODUCTION

PDM activities detect changes in the physical condition of equipment or give signs of failure. Accordingly, appropriate maintenance work is done for maximizing the service life of equipment without increasing the risk of failure.

There are two kinds of PDM:

(*i*) Condition-based predictive maintenance.

(*ii*) Statistical-based predictive maintenance.

Condition-based predictive maintenance is based on continuous or periodic condition monitoring equipment. Statistical-based predictive maintenance is based on statistical data from the recording of the stoppages of the in-plant items and components.

Disadvantage of predictive maintenance is that it has too much dependence on information and the correct interpretation of the information. The predictive maintenance monitors the condition of machines or equipment to determine the actual mean time to failure whereas preventive maintenance depends on industrial average life statistics.

Predictive maintenance techniques help in determining the condition of equipment to predict when maintenance should be done. This helps cost savings over routine or time-based preventive maintenance as these are performed only when necessary. Predictive maintenance leads to less maintenance. It is important to know that the right information in the right time is the key. Knowledge of equipment needing maintenance, the maintenance work can be better planned.

It gives increased equipment life time, increased plant safety, less accident with negative impact on environment, an optimised spare parts handling etc.

Predictive or condition-based maintenance evaluates the condition of equipment by performing periodic or continuous equipment condition monitoring. The ultimate goal is to perform maintenance at a scheduled point in time when the maintenance activity is most cost-effective and before the equipment loses performance within a threshold.

Predictive maintenance uses principles of statistical process control to determine at what point in the future maintenance activities will be appropriate. Predictive inspections are performed while equipment is in service which minimizes disruption of normal system operations. PDM results in substantial cost savings and higher system reliability.

Reliability-control maintenance (RCM) use the predictive maintenance technique in addition to traditional preventive measures. For evaluation of equipment condition, predictive maintenance uses non-destructive testing technologies such as infrared, acoustic, corona detection, vibration analysis, sound level measurements, oil analysis and other specific on-line tests.

These utilize measurements on the actual equipment in combination with measurement of process performance, measured by other devices, to trigger maintenance conditions.

Vibration analysis is used on high-speed rotating equipment. This can be most expensive. Vibration analysis helps to evaluate the condition of equipment and avoid failures. Acoustic analysis is done on a sonic or ultrasonic level. Ultrasonic techniques make it possible to hear friction and stress in rotating machinery and this predicts deterioration earlier than conventional techniques. Ultrasonic technology is sensitive to high-frequency sounds that are in inaudible to the human ear and separates them from lower-frequency sounds and mechanical vibration. Machine friction and stress waves give distinctive sounds in the upper ultrasonic range. Any changes in these friction and stress waves indicate deteriorating conditions. Ultrasonic measurement and analysis helps to differentiate normal wear from abnormal wear, physical damage, imbalance conditions, and lubrication problems.

Sonic monitoring equipment is less expensive but are limited as compared to ultrasonic technologies. Sonic technology is useful only on mechanical equipment. The ultrasonic equipment can detect electrical problems and is more flexible and reliable in detecting mechanical problems.

The widest range of application is that of infrared monitoring and analysis. It is used for detecting both mechanical and electrical failures. Currently, this is the most cost-effective technology.

Oil analysis technique can be classified into two categories: used oil analysis and wear particle analysis. Used oil analysis determines the condition of the lubricant. The quality of the lubricant, and checks its suitability for continued use. Wear particle analysis determine the mechanical condition of machine components which are lubricated. Wear particle analysis, helps to identify the composition of the solid material present and evaluate particle type, size, concentration, distribution and morphology.

The equipment or machinery wear records are system recording of critical components such as bearings, machine slides or elements which deteriorate due to usage of the machine or equipment. Records are maintained for each such item similar to equipment maintenance history formats already discussed. Such records will be very handy in planning and executing predictive maintenance.

There are several international standards for condition monitoring techniques. These can be for vibration, noise, wear, etc. These standards help in evaluation and predicting timing and frequency of maintenance. The equipments used for condition monitoring are basically equipments of predictive maintenance. These can be vibration sensors analyzer of vibration, noise, wear, etc.

17.2. COMPUTERIZED MAINTENANCE

Computerized maintenance software package is used to maintain a computer database of information about an organization's maintenance operations. This information help to do maintenance more effectively such as determining which machines require maintenance and which store rooms contain the spare parts they used. Management, make decisions by calculating the cost of machine breakdown repair versus preventive maintenance for each machine. This leads to better allocation of resources.

Computerized maintenance management system (CMMS) can produce status reports and documents giving details or summaries of maintenance activities. Many CMMS can be either web-based or LAN based. CMMS helps in scheduling jobs, assigning personnel, reserving materials, recording costs, and tracking relevant information such as the cause of the problem if any, down time involved if any, and recommendations for future action. The CMMS schedules preventive maintenance automatically based on maintenance plans. Different techniques are used for repairing when a job should be performed. These can be:

(*i*) **Inspection:** CMMS track preventive maintenance jobs including step-by-step instructions or check-lists, lists of materials required, etc.

(*ii*) **Machine Breakdown:** Data of machine breakdown, details of repairs completed and repairs to be done are kept. Preventive maintenance tasks are generally undertaken during breakdown repair as such preventive maintence requires rescheduling.

CMMS helps in recording data about equipment and property including maintenance activities, specifications, purchase date, expected lifetime, warranty information, service contracts, service history, spare parts etc. The CMMS may also generate matrices to measure effectiveness of asset management. CMMS helps in management of spare parts, tools, and other materials including the reservation of materials for particular jobs, recording where materials, are stored, determining when more materials should be purchased, tracking shipment receipts, and taking inventory.

CMMS also helps in management of permits and other documentation required for the processing of safety requirements. These safety requirements can include electrical safety, and others. The advantage of a computerized system are the following:

(*i*) Faster data processing and information retrieval with accuracy.

(*ii*) As soon as feedback is processed, line data is available.

(*iii*) Reporting can be selective and ranked.

(*iv*) The maintenance planning department has less paper work as such much more time is available for planning and follow-up activities.

17.3. METHODS OF CONDITION MONITORING

Symptoms of imminent failure often appear before the occurrence of failure of a component, but these symptoms often go unnoticed. The form of deterioration of a critical parameter of a component indicate these symptoms. The basic principle of the condition monitoring is to monitor, either continuously or periodically, a selected parameter of the component to assess machine condition. Condition monitoring techniques may be broadly classified under four basic heads namely visual monitoring, performance monitoring, vibration and noise monitoring and wear debris monitoring.

(*i*) Visual Monitoring

The visual inspection of equipments and machines gives a direct indication of component condition by using past experience as the basis of assessment. However, it is generally limited to directly visible and stationary components and parts of machines. The extension of this method may be aided by magnifying glasses and low power microscopes, light probe, boroscopes and endoscopes.

(*ii*) Performance Monitoring

Performance monitoring normally is measurement of output, or sometimes the comparison of input and output which is assessment of the relationship between the input and output of the machine. The monitoring of equipment/machine relate performance deterioration to the possible cause. A monitoring system is normally the recording of the relevant performance parameters. Important parameters and their derivates with reference information are compared which provides an assessment of the extent of machine deterioration. The parameters selected can be primary parameters such as pressure, force, electric charge etc. It can be secondary parameters such as impulse, torque, rate of fuel flow etc.

(*iii*) Vibration and Noise Monitoring

Vibration and noise monitoring have become important techniques of condition monitoring. Rotating machines such as engines, turbines, and turbo generators, industrial fans, etc., produce a sound or vibration pattern. These repeat periodically with a certain time interval. This pattern is a signature which is analyzed for condition monitoring. The bearings of a machine, or an equipment are used for measuring vibration because the bearings take the basic dynamic loads and forces of the machine.

Experienced workers develop a knock of detecting defects of sudden changes of sound by their ears close to the equipment or by a screw driver test. Microphones are used for noise monitoring in practice. Decibel (dB) is used as the unit of measure of sound pressure level.

(*iv*) Wear Debris Monitoring

Wear debris monitoring is used for the analysis of lubricating oils. The debris is generated by the wear of machine components. It helps to assess the level of degradation of the oil itself and to evaluate the degree of wear and wear rate of components. Various methods are used for wear measurements such as infra-red and ultra-violet spectroscopy, mass spectrometry, gas chromatography and thermal analysis.

17.4. NON-DESTRUCTIVE TESTING

Non-destructive testing is a wide group of analysis techniques used to evaluate the properties of a materials, component or system without causing damage. The term Non-destructive Testing (NDT), Non-destructive Examination (NDE), Non-destructive Inspection (NDI) and Non-destructive Evaluation (NDE) are also used for the same purpose. It is a highly valuable technique which saves both money and time in product evaluation, troubleshooting, and research. NDT methods also use ultrasonic, magnetic-particle, liquid penetrant, radiographic, Remote Visual Inspection (RVI), eddy current testing, etc. It is also used in forensic science, medicine and art.

NDT methods are based upon one of electromagnetic radiation, sound, and inherent properties of materials to examine samples. Microscopy principle may be used to examine external surfaces in details. However, sample preparation techniques for metallography, optical mircorscopy and electron microscopy are normally destructive as the surfaces are made smooth through polishing liquid penetrant testing uses dyes, fluorescent or non-fluorescing, in fluids for non-magnetic materials normally metals. Use of a liquid suspension of fire iron particles applied to a part while it is in an externally applied magnetic field is another commonly used method for magnetic materials. Thermal properties of an alloy to quickly and easily characterize many alloy are used in thermoelectric effect or use of the seeback effect.

17.4.1. Applications of NDT

(*i*) Weld Verification

Welding are done to join two or more metal surfaces. Due to loads and fatigue there is a chance that product may fail if welding is not as per specifications. In welding, the base metal must reach a certain temperature during the process, must cool at a specific rate, and must be welded with compatible materials, for strong joints. Cracks may develop in the weld causes it to fail. Lack of fusion of the weld to the base metal, cracks and variations in weld density are causes of defective welding.

NDT techniques may be used for testing welds. Industrial radiography using X-rays or gamma rays, ultrasonic testing, liquid penetrant testing or via eddy current are techniques used. These tests indicate a lack of cracks in the radiography, show clear passage of sound through the weld and back, or indicate a clear surface without penetrant captured in cracks.

(*ii*) Structural Mechanics

Complex structures undergo different loads during their lifetime. For example, the turbo-machinery in a liquid-fuel rocket. The structure undergoes a dynamic input in NDT, similar to a controlled impulse. Displacement or acceleration at different points of the structure are measured. This output is recorded and compared with known input. Differences may indicate an inappropriate model, failed components or an inadequate control system.

(*iii*) Radiography in Medicine

Elements of the body such as bones or molecules have a known response to certain radiographic inputs such as X-rays. A known element such as radiography can be used to image parts or function of the body by measuring and interpreting the response to the radiographic input. Many bone fractures and diseases may be detected and localized in preparation for treatment in this manner. NDT technique also used X-rays to examine the interior of mechanical systems in manufacturing.

Applications

NDT is used in a variety of settings that covers a wide range of industrial activity such as engine parts, automotive frames, airframes, spare frames, propellers, gas turbine engines, bridges, machine parts, boilers, turbine bores, rail inspection, leak detection, submarines and other naval workships, and medical imaging applications, etc.

17.5. LIQUID PENETRATE

Liquid penetrate is also called dye penetrate. Liquid penetrate is Liquid Penetrant Inspection (LPI) or Penetrant Testing (PT). It is a low cost inspection method used to locate surface-breaking defects in metals, plastics, or ceramics. LPI is used to detect casting, forging and welding surface defects such as cracks, surface porosities, and leaks in new products, and fatigue cracks on in-service components.

LPI is based upon capillary action. Low surface tension fluid penetrates into clean and dry surface-breaking discontinuities. After adequate penetration time, the excess penetrant is removed and a developer is applied. The developer draws penetrant out of the flow and defects becomes visible. Inspection is performed under ultraviolet or white light, depending upon the type of dye used fluorescent or non-fluorescent.

The basic stages of liquid penetrant inspection are shown in Fig. 17.1.

First the surface is cleaned thoroughly to remove all traces of dirt and grease. Then, a liquid penetrant which is a bright coloured or fluorescent liquid is applied to the surface. The liquid is allowed to soak into the material's surface for about 20 minutes. Subsequently, the excess liquid penetrant is wiped from the surface and a developer is applied. The developer is a dry white powder. It draws penetrant out of any cracks by reserve capillary action to produce indication on the surface. These coloured indications are broader then the actual flow and are therefore quiet visible. Fluorescent penetrants are normally used when the maximum flow sensibility is required. However, it requires darkened conditions with a UV lamp. The most commonly used systems are solvent removable, or water washable. These systems are often used to check weld quality during fabrication.

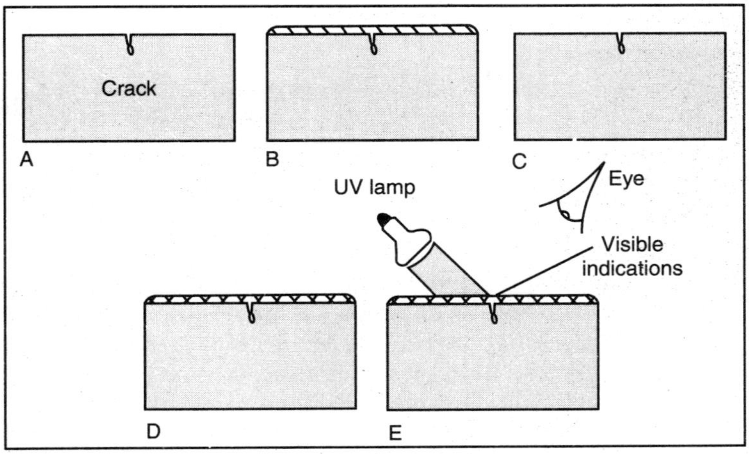

A: Sample before testing
B: Liquid penetrant applied.
C: Surplus wiped off leaving penetrant in crack.
D: Developer powder applied, dye soaks into powder.
E: UV lamp shows up fluorescent indications.

Fig. 17.1. Stages of liquid penetrant inspection

Test surfaces must be cleared adequately. The excess penetrant should be removed completely. Thus, personnel must be properly trained and qualified under certification scheme or equivalent scheme.

Liquid penetrant testing is a fast, cheap and relatively simple means of surface inspection. The aerospace industry uses automated fluorescent penetrant testing to look for fatigue cracking in turbine blades. The construction industry uses dye penetrant testing as a quick and simple method for checking that welds and other susceptible areas are free from surface-breaking flaws.

17.6. MAGNETIC PARTICLES

The particles which are used for magnetic particle inspection plays, a vital role as they form the indication that alert the inspector to defects. Particles are very tiny piece of iron or iron oxide. A paint like pigment is bonded to their surfaces to give the particles colour. The metal used for the particles has high magnetic permeability and low retentivity. High magnetic permeability makes the particles attract easily to small magnetic leakage fields from discontinuities such as flaws. Low retentivity helps the particles themselves in not becoming strongly magnetized and these particles do not stick to each other or the surface of the part. Particles can be in a dry or a wet form. Dry magnetic particles colour can be in red, black, gray, yellow etc., for a high level of contrast between the particles and the part being inspected. These particles are around 50 μm in size, and are about three times smaller in diameter and more than 20 times lighter than the coarse particles.

Magnetic particles are also supplied in water or oil. The particles used with the wet method are smaller in size than those used in the dry method. The particles are typically 10 μm and smaller. The synthetic iron oxides have particles diameters around 0.1 μm.

(*i*) Dry Particle Inspection

Dry particles are dusted onto the surface of the test object as the item is magnetized. When an electromagnetic yoke is used, the AC or half wave DC current cracks a pulsating magnetic force that provides mobility to the powder. The dry powders are used for unground welds and rough as cast surfaces, subsurface cracks.

The part surface is cleaned. The surface must be free of grease, oil or other moisture. The magnetizing force is established for the necessary magnetic flux. Magnetic dust particles are applied. The excess powder is removed. The air needs to be strong enough to remove the excess particles but not strong enough to dislodge particles hold by a magnetic flux leakage field. The areas where the magnetic particles are clustered are to be identified for cracks.

(*ii*) Wet Suspension Inspection

It involves applying the particles while they are suspended in a liquid carrier. It is performed using a stationary, wet and horizontal inspection. In this first, the surfaces of the components are covered with uniform layer of particles. The liquid carrier provides mobility to the particles for an extended period of time making particles to float to small leakage fields to form a visible indication. Wet inspection is best for detecting very small discontinuities on smooth surfaces.

The surface is cleared and free of grease, oil and moisture. The suspension is gently sprayed over the surface of the part. The magnetizing force is applied immediately after applying the suspension of magnetic particles. The magnetic particle clustered is located for crack detection.

17.7. ULTRASONIC TESTING

Ultrasonic testing (UT) uses very short ultrasonic pulse-waves with center frequencies ranging from 0.1-15 MHz and sometimes upto 50 MHz. UT waves are launched into materials to detect internal flaws. This technique is also used to determine the thickness of the test object. Generally, UT is done on metals and alloys but it can also be used on concrete, wood and composites.

An ultrasound transducer is connected to a diagnostic machine. It is passed over the object being inspected. The transducer is separated from the test object by a couplant such as oil or by water.

Reflection and attenuation are the two methods of receiving the ultrasound waveform. The transducer performs both the sending and the receiving of the pulsed waves as the sound is reflected back to the device in case of reflection mode. Reflection ultrasound comes from an interface or an imperfection within the object.

The diagnostic machine displays these results in the form of a signal. The signal amplitude represents the intensity of the reflection and the distance represents the arrival time of the reflection.

Consider now, attenuation mode. A transmitter sends ultrasound through one surface. A separate receiver detects the amount that has reached it on another surface after travelling through the medium. The amount of sound transmitted is reduced due to imperfections. This reduction reveals the imperfection.

Ultrasonic waves have high penetrating power per detection of flaws deep in the part. It has high sensitivity for the detection of extremely small flaws. The size, orientation, shape and nature of defects can also be estimated. It is portable and also useful in highly automated operation.

Experienced techniques are required for manual operation. The development of inspection procedures requires extensive technical knowledge. Rough, irregular in shape, very small or thin or not homogeneous parts are difficult to inspect. Surface preparation is essential by cleaning and removing loose scale, paint, etc. However, the paint that is properly boonded to a surface need not be removed. Couplants are essential for effective transfer of ultrasonic wave energy between transducers and parts being inspected. When using water based couplants the inspected item must be water resistant so no rusting takes place.

17.8. OIL ANALYSIS

The laboratory analysis of a lubricant's properties, suspended contaminants, and wear debris is known as oil analysis (OA). OA is conducted during routine preventive maintenance which gives meaningful and accurate information on lubricant and machine condition. Oil analysis provides the life of a particular machine and a trend to eliminate costly repairs. OA can be divided into three categories namely analysis of oil properties including those of the base oil and its additives, analysis of contaminants and analysis of wear debris from machinery.

Oil sampling is a procedure for collecting a volume of fluid from lubricated or hydraulic machinery for the purpose of oil analysis. One has to ensure that procedures are used to minimize disturbance of the sample during and after the sampling process. Oil samples drawn in small quality in a clean bottle are sealed and sent to a laboratory for analysis.

It is established that increases in component wear can be confirmed by detecting increases in the wear metal content of the lubricating oil. OA includes the analysis of the additives in oils to determine if an extended drain interval may be used.

OA helps to determine the remaining useful life of additives in the oil. Comparison of the OA results of new and used oil, can determine when an oil must be replaced. Careful analysis might even allow the oil go to its original additive levels by either adding fresh oil or replenishing additives that were depleted.

17.9. RADIOGRAPHIC TESTING (RT)

RT is a non-destructive Testing (NDT). It involves the use of penetrating X-rays or gamma radiation, to examine parts and products for imperfections. The source of radiation used is an X-ray machine or radioactive isotope.

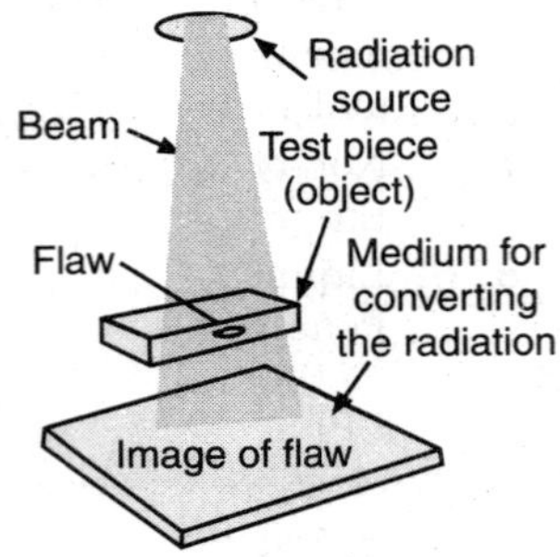

Fig. 17.2. Radiographic Testing (RT)

Radiation is directed through a part and onto film or other media as shown in Fig. 17.2. The internal soundness of the part is seen in the resulting shadowgraph. Density changes in the film indicates the possible imperfections.

The radiographic applications evaluate the material properties such as composition, density uniformity and cell or particle size. It also evaluates the manufacturing and assembly properties such as voids, cracks, bond integrity of welds etc. In addition verification of proper assembly of component pieces can also be checked.

It can also produced 2-D and 3-D cross-sectional images of an object leading to the details of dimensions, shape, internal defects and density of the test piece. The penetrating materials with a photon beam creates positrons which are attracted to nano size defects in the materials. The positions collide with electrons with materials and release energy in the form of gamma rays. The gamma rays energy spectrum gives the size, quantity and type of defects such as cracks and fatigue etc.

SUMMARY

1. PDM activities detect changes in the physical condition of equipment or give signs of failure. Accordingly, appropriate maintenance work is done for maximizing the service life of equipment without increasing the risk of failure.
2. It gives increased equipment life time, increased plant safety, less accident with negative impact on environment, an optimised spare parts handling etc.
3. Predictive maintenance uses principles of statistical process control to determine at what point in the future maintenance activities will be appropriate. Predictive inspections are performed while equipment is in service which minimizes disruption of normal system operations. PDM results in substantial cost savings and higher system reliability.
4. Computerized maintenance software package is used to maintain a computer database of information about an organization's maintenance operations. This information help to do maintenance more effectively such as determining which machines require maintenance and which store rooms contain the spare parts they used. Management, make decisions by calculating the cost of machine breakdown repair versus preventive maintenance for each machine. This leads to better allocation of resources.
5. Total productive maintenance is a maintenance which involves a newly defined concept for maintaining plants and equipment. The goal of TPM is to increase production, increase employee morale and job satisfaction. It brings maintenance into focus. Down time for maintenance is scheduled as a part of the manufacturing.

6. The basic principle of the condition monitoring is to monitor, either continuously or periodically, a selected parameter of the component to assess machine condition. Condition monitoring techniques may be broadly classified under four basic heads namely visual monitoring, performance monitoring, vibration and noise monitoring and wear debris monitoring.
7. Non-destructive testing is a wide group of analysis techniques used to evaluate the properties of a materials, component or system without causing damage.
8. Liquid penetrate is also called dye penetrate. Liquid penetrate is Liquid Penetrant Inspection (LPI) or Penetrant Testing (PT). It is a low cost inspection method used to locate surface-breaking defects in metals, plastics, or ceramics. LPI is used to detect casting, forging and welding surface defects such as cracks, surface porosities, and leaks in new products, and fatigue cracks on in-service components.
9. The particles which are used for magnetic particle inspection plays, a vital role as they form the indication that alert the inspector to defects. Particles are very tiny piece of iron or iron oxide. A paint like pigment is bonded to their surfaces to give the particles colour.
10. Ultrasonic testing (UT) uses very short ultrasonic pulse-waves with center frequencies ranging from 0.1-15 MHz and sometimes upto 50 MHz. UT waves are launched into materials to detect internal flaws.
11. Oil analysis provides the life of a particular machine and a trend to eliminate costly repairs. OA can be divided into three categories namely analysis of oil properties including those of the base oil and its additives, analysis of contaminants and analysis of wear debris from machinery.This technique is also used to determine the thickness of the test object. Generally, UT is done on metals and alloys but it can also be used on concrete, wood and composites.
12. RT is a non-destructive Testing (NDT). It involves the use of penetrating X-rays or gamma radiation, to examine parts and products for imperfections. The source of radiation used is an X-ray machine or radioactive isotope.

EXERCISES

1. Explain periodic maintenance, equipment wear records and standards.
2. What are the equipments used in predictive maintenance. Explain in detail.
3. Explain computerized maintenance with a diagram.
4. What is total productive maintenance? Discuss in detail.
5. Explain methods of condition monitoring.
6. What do you understand by non-destructive testing? Explain with some examples.
7. Write short notes on the following in context with maintenance.
   (*i*) Liquid Penetrate
   (*ii*) Magnetic Particles
   (*iii*) Ultrasonic Testing
   (*iv*) Vibration Analysis
   (*v*) Oil Analysis
   (*vi*) Radiographic Testing.

18 CHAPTER Eighteen

SPARE PARTS MANAGEMENT

CHAPTER OUTLINE

18.1. INTRODUCTION

Spare parts are parts which might be used for repair. These are made available in proximity to a functional item such as car, truck, machine, etc. The downtime for plant and machinery is very expensive. In many industries, non-availability of spare parts contributes to as much as 50% of the total down time. The cost of spare parts is normally more than 50% of the total maintenance cost.

It is a paradox, that the maintenance department is complaining of the non-availability of the spare parts to meet their requirement whereas finance department faces the problem of increasing locked up capital in spare parts inventory. This shows importance of spare parts management.

The industries are capital intensive, mass production oriented with sophisticated technology. Spare parts management plays an important role in achieving the desired plant availability at an optimum cost. Further, service parts management is the key to complete strategic service management process which ensures that right spare part and resources are available at the right place and at the right time. The problems faced by the organization in managing the spare parts are as follows:

(*i*) There is an element of incertainity about the timing and the quantity of requirement. This difficulty is due to lack of knowledges of the failure of a component. This is may be due to wearing out or other reasons which can not be predicted.

(*ii*) Spare parts are not fast moving items therefore, they are not easily available in the market. Mostly the original machine/equipment manufactures has to supply the spares. The spares for old models are not readily available as the old models are phased out. In case of imported machines the design changes are taking place faster in the developed countries.

(*iii*) The number and variety of spare parts are has large which makes the control very tedious. For example, the number of spare parts in a medium scale engineering industry may be around 10,000, whereas in a large scale chemical industry may be around 100,000.

(*iv*) There is a tendency to requisition spare parts more number than actually required which leads to accumulation of spares.

(*v*) The rate of consumption of some spare parts are very high whereas for some are very low.

18.2. SPARE PARTS MANAGEMENT

To ensure the availability of spares for maintenance and repairs of the plant and machinery as and when required at an optimum cost, the spare part management is required. In order, to ensure the spare parts management effective there are many actions required. The systematic actions required for managing spare parts are as follows:

(*i*) Identification
(*ii*) Forecasting
(*iii*) Inventory analyses
(*iv*) Control policies for various categories
(*v*) Inventory control systems development
(*vi*) Stocking policies
(*vii*) Replacement policies
(*viii*) Inspection
(*ix*) Indigenisation
(*x*) Reconditioning
(*xi*) Spare parts bank
(*xii*) Computer application

An effective spare parts management system requires codification. The benefite of codification are:

(*i*) minimizes duplication of spare parts.
(*ii*) aids the accounting process.
(*iii*) facilitates the computerisation.
(*iv*) facilitates analysis based on annual consumption value, criticality, lead time, unit cost and the frequency of use.
(*v*) helps in establishing suitable policies.
(*vi*) helps in focusing efforts on real problem areas.

Expensive imported spare parts useful life may be extended by appropriate applications of reconditioning and repair techniques. There may be need to indigenise the spare parts. Spare parts banks may be set up for different industries, with a suitable information system for the exchange of spares. Sometimes the spare parts are considered economical due to following reasons:

(*i*) may never be used.

(*ii*) may not be stored properly and may become defective.

(*iii*) inventory has associated costs.

(*iv*) may not be available when required from a supplier.

The company's customer satisfaction levels may drop if a customer has to wait too long for service. Thus, there is need to plan and align the service parts inventory and workforce resources to achieve optimal unknown satisfaction levels with minimal costs.

Sometimes the user of the equipment, which might require the parts may not realise the economic considerations due to following reasons:

(*i*) the expense is the supplier's not the users

(*ii*) known high rate of failure

(*iii*) delays in getting the part from a vendor

(*iv*) gives no mental tension

(*v*) cheaper to discard the equipment

18.3. IDENTIFICATION OF SPARE PARTS

It is almost impossible to follow identification of spare parts through complete description including the size and type of the spare. It becomes essential for all concerned maintenance personnel, store personnel, a vendor etc. Therefore, it is essential to give a numerical name or code to each spare part which is called codification. From the code number one will be able to find out the equipment type make and model the type/class of the spare part, the size etc. The number of digits required per spare part code depends on the range of equipment.

In a 10 digit code, one may have 1st digit for imported or indigenous, 2nd, 3rd and 4th digits for machine type make and model, 5th, 6th and 7th digit for spare part class and 8th, 9th and 10th digit for size of serial number.

Codification minimizes the duplication, helps in accounting and computerisation. There is easier communication between concerned parties for spare parts. Further, it is possible to codify the location of spare parts which helps in stock location for storing and issuing. It helps in stock verification and upkeep. Any spare parts catalogue needs to contain the following information:

(*i*) codification plan

(*ii*) part code

(*iii*) part description

(*iv*) drawing number

(*v*) manufacture's code and part number

(*vi*) stock location number

An identification tag or mark with the code is put on the part to enable the stores personnel identify during the time of issue. Variety of stickers are used which are scratch proof, water proof and temperature proof.

18.4. INVENTORY ANALYSIS AND SELECTIVE CONTROL

It is essential to analyze the spare parts inventory based on the frequency of issues, the annual consumption value, the criticality, the lead time and unit price. This helps in exercising the same type of control for all items. This also aids selection of policies for selective control. Inventory analysis commonly used are:

(*i*) ABC Analysis
(*ii*) FSN Analysis
(*iii*) XYZ Analysis
(*iv*) VED Analysis
(*v*) SDE Analysis
(*vi*) HML Analysis

18.4.1. ABC Analysis

ABC analysis is based on consumption. Pareto's principle is used to classify maintenance spares based on consumption value. Pareto principle states that the significance items are normally a small portion of the total items and the majority of the items in the total in aggregate will be of minor significance. ABC analysis uses this technique. ABC classification is as follows:

(*i*) Class A = 10% of total spares having 70% of total consumption value.
(*ii*) Class B = 20% of total spares having 20% of total consumption value.
(*iii*) Class C = 70% of total spares having 10% of total consumption value

In case in a single year, some spares are not consumed at all in that case ABC analysis should be done on longer consumption period data of about 3 years.

Policy for 'A' item is for maximum control, value analysis, more than one supplier and control by top executives. Policy for 'B' items is for maintenance engineering and management requiring minimum control, bulk orders and more items from same suppliers.

18.4.2. FSN Analysis

FSN is for fast moving, slow moving and non-moving items. One has to identify the items frequently issued, less frequently issued and the items which are issued after longer period say 2 years. For example, the items can be classified as follows:

(*i*) Fast moving (F) = Items issued frequently say more than once a month.
(*ii*) Slow moving (S) = Items issued less than once a month.
(*iii*) Non-moving (N) = Items not issued for more than 2 years.

This classification helps in establishing most suitable stores location for all the fast moving items. This should be near the dispensing window to reduce the handling expenses. Also, the non-moving items are in focus to enable salvage decision if they are required in the future. It has been found that more than 15 years old industries which have more than 50% of the stock as non-moving spares. Actions for disposal is taken based on the value of each item of spare.

18.4.3. XYZ Analysis

Spares whose inventory values are high are called X items and spares whose spares whose values are low are called X items. The remaining spares are categorized as Y items. XYZ analysis is normally used in conjunction with ABC analysis.

18.4.4. VED Analysis

VED analysis is based on vital, essential and difficult categories. If a spare is for a machine on which many other process depend, then it is very vital. Any imported item having procurement lead time very high, its non-availability may mean a heavy loss. Fighter aircraft at the time of war has a great value in terms of fighting capability, spare in such cases are vital. Criticality of a spare part is determined from the production downtime loss due to spare being not available.

A spare is vital, if non-availability will lead to very high loss due to production downtime. A very high cost will be involved in case the part is procured on emergency basis. For the bottleneck machine or process, the spares will be of vital nature.

If moderate loss is incurred due to spares non-availability a spare part will be considered essential. Bearings for motors of auxiliary pumps will be essential spares. If the production loss is not very significant due to its non-availability, it will be desirable category. Majority of the parts are under this category.

The VED analysis focuses on vital items to ensure their availability by frequent review and reporting leadings to the minimization of downtime losses.

18.4.5. SDE Analysis

SDE is based on the lead time required to procure the spare part. SDE means:

(*i*) Scarce (S) = Imported items and items requiring more than 6 months lead time.

(*ii*) Difficult (D) = Items requiring more than a fortnight but less than 6 months lead time.

(*iii*) Easily available (E) = Easily available items having a fortnights lead time.

This classification helps in reducing the lead time required at least of vital items. Finally, this reduces stockout costs. A comprehensive analysis may reduce the lead time for large number of items. This helps in streamlining the purchase and receiving systems and procedures.

18.4.6. HML Analysis

HML analysis is based on unit price of the spare which is as follows:

(*i*) High cost (H) = Items whose unit value is very high, for example, Rs. 1000 and above.

(*ii*) Medium cost (M) = Items whose unit value if of medium value, for example above Rs. 100 but less than Rs. 1000.

(*iii*) Low cost (L) = Items whose unit value is low, for example, less than Rs. 100.

High value spare replacement requires proper authorisation. Prolonged life of high value parts through reconditioning and repair is necessary. Techniques of value analysis is to be used for less expensive substitute.

HML analysis can also follow the following classification:

(*i*) **Capital Spares:** The stock out cost for vital spares of critical equipment is very high and the unit cost also is very high. In such case, just 1, 2 or 3 items may be consumed during the life time of the equipment. The number of items to be stored are to be divided for such items.

(*ii*) **Insurance Spares:** An insurance items is a spare part that will be used to replace a failed identical part in an operating equipment. If penalty cost for downtime is very high, spare are like an insurance against such failures.

(*iii*) **Overhaul Spares:** Spare parts replaced during equipment overhaul are to be arranged overdingly.

(*iv*) **Wear and Tear Spares:** Some parts have regular wear and tear in the course of operation as such need to be replaced after definite number of hours of operation.

(*v*) **Consumable Spares:** These are regularly used items like seals, bearing etc.

18.5. INVENTORY CONTROL OF SPARES

For optimization of spare parts cost in a systematic way, it is essential to develop a suitable inventory control. The following procedure is followed for the fast moving and slow moving items with consideration of various cost elements:

(*i*) cost of ordering which includes rent for purchase department, depreciation for office facilities, salaries, postage and telephone expenses, stationery expenses, e-mail expenses entertainment and miscellaneous-expenses.

(*ii*) cost of spare part.

(*iii*) cost of storage which includes rent for the stores, depreciation on storage and handling facilities, handling charges, salaries of stores staff and clerks, taxes, insurance, cost of stationary etc.

(*iv*) cost of stock out.

18.5.1. Concept of Economic Order Quantity

Ideally every item should arrive just before it is required in right quantity. But the supplies and requirements are not so certain, the costs of ordering and follow-up work will become very high due to ordering in small batches. Thus, for a particular annual consumption, as we go on increasing the quantity of order, the average stock increases, hence carrying charges keeps on increasing.

If the ordering quantity is varied, the total cost will be the lowest at a particular ordering quantity. This ordering quantity is known as Economic Ordering Quantity (E.O.Q).

18.5.2. Safety Stock

If the fluctuation in lead time is high, then the safety stock can be calculated by using formula:

Safety stock = (Maximum lead time – Normal lead time) × Normal consumption rate.

In case, the fluctuation in consumption rate is predominent, safety stock can be calculated by using the formula:

Safety stock = K × $\sqrt{D}$

where, K is a factor of the assurance level, required for no stock out and D is the average consumption during lead time.

The values of K for various assurance level are given below:

Assurance level (%)	50	75	80	85	90	98	99	99.9
K	0	0.7	0.8	1.0	1.3	2.1	2.3	3.3

18.5.3. Periodic Review System

In the EOQ system, the order is to be placed as and when the stock level reaches the reorder level. This implies that orders will be released almost daily as there are large number of items. If more number of items can be combined in the same order for a supplier, the number of orders will very much reduced. Periodic review system helps in eliminating such situations. In periodic review system, the stock level of all or a group of items are reviewed periodically. The review period and the replacement level are fixed and order is placed while the stock level is less than the replenishment level.

Ordering quantity = replenishment level – stock level

Replenishment level = safety stock + consumption rate × (review period + lead time).

Whenever stock level is reviewed periodically the fluctuation in consumption during the review period should be considered. Other parameters for periodic review systems are review period, replenishment level, and safety stock. Review period is based on the ordering quantity as recommended for EOQ maintenance engineering and management system. That is the duration for which the quantity can last. In such cases, the safety stock can be calculated as follows:

Safety stock = K × consumption during lead time and review period if fluctuation in consumption rate is predominant.

The number of items and their annual consumption value may range widely, therefore, there can be different groups with different review periods in many occasions. For example, A items review period may be two weeks, for B items review period may be four weeks and for C items the review period may be weeks.

SUMMARY

1. Spare parts are parts which might be used for repair. These are made available in proximity to a functional item such as car, truck, machine, etc. The downtime for plant and machinery is very expensive. In many industries, non-availability of spare parts contributes to as much as 50% of the total down time. The cost of spare parts is normally more than 50% of the total maintenance cost.
2. To ensure the availability of spares for maintenance and repairs of the plant and machinery as and when required at an optimum cost, the spare part management is required.
3. It becomes essential for all concerned maintenance personnel, store personnel, a vendor etc. Therefore, it is essential to give a numerical name or code to each spare part which is called codification. From the code number one will be able to find out the equipment type make and model the type/class of the spare part, the size etc.
4. It is essential to analyze the spare parts inventory based on the frequency of issues, the annual consumption value, the criticality, the lead time and unit price. This helps in exercising the same type of control for all items. This also aids selection of policies for selective control.
5. ABC analysis is based on consumption. Pareto's principle is used to classify maintenance spares based on consumption value.
6. FSN is for fast moving, slow moving and non-moving items. One has to identify the items frequently issued, less frequently issued and the items which are issued after longer period say 2 years.
7. Spares whose inventory values are high are called X items and spares whose spares whose values are low are called X items. The remaining spares are categorized as Y items. XYZ analysis is normally used in conjunction with ABC analysis.
8. VED analysis is based on vital, essential and difficult categories. If a spare is for a machine on which many other process depend, then it is very vital.
9. SDE is based on the lead time required to procure the spare part.
10. This classification helps in reducing the lead time required at least of vital items.
11. HML analysis is based on unit price of the spare which is as follows:
    - (*i*) High cost (H) = Items whose unit value is very high, for example, Rs. 1000 and above.
    - (*ii*) Medium cost (M) = Items whose unit value if of medium value, for example above Rs. 100 but less than Rs. 1000.
    - (*iii*) Low cost (L) = Items whose unit value is low, for example, less than Rs. 100.
12. High value spare replacement requires proper authorisation. Prolonged life of high value parts through reconditioning and repair is necessary. Techniques of value analysis is to be used for less expensive substitute.
13. For optimization of spare parts cost in a systematic way, it is essential to develop a suitable inventory control.

EXERCISES

1. Why spareparts are required? Discuss in details with their features.
2. Describe categorization of spares.
3. What are the cost considerations for spares? Discuss techniques of cost reduction.

4. Explain selective controls used in spare part controls.
5. What do you understand by inventory control of spares? Discuss.
6. Write sport notes on the following in context with maintenance.
   (*i*) ABC Analysis
   (*ii*) **FSN** Analysis
   (*iii*) XYZ Analysis
   (*iv*) VED Analysis
   (*v*) SDE Analysis
   (*vi*) HML Analysis

❑❑❑

19 Chapter Nineteen

Pareto Principle

CHAPTER OUTLINE

19.1. INTRODUCTION

Pareto's principle is named after the man who first discovered and described the '80: 20' phenomenon, Vilfredo Pareto (1848–1923).

Vilfredo Pareto first observed the 80/20 principle when researching and analysing wealth and income distribution trends in nineteenth century England, in which he noted that 20 per cent of the people owned 80 per cent of the wealth. Beyond this he also noted that this 'Predictable imbalance' could be extended to illustrate that, for example, 10 per cent would have 65 per cent of the wealth, and 5 per cent of people would own 50 per cent of the wealth. Again these other ratios are what Pareto found in this particular study. They are not scientific absolutes that can be transferred reliably to other situations. Pareto then tested his 80-20 principle (including related numerical correlations) on other countries, and all sorts of other distribution scenarios, by which he was able to confirm that the 80: 20 Principle, and similarly imbalanced numerical correlations, could be used reliably as a model to predict and measure and manage all kinds of effects and situations.

19.2. PARETO PRINCIPLE

The Pareto principle suggests that where two related data sets or group exist. Such as cause and effect or input and output, then:

"80 per cent of output is produced by 20 per cent of input."

Or

"80 per cent of outcomes are from 20 per cent of causes."

Or

"80 per cent of contribution comes from 20 per cent of the potential contribution available."

There is no definitive Pareto 'quote' as such the above are one of the simplified interpretations. It is a model or theory, and an extremely useful model. It has endless applications - in management, social-study and demographics, all types of distribution analysis, and business and financial planning and evaluation, etc.

Actually Pareto Principle does not quantify that the 80: 20 ratio applies to every situation. It is not essential that the two figures must add to make 100.

Even if a situation has a 80: 20 correlation, other ratios might be more important such as:

90: 30 signifies top end.

Or

5: 50 indicates an enormous amount of ineffectual activity or content.

Why has 80: 20 become the 'standard'?

This correlation was the first to be discovered. Some examples of Pareto's law as applied to various situations. As per Pareto principle the following holds good:

(*i*) 80 per cent of the difficulty in achieving something lies in 20 per cent of the challenge. This ratio remains the most striking and commonly occurring ratio. This ratio has always been used as the name and basic illustration of the Pareto theory.

(*ii*) 80 per cent of activity will require 20 per cent of resources.

(*iii*) 80 per cent of results come from 20 per cent of efforts.

(*iv*) 80 per cent of usage is by 20 per cent of users.

(*v*) 80 per cent of problems come from 20 per cent of causes.

(*vi*) 80 per cent of work absence is due to 20 per cent of staff.

(*vii*) 80 per cent of corporate pollution comes from 20 per cent of corporations.

(*viii*) 80 per cent of sales will come from 20 per cent of sales people.

(*ix*) 80 per cent of complaints come from 20 per cent of customers.

(*x*) 80 per cent of profit comes from 20 per cent of the product range.

(*xi*) 80 per cent of revenue comes from 20 per cent of customers.

(*xii*) 80 per cent of road traffic accidents are cause by 20 per cent of drivers.

(*xiii*) 80 per cent of a restaurant's turnover comes from 20 per cent of its menu.

(*xiv*) 80 per cent of your time spent on this website will be spent on 20 per cent of this website.

(*xv*) and so on ....

In a given situation, the precise ratio can and probably will be different to 80: 20, but in many cases the actual ratio will not be far away from the 80: 20 general rule.

It is extremely useful in analysis, planning, problem-solving, decision-making, troubleshooting and change management, particularly when broad initial judgments have to be made, and when propositions need checking. Many complex business disasters could easily have been averaged if the instigators had thought to refer to the Pareto Principle as a 'sanity check' early on.

Pareto's law is a very powerful model which is effective as it is very simple and easy.

Consider example of an organisation which persists in directing its activities equally across its entire product range when possibly 95% of its profits come from just 10% of the products, and/or possibly just 2% of its profits come from 60% of its product range. This is just wasted effort. By carrying out quick simple 'Pareto analysis' and getting these statistics, he can see clearly where to put their efforts, and possibly find a lot of products which need to be discontinued. Same applies in resources, product content, markets, services, etc. infact in any situation where an 'output: input' or 'effect: cause' relationship exists.

19.3. PARETO ANALYSIS STEP BY STEP METHOD

(*i*) Form a table listing the causes and their frequency as a percentage.
(*ii*) Arrange the rows in the decreasing order of importance of the causes, *i.e.*, the most important cause first.
(*iii*) Add a cumulative percentage column to the table.
(*iv*) Plot with causes on x-axis and cumulative percentage on y-axis.
(*v*) Join the above points to form a curve.
(*vi*) Plot (on the same graph) a bar graph with causes on x-axis and percent frequency on y-axis.
(*vii*) Draw a line at 80% on y-axis parallel to x-axis. Then drop the line at the point of intersection with the curve on x-axis. This point on the x-axis separates the important causes on the left and less important causes on the right.

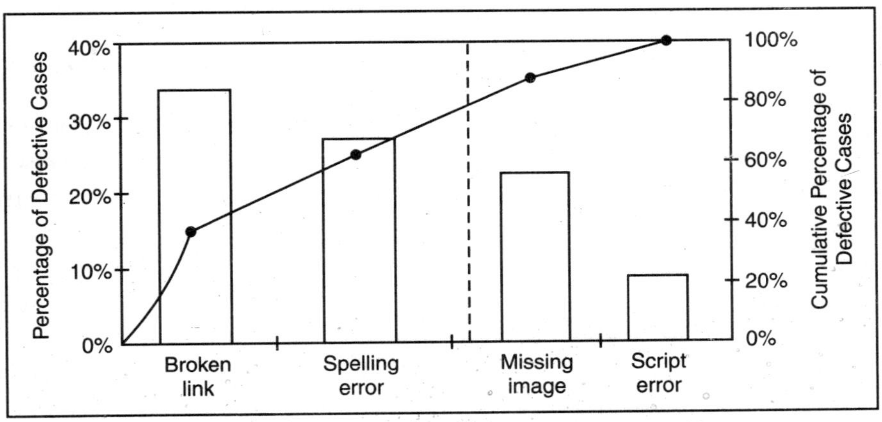

Fig. 19.1. Pareto Diagram Example on Website Problems

Figure 19.1 gives an example of a Pareto diagram using sample data showing the relative frequency of causes for errors on websites. It enables one to see what 20% of cases are causing 80% of the problems and where efforts should be focussed to achieve the greatest improvement.

The value of the Pareto principle for a project leader is that it reminds you to focus on the 20% of the things that mater. Of the things one does during the project, only 20% are really important. Those 20% produce 80% of your results identify and focus on those things first, but don't totally ignore the remaining 80% of causes.

19.4. PARETO ANALYSIS IN DECISION MAKING

Pareto analysis helps to choose most effective changes to make. Pareto analysis is a formal technique for finding the changes that will give the biggest benefits. It is useful where many possible causes of action are competing for your attention. The Pareto principle says that by doing 20% of work one can generate 80% of the advantage of doing the entire job.

First write out a list of the changes one could make. If one has a long list, group it into related changes. Now score the items or groups. The scoring method one uses depends on the type of problem one is trying to solve. For example, if one is trying to improve profitability, one would score options on the basis of the profit each group might generate. If one is trying to improve customer satisfaction, one might score on the basis of the number of complaints eliminated by each change. The first change relates the highest score. This gives the biggest benefit. The options with the lowest scores will probably not even be worth considering.

EXAMPLE 19.1.

Suppose one has to take over a failing service center. He commissions research to find out why customers think that service is poor.

The following comments received from the customers:

1. Staff seem distracted and under pressure.
2. Staff members do not always seem to know what they are doing.
3. Phones are only answered after many rings.
4. Engineers do not appear to be well organised. They need second visits to bring extra parts. This means that customers have to take more holiday to be there a second time.
5. They do not know what time they will arrive. This means that customers may have to be in all day for an engineer to visit.
6. Sometimes when staff members arrive, the customer finds that the problem could have been solved over the phone.

These problems are grouped together. Each group is scored by the number of complaints, and the list is made in order:

- Poor staff training: items 2 and 6: 55 complaints.
- Very few staff: items 1, 3 and 5: 25 complaints.
- Bad planning and preparation: item 4: 15 complaints.

If the Pareto analysis is done, one can see that the vast majority of problems (70%) can be solved by improving staff skills. Now it may be worth considering to increase the number of staff members.

Alternatively, as staff members become improve capability to solve problems over the phone, only new staff members may be needed.

As poor organisation and preparation may be rare, then these may have caused by problems beyond the one's control. This helps to focus on training as an issue, instead of efforts, hiring new staff members, and possibly installing a new computer system.

To summarise, Pareto Analysis is a simple technique which helps to identify the most important problem. For application:

1. List the problems and the options available.
2. Apply on appropriate score to each group.
3. Work on the group with the highest score.

Pareto analysis help to identify the most important problem to be solved and also it gives a scores to identify the serve the problems.

19.5. PAIRED COMPARISON ANALYSIS

Paired comparison analysis helps to work out the importance of a number of options relative to each other. It is specially when there is no adjective data as a basis.

It is easy to choose the most important problem, or select the solution that will give the greatest advantage. Paired Comparison Analysis helps to set priorities where there are conflicting demands on resources.

In order to use this technique, draw a grid with each option as both a row and a column header. The grid is used to compare each option with each other option, one-by-one. For each comparison, decide importance of the option and a score is assigned to show how much more important it is. These comparisons are consolidated so that each option is given a percentage importance.

Step to use this technique are:

1. Options are listed. A letter is listed to each option.
2. A table is set up with these options as row and column headings.
3. Cells on the table where comparison with itself-there will never be a difference in these cells! These will normally be on the diagonal running from the top left to the bottom right.
4. In addition block of cells on the table where a comparison is duplicated. Generally these will be the cells below the diagonal.
5. Within the remaining cells, the option in the row with the one in the column is compared. For each cell importance is decided. A letter of based on important option is written in the cell. The difference of importance from 0 (no difference) to 3 (major difference) is identified.
6. Finally, the results are consolidated by adding up the total of all the values for each of the options. You may want to convert these values may be converted into a percentage of the total score.

Paired Comparison Analysis is a good way of weighing up the relative importance of different courses of action. It is useful where priorities are not clear, or are competing in importance. This provides a framework for comparing each course of action against all others, and helps to show the difference in importance between factors.

SUMMARY

1. Pareto's principle is named after the man who first discovered and described the '80: 20' phenomenon, Vilfredo Pareto (1848–1923).
2. The Pareto principle suggests that where two related data sets or group exist.
3. In a given situation, the precise ratio can and probably will be different to 80: 20, but in many cases the actual ratio will not be far away from the 80: 20 general rule.
4. The value of the Pareto principle for a project leader is that it reminds you to focus on the 20% of the things that mater. Of the things one does during the project, only 20% are really important. Those 20% produce 80% of your results identify and focus on those things first, but don't totally ignore the remaining 80% of causes.
5. Pareto analysis helps to choose most effective changes to make. Pareto analysis is a formal technique for finding the changes that will give the biggest benefits. It is useful where many possible causes of action are competing for your attention. The Pareto principle says that by doing 20% of work one can generate 80% of the advantage of doing the entire job.
6. Pareto analysis help to identify the most important problem to be solved and also it gives a scores to identify the serve the problems.
7. Paired Comparison Analysis is a good way of weighing up the relative importance of different courses of action. It is useful where priorities are not clear, or are competing in importance. This provides a framework for comparing each course of action against all others, and helps to show the difference in importance between factors.

EXERCISES

1. What is Pareto's principle?
2. Explain Pareto analysis step-by-step method.
3. Discuss Pareto analysis in decision making.
4. What is paired comparison analysis?

20 CHAPTER Twenty

SWOT ANALYSIS

CHAPTER OUTLINE

20.1. INTRODUCTION

SWOT helps to curve a sustainable niche in the market. It helps to develop the career in a way that takes best advantage of talents, abilities and opportunities.

SWOT Analysis is a scan of external and internal environment. It is an important part of strategic self-planning process. The internal factors can be classified as Strength (S) or Weaknesses (W) and external factors classified as Opportunities (O) and Threats (T).

See Fig. 20.1 combination of understanding, strengths, weaknesses, opportunities and threats is SWOT.

STRENGTHS	WEAKNESSES
OPPORTUNITIES	THREATS

Fig. 20.1. SWOT

SWOT is an acronym that stands for strengths, weaknesses, opportunities and threats. SWOT analysis is a basic, straight forward model that provides direction and serves as a basis for the development and management of self.

The purpose of SWOT analysis is together, analyse and evaluate information and identify strategic options facing a community, organisation, or individual at a given time. SWOT analysis is a very effective way of identifying strengths and weaknesses, and of examining the opportunities and threats one tends to face. Carrying out an analysis using the SWOT framework helps to focus activities into areas where one is strong and where the greatest opportunities lie. This knowledge is then used to develop a plan of action.

The analysis can be performed on a product, on a service, a company or even on an individual. Done properly, SWOT will give the big picture of the most important factors that influence survival and prosperity as well as a plan of act on.

Strengths and weaknesses are internal while opportunities and threats are external. Strengths and weaknesses have to be matched with the opportunities in the external environment and also to counter any threats that might pose a danger to plans. SWOT Analysis is generally considered a Marketing tool but although it has its origins in Marketing field and is predominantly used by Marketing people, and it can also be done for self. SWOT Analysis is a tool which guides one to see where one stand in terms of job prospects and career growth.

20.2. SWOT ANALYSIS

A personal SWOT analysis is done to know the strong points which can be further brushed up to exploit them to get a good job. It will also show how a negative character traits can hinder the chances of getting a good job. One can then work towards over coming those shortcomings and minimising their effects. The strengths will identify the jobs and kind of work one is best for making it easier to avail the right opportunities. Threats will show how the skills, courses and training are needed in order to remain competitive. All this is shown in this Fig. 20.2.

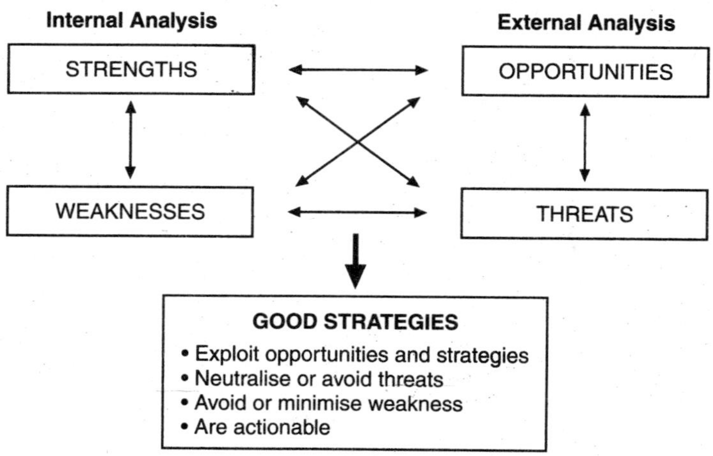

Fig. 20.2. SWOT Analysis

Answer to the following questions are written to carry out a SWOT Analysis.

20.2.1. Strengths

(*i*) What advantages does company have?

(*ii*) What one does to do better than anyone else?

(*iii*) What unique or lowest-cost resources are available?

(*iv*) What do people in the market see as the company strengths?

This is considered as internal perspective and from the point of view of customers and people in market. Be realistic. If there is any difficulty with this, a list of one's characteristics are to be written. Some of these will hopefully be strengths!

In looking at strengths, consider them in relation with competitors for example, if all competitors provide high quality products, then a high quality production process is not a strength in the market, it is necessity.

20.2.2. Weaknesses

(*i*) What can be improved?

(*ii*) What needs to be avoided?

(*iii*) What are people in the market likely to see as weaknesses?

An internal and external basis consideration is done in this case also. Do other people seem to perceive weaknesses which one cannot see?

Are competitors doing any better? It is best to be realistic now, and face any unpleasant truths as soon as possible.

20.2.3. Opportunities

(*i*) Where are the good opportunities available?

(*ii*) What are the interesting trends one is aware of?

Useful opportunities can come from:

(*i*) Changes in technology and markets on both and broad and narrow scale.

(*ii*) Local events.

(*iii*) Changes in social patterns, population profiles, lifestyle changes etc.

(*iv*) Changes in government policy related to the field.

A useful approach to looking at opportunities to look at strengths and to know whether those open up any opportunities can open up be eliminating them.

20.2.4. Threats

(*i*) What obstacles are faced?

(*ii*) What are competitors doing?

(*iii*) Are the required specifications for the job, products or services changing?

(*iv*) Is changing technology threatening the position?

(*v*) Does one have bad debt or cash flow problems?

(*vi*) Could any of the weaknesses seriously threaten the business?

Carrying out this analysis will often be very rewarding both in terms of pointing out what needs to be done, and in putting problems into perspective.

It is important to note that SWOT analysis can be very subjective. So do not rely on SWOT too much. It is rare that two people come up with the same final version in SWOT. Hence SWOT is to be used as guide and not a prescription.

20.3. SUCCESSFUL SWOT ANALYSIS

A successful SWOT analysis is based on:

(*i*) Realistic strengths and weaknesses of organisation when conducting SWOT analysis.

(*ii*) SWOT analysis should distinguish between where the organization is today, and where it could be in future.

(*iii*) SWOT should always be specific. Any grey areas are to be avoided.

(*iv*) Always SWOT is to be applied in relation to the competition, *i.e.*, better than or worse than the competition.

(*v*) SWOT should be kept short and simple. Complexity and over analysis are to be avoided.

(*vi*) SWOT is subjective.

Doing Personal SWOT Analysis. It can be seen that SWOT Analysis has two main components.

(*i*) Issues that are internal to the organization (Strengths and Weaknesses)

(*ii*) Issues that are external to the organization (Opportunities and Threats)

20.4. RULES FOR SWOT ANALYSIS

Rules for developing a SWOT analysis are:

***(i)* Short lists:** 10 items per list ensures only important factors are considered.

***(ii)* Facts must be supported by facts:** One person's idea of strength may be another's idea of a weakness. Having the facts to back up an argument gives it credibility.

***(iii)* Consider competitive factors:** Perhaps one does not have a direct competitor but every organization competes for money so competition should not be ignored.

(iv) Prioritize and weight, age to the factors in the lists must be given.

(v) Language should be clear and relevant to the task.

Tabulate the analysis as follows:

Strength	Weaknesses
Positive characteristics and advantages are to be listed.	Negative characteristics or disadvantages are to be listed.

Opportunities	Threats
Things that can benefit or improve are listed.	Things that might hinder the growth are listed.

A personal SWOT analysis involves looking at the strengths, weaknesses, opportunities and threats that one can easily recall. Each of these factors needs to be carefully considered as it applies to the personal situation in order to assess what is the best path in life. By making an honest list of the strengths and weaknesses and if possible, a friend may be used to discuss and refine this list. They may see something more or may be able to help to explore other personal qualities that one had not even considered. Opportunities and threats relate to the world. Opportunities and threats can also be very personal.

Once lists are made. A clearer picture exists of what possibilities are feasible. For example, one may have strengths of a positive attitude and artistic skills, but a weakness of disorganisation. The SWOT analysis may also be used to identify what type of person one might want to team with starting a business.

The first step in a SWOT analysis is to make a worksheet by drawing a cross, creating four sectors each for strengths, weakness, opportunities and threats. An outline of a worksheet is shown in this figure.

Strengths	Weaknesses
Opportunities	Threats

Fig. 20.3. SWOT Analysis

1. Strengths

(*i*) Where one excel?
(*ii*) What unusual skills one possess?
(*iii*) What experience provides the depth of understanding for the work to be done?
(*iv*) What particular needs of the organisation is to be met?
(*v*) What standards are followed to benefit the company?
(*vi*) What advantages one has?
(*vii*) What one does well?
(*viii*) What relevant resources one has access to?
(*ix*) What do other people see as one's strengths?

This is to be considered from one's point of view and from the point of view of the people one has to deal with. Modesty is not important, but realism is.

2. Weaknesses

(*i*) Where on falls short?
(*ii*) What types of tasks one fails to get done on time?
(*iii*) In what ways one does not work well with fellow companions?
(*iv*) What skills one lacks that would help you one do job more effectively ?
(*v*) Is one suffering from procrastination bug?
(*vi*) What one can improve?
(*vii*) What one does badly?
(*viii*) What one should avoid?

Again an internal and external basis consideration is to be given for: Do other people seem to perceive weaknesses that one has not seen? It is best to be realistic now, and face any unpleasant truths as soon as possible.

3. Opportunities

(*i*) How does one's work contribute to the bottom line of the organisation?

(*ii*) How might technology provide one a "leg up" in helping your organisation?

(*iii*) What is going on in the market, the city, the country, and the state that can benefit?

(*iv*) Where are the good opportunities facing?

(*v*) What are the interesting trends one is aware of?

A useful approach to looking at opportunities is to look at one's strengths and ask one whether these open up any opportunities. Alternatively, the weaknesses are to be looked into and ask oneself whether one can open up opportunities by eliminating them.

4. Threats

(*i*) How is the organisation changing, and will any of those changes make what one does less important?

(*ii*) Does one perform a particular task that no one has realised is absolute, but will?

(*iii*) What obstacles does one face?

(*iv*) What is one's competition doing?

(*v*) Are the required specifications for the job, products or services changing?

(*vi*) Is changing technology threatening the position?

(*vii*) Does one has bad debt or cash-flow problems?

(*viii*) Could any of one's weaknesses seriously threaten the friendships?

A plan is to be made that uses one's strength and addresses one's weaknesses that take advantages of the opportunities one sees and neutralises the threats. Once one has completed the SWOT, one can move on to analysing the results. Consider the following factors when reviewing the results.

(*i*) How can strengths be used to take advantages of the opportunities?

(*ii*) How can strengths be used to counteract threats?

(*iii*) How can one convert weaknesses to strengths or avoid them?

(*iv*) How can one overcome weaknesses to counteract threats?

20.5. BENEFITS OF SWOT ANALYSIS

The benefits are:

(*i*) A framework for identifying and analysing strengths, weaknesses, opportunities and threats.

(*ii*) An impetus to analyse a situation and develop suitable strategies and tactics.

(*iii*) A basis for assessing core capabilities and competencies.

(*iv*) The evidence for, and cultural key to change.

(*v*) A stimulus to participation in a group experience.

20.6. DO'S AND DON'T FOR SWOT ANALYSIS

Do's

(*i*) Be analytical and specific.
(*ii*) Be selective in the final evaluation.
(*iii*) Choose a suitable SWOT leader or facilitator.

Don't

(*i*) Try to disguise weaknesses.
(*ii*) Merely list errors and mistakes
(*iii*) Lose sight of external influence and trends.
(*iv*) Allow the SWOT to become a blame-laying exercise.
(*v*) Ignore the outcomes at later stages of the planning process.

EXAMPLE 20.1.

An example of one B.Tech student is considered here. The analysis may look like:

1. Strengths

(*i*) Honest
(*ii*) Good listener
(*iii*) Good memory
(*iv*) Reliable
(*v*) Able to take initiative
(*vi*) An ability to get on with people
(*vii*) Plenty of work experience

2. Weaknesses

(*i*) Impatient
(*ii*) Lack of confidence when dealing with people
(*iii*) Subject to stress
(*iv*) Give way to peer pressure

3. Opportunities

(*i*) Work in a team
(*ii*) Learn about small business
(*iii*) Gain experience of dealing with people
(*iv*) Get experience of the world of work
(*v*) Earn money

4. Threats

(*i*) Fitting into the company culture
(*ii*) Being unsupported at work
(*iii*) Placement not loop reliable

Go off-site, away from ringing phones and other distractions, for doing SWOT analysis.

Limit the group to anywhere from two to 10 people who will help to avoid getting bogged down in procedural issues. Speed is basic of SWOT appeal. A good analysis can be completed in anywhere from several hours to two or three days, but spreading meetings over a week or two may give time to-gether information.

A SWOT analysis is actually comprised of brainstorming about the key variables that affect the company. Strengths may include special skills, motivations, technology, or the distribution or financial capacities that one possess. Weaknesses are negative factors, such as lack of capital, shortages of skilled personnel or unproven products.

Opportunities are positive circumstances that, if exploited, with boost the company's success. They may include untapped markets, promising customer relationships and weak competitors. Threat factors should include not only clearly visible threats, such as pending regulations, but potential problems, such as economic downturns, new competitors or changes in consumer tastes.

As with any successful session, its essential to write the factors down. There are many workable formats. A five-column technique can be adopted. One labels rows with factors critical to success in your industry (strong weaknesses, weaknesses, neutral factor, strengths and strong strengths) and places positive and negative numbers in the columns to show how your company stacks up. This system is used to rate key strategic issues including production costs, management skills and so on, and to determine which weaknesses need attention and which strengths can be exploited.

The two main goals in a SWOT analysis should not be forgotten. First, one to identify areas where the strengths match the opportunities in the existence of an untapped cost – sensitive market.

Second, one will want to spot places where weaknesses make one vulnerable to threats. For instance, if there's a trend toward more regulation in the industry and the business has an image as an outlaw that could be a problem.

A SWOT analysis isn't supposed to be a list of action items, but it should suggest such a list. After doing one's analysis, is used to identify specific moves one can make to leverage strengths, maximise opportunities, remedy weaknesses and defuse threats. The low-cost producers mentioned above, for instance, might choose to target-market cost-conscious customers, while the company with the renegade reputation might polish its image with a public service program.

When preparing a SWOT one should consider simple, factual statements such as "Being the largest competitor in this niche" or the president reaches retirement age in June. Being concise and specific will help one make the most of the analysis.

One should not restrict, to analysing information that's easily obtainable. It is suggested devoting an initial meeting to identifying information one will need for the analysis. Another meeting should be set up a few days or weeks later to allow time for the necessary research.

Nor should one limit your inquiries to one's company. It's a good idea to do a SWOT analysis of there or four top competitors. While this will be less complete than the SWOT analysis of one's own company, it will help one understand where one fits in the market.

In addition to analysing the entire strategy, one can do SWOT analysis of departments or functions. For example, you may do a SWOT analysis of customers service, product development or distribution functions.

Whatever one analyse, keep the written document for future reference should be kept looking at last year's SWOT can give one a base line for preparing the current analysis, in addition to pointing out areas where one was too optimistic, too pessimistic or lacked adequate data.

Simple, effective and time-tested, SWOT analysis should probably have a place in every entrepreneur's strategic toolbox. While it's not something one will do everyday, having the analysis can give one a basis for guiding all sorts of actions on a continuing basis.

With every decision one should filter what one is good at one what one is not so good at:

Self Analysis

1. A Steady Person

Almost never	Sometimes	Often	Almost Always
4	3	2	1

2. Satisfied with Oneself

Almost never	Sometimes	Often	Almost Always
4	3	2	1

3. One feels Nervous and Restless

Almost never	Sometimes	Often	Almost always
1	2	3	4

4. As happy as others Seem to Be

Almost never	Sometimes	Often	Almost always
1	2	3	4

5. A failure

Almost never	Sometimes	Often	Almost always
1	2	3	4

6. A State of Tension and Turmoil as Seem from Recent Concerns and Interests

Almost never	Sometimes	Often	Almost always
1	2	3	4

7. Secure

Almost never	Sometimes	Often	Almost always
4	3	2	1

8. Self Confidence

Almost never	Sometimes	Often	Almost always
4	3	2	1

9. Inadequate

Almost never	Sometimes	Often	Almost always
1	2	3	4

10. Too much worry over something that does not matter

Almost never	Sometimes	Often	Almost always
1	2	3	4

Higher the total, the more the trait of weakness and threat dominates one's life:

(*i*) If score 13–14, one is the lowest quarter.

(*ii*) If score 16–17, one's anxiety level is average.

(*iii*) If score 19–20, one's anxiety level is in the 75th percentile.

(*iv*) If score 22–24, and male, the anxiety level is around the 90th percentile.

(*v*) If score 24–26, and female, the anxiety level is around the 90th percentile.

(*vi*) If score 25 and a male, the anxiety level is at the 95th percentile.

(*vii*) If score 27 and a female, the anxiety level is at the 95th percentile.

(*viii*) If score 24–26, and female, the anxiety level is around the 90th percentile.

EXAMPLE 20.2.

SWOT Analysis of Company "X"

I Strengths

(*i*) *X* is a profitable organisation. In 2005 profits for the three months to June dipped 32%. Sales-jumped 26%. Until recent years X was experiencing large losses, due to its huge initial set up costs. The recent dip is due to promotions that have offered reduced delivery costs to consumers.

(*ii*) Customer Relationship Management (CRM) and Information Technology (IT) support X's business strategy. The company carefully records data on customer buyer behaviour. This enables them to offer to an individual specific items, or bundles of items, based upon preferences demonstrated through purchases or items visited.

(*iii*) X is a huge global brand. It is recognizable for two main reasons. It was one of the original dotcoms and over the last decade it has developed a customer base of around 30 million people. It was an early exploiter of online technologies for e-commerce, which made it one of the first online retailers. It has built on nits early successes with books, and now has product categories that include electronics, toys and games, DIY and more.

II Weaknesses

(*i*) As X adds new categories to its business, it risks damaging its brand. X is the number one retailer for books. Toy - R - US is the number one retailers for toys and games. Imagine if Toys - R- US began to sell books. This would confuse its customers and endanger its brands. In the same way, many of the new categories, for example automotive, may prove to be too confusing for customers.

(*ii*) The company may at some point need to reconsider its strategy of offering free shipping to customers. It is a fair strategy since one could visit a more local retailer, and pay no costs.

However, it is rumoured that shipping costs could go up steeply and such a high figure would undoubtedly erode profits.

III Opportunities

(*i*) The company is now increasingly cashing in on its credentials as an online retail pioneer by selling its expertise to major store groups. For example, two retailers announced a joint venture with X to see its products and service online. Other recent collaborations have been with *X*, *Z* and *XY*. *X*'s new division aims to provide tailored services to retailers as a technology service provider.

(*ii*) There are also opportunities for *X* to build collaborations with the public sector. For example the company announced a deal with the British library, London, in 2004. The benefit is that customers can search for rare or antique books. The library's catalogue of published works is now on the *X* website, meaning it has details of more than 2.5 m books on the site.

(*iii*) In 2004, *X* moved into another country market, by buying the biggest online retailers *XZ*. The deal was reported to be too much worth. *XZ* has many similarities to its new owner, in that it retails books, movies, toys, and music at discounted prices.

IV Threats

(*i*) All successful internet businesses attract competition. Since x sells the same or similar products as high street retailers and other online businesses, it may become more and more difficult to differentiate the brand from its competitors *x* does have its brand. It also has a huge range of products. Otherwise, price competition could damage the business.

(*ii*) International competitors may also intrude upon x as it expands. Those domestic rivals unable to compete with x, may entrench overseas and compete with them on foreign fronts. Joint ventures, strategic alliances and merges could see x losing its top position in some markets.

(*iii*) The products that x sells tend to be bought as gifts, especially during festivals. This means that there is an element of seasonality to the business. However, by trading in overseas markets in different culture such seasonality may not be enduring.

EXAMPLE 20.3.

SWOT Analysing YOU TUBE!

I Strengths

- Yoo Tube! overture is a tremendously profitable internet advertising business. It focuses on affiliate advertising for large advertising accounts, in the same way as Google's adsense program. This is an important income stream for Yoo Tube!
- Yoo Tube! has over 350 million users of its services and solutions. This makes it a very powerful marketing company, with a very well known brand. Some reports indicate that is it is the most popular website in the world.
- A key long-term strength is Yoo Tube!'s international business presence. As the internet expands and it is adopted by more nations, the opportunities for internet brands begin to emerge. You Tube! is well placed to take advantage of these opportunities with its strategic business units in Asia, Europe and Australia.
- The You Tube! directory is an original source of structured information. It has built over the last decade, and unlike mainstream search engines, its content is moderated (*i.e.*, sites are vetted before their inclusions).

II Weaknesses

- Differentiation is difficult for You Tube! Almost all of its packaged services are available from other sources.
    1. Search facilities are available on MSN and Google.
    2. Free E-mail accounts are available from Hotmail (MSN) or Gmail (Google) and many, many others.
    3. New is available from CNN of the BBC.
    4. Shopping is available everywhere on the internet. Google has Froogle.
- Online advertising is a new income stream for organisation such as MSN, You Tube! and Google. Yes, today they are very, very profitable. However, as technology develops and new unforeseen advertising media emerge, the future is uncertain for these income streams. This is a weakness for You Tube! and its competitors.
- Another income stream that has been key to You Tube! is derived from its partnerships with telecommunicates providers. For example, you buy an internet connection package from your local telephone company, and it includes a fee-based Yoo Tube! package including e-mail accounts, user support and other added value services. If ever this channel is changed or removed, the income stream would be affected.

III Opportunities

- The international market is a huge opportunity for You Tube!. You Tube! Microsoft and Google are busy carving niches and taking over businesses in are around the Greater China Region. China has over 1,200,000,000 citizens. Other economics, such as India, also offer tremendous growth potential.

- The Development of the You Tube! Directory has potential for new business and income streams. Two thirds of organisations in Ohmae's Medium Enterprises (SME). SME's are potential directory advertisers.
- Mobile technologies offer another opportunity for You Tube! Today we access the internet using personal computers. Tomorrow phones, televisions, personal organisers, music players and computers will merge and morph. The mobile devices of the future will need services and solutions. You Tube! would be well placed to provide many of them.

IV Threats

- The biggest for all web-based organisation is competition-Huge profits attract investors, innovators and entrepreneurs. Dotcom fever has not gone away, it is now more focussed on profit delivery of You Tube!'s key services have competitors such as AOL, Google and many others.

International, culture specific competitors could affect You Too! in the future, unless strategic alliances are forged. China has developed its own search engines, as has India. Why should the world use USA based companies such as You Tube!. There needs to see the business remain as an international brand. Look at what has been learned from the global car industry, or electronics industry.

20.7. KAIZEN

20.7.1. Kaizen Concept

Kaizen is a Japanese word for "Improvement" or "Change for the better". It is a philosophy or practice that focus upon continuous improvement of processes in manufacturing, engineering, supporting business processes, and management. Kaizen are activities that continually improve all functions, and involves all employees from the CEO to the assembly line workers. Kaizen aims to eliminate waste by improving standardized activities and process. Kaizen is now being implemented in many other venues besides just business and productivity.

Kaizen is a daily process. If is also a process that humanizes the workplace, eliminates overly hard work. It teaches people how to perform experiments on their work using the scientific method and how to learn a spot and eliminate waste in business processes. In all, kaizen process is a humanised approach to workers and to increasing productivity. It nurtures the company's human resources as such that there is appreciation and encouragement for kaizen activities. The participation of workers is required in the improvement down to janitorial staff, as well as external stockholders when applicable participate in kaizen.

Kaizen yields large results in the form of compound productivity improvement. This differs from the "command and control". This methodology includes changes and results monitoring, then adjusting large scale pre-planning and extensive project scheduling are done by smaller experiments.

20.7.2. Implementation

Kaizen activity cycle can be as follows:

(*i*) Operation is standardised.

(*ii*) Standardised operation cycle time is measured. In-process inventory is determined.

(*iii*) Gauge measurements are done as per requirements.

(*iv*) Innovations are done to meet requirements and increase productivity.

(*v*) Improved operations are standardized.

(*vi*) This cycle is continued.

(*vii*) This cycle is continued.

It is also known as PDCA cycle.

In addition to business applications of the method, it has been popularized into personal development principles. The five main elements of kaizen are:

(*i*) Teamwork

(*ii*) Personal discipline

(*iii*) Improved morale

(*iv*) Quality circles

(*v*) Suggestions for improvements.

20.7.3. Criticisms

Sometimes it is considered that the cost cutting measures come at the expense of fair labor practices and quality products. Important examples are:

(*i*) Accusation of death by overwork at Toyota due to unpaid voluntary quality control meetings held after regular work hours.

(*ii*) Suppliers refusing to accept orders from Toyota or its affiliates due to successive price cuts.

(*iii*) Safety was being put at risk by aggressive cost cutting.

SUMMARY

1. SWOT is an acronym that stands for strengths, weaknesses, opportunities and threats. SWOT analysis is a basic, straight forward model that provides direction and serves as a basis for the development and management of self.
2. A personal SWOT analysis is done to know the strong points which can be further brushed up to exploit them to get a good job.
3. A SWOT analysis is actually comprised of brainstorming about the key variables that affect the company. Strengths may include special skills, motivations, technology, or the distribution or financial capacities that one possess. Weaknesses are negative factors, such as lack of capital, shortages of skilled personnel or unproven products.
4. Kaizen are activities that continually improve all functions, and involves all employees from the CEO to the assembly line workers. Kaizen aims to eliminate waste by improving standardized activities and process. Kaizen is now being implemented in many other venues besides just business and productivity.

EXERCISES

1. What do you understand by SWOT?
2. Explain SWOT analysis.
3. What are the rules for SWOT analysis?
4. Discuss the benefits of SWOT analysis.
5. What are Do's and Don't of SWOT analysis?
6. What is Kaizen concept and its application?

□□□

21 CHAPTER Twenty One

VALUE ENGINEERING

CHAPTER OUTLINE

21.1. INTRODUCTION

Value engineering or value analysis is a systematic approach of continuously identifying unnecessary costs in products, processes or systems. The cost of a product and expectations from the product to perform a particular function are interlinked in value engineering. Value engineering is used on identifying a series of step-by-step techniques to identify the unnecessary costs and to eliminate them. It concentrates on their functions and their cost for this purpose.

The required or needed performance at minimum cost is the value. Needed performance is what the customer expects. If it is less than the desired performance, one should focus on eliminating the waste curved due to overdesign, such as costly material, high factor of safety in design, etc. Value in general is defined as the ratio of function and cost. This is given by:

$$\text{Value} = \frac{\text{Function}}{\text{Cost}}$$

Here, function is expressed as units of performance and cost is expressed as a monetary unit, such as ₹ related to expenditure of resources. Hence, value is expressed as a relationship of what a product or service/system achieves and at what cost.

For example, if there is need to travel between two cities. The distance between these cities is 1000 km. The air ticket costs ₹ 4000, the train ticket costs ₹ 3000 and the road-but costs ₹ 2000. Then by air way the value of air travel is 1000 km/₹ 4000 or 0.25 km/₹ Similarly, value of train is 1000/300 or 0.33 km/₹ and the value of road travel is 0.5 km/₹ In this example, we have ignored any function other than distance travelled. If comfort, quality of services or time to travel are also important, then the numerator in value or function is a combination of the needed performance measure. However, it is important to mole in this example that value gives a quantitative measure of what we are getting at every spent rupee.

21.2. COST AND WORTH

Cost is the amount that is actually paid for a product or services. Worth is what is to be paid for the functions which is needed. So, the reasonable cost of the functions required is the worth of product. The objective of value engineering is to make cost close to worth.

$$\text{If cost} \sim \text{worth, then Value} = \frac{\text{worth or intended function}}{\text{Cost} \approx 1}$$

The objective of cost worth analysis is to select areas of higher cost worth ratio because these areas are the potential areas of improvement through value and analysis. Worth is applicable for the function and not for the product, services or system.

21.3. OBJECTIVE AND DEFINITION OF VALUE ENGINEERING

21.3.1. Objective

Basic ideas related to value engineering were developed by Lawrence D. Miles in 1947 after World War II.

There was overall recession during that period. Material was in short supply. Trained and skill manpower was insufficient. There was need of alternative material, improved processing requirements, better designs, low cost operation, and efficient system. Cost cutting was the objective while retaining the desired function of the product.

Value engineering is the systematic application of recognized techniques to identify the function of a product or service; establish norms for the function; and provide the necessory function related to these at lowest cost.

21.3.2. Definition

Value engineering is a technique for analyzing the functions of a product, service or a system with the purpose of achieving the required functions at the lowest overall cost consistent. All the requirements, which comprise its value, such as performance, reliability, maintainability appearance etc. are to be kept intact. Figure 21.1 describes the value engineering.

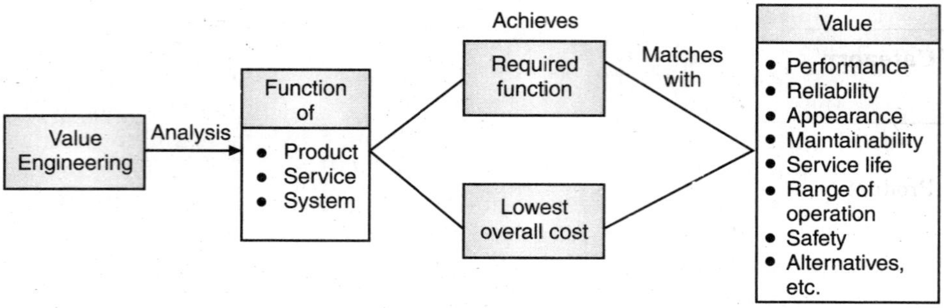

Fig. 21.1. Value Engineering

21.4. VALUE ENGINEERING TERMS

Value engineering, value analysis, value management, value performance, value control etc. are the many terms used for same meaning. However, we would generally stick to the term- value engineering or *VE*.

Value is categorized into following types:

(*i*) **Use value:** It is defined as the qualities and properties needed to accomplish a service, product or work.

(*ii*) **Esteem value:** It provides properties, features and attractiveness to a service, product or work which makes the ownership desirable.

(*iii*) **Scrap value:** It is the money which can be recovered when the item is not needed. It is the scrap value.

(*iv*) **Cost value:** It is the total cost of material, labour, overhead and services to produce an item (or to deliver a service).

(*v*) **Exchange value:** It is the property and qualities which enable to exchange (or trade) a product (or service) for something else, which is needed by the exchanger.

A customer purchases an item when the exchange value is less than the perceived (or guessed) use and esteem value. The most important issue, which a customer generally looks into a product, is a combination of use and esteem value. Value is a relative concept. Normally, it increases when product cost decreases.

Value can change with time and place. So, it has its own dynamics. It can be computed performance wise. Value is a relationship between want and willingness to pay for it.

21.5. FUNCTION

Function of an item is its utilization. For every product or services, there must be reasons to justify why it is of any use. For example, pen makes a mark where pen is a product and mark is a function. Other examples are shown in Table 21.1.

Table 21.1. Function and Value

Category	Item	Function		Value
		Verb	Noun	
Product	Book	Enriches	Knowledge	Use
	Cloth	Cowers	Body	Use
	Shaft	Transmits	Power	Use
	Old stamp	Exchange	Money	Exchange
	Pillers	Hold	Roof	Use
Services	Firebrigade	Extinguishes	Fire	Use
	Consultant	Consults	Client	Use
System	Organization	Facilitates	Management	Use

21.6. EFFECT OF FUNCTION AND COST ON VALUE

Change in function and cost the value is affected. It increases with increase in function while cost is same and with decrease in cost while function is same. When increased function is associated with increased cost, value would increase if improvement in function is more then increased cost. Similarly when cost is less while some functions are compromised, value would increase if and only if cost reduction is more than less in function. Table 21.2. shows the method to increase value

Table 21.2. Method to Increase Value

S. No.	Cost (C)	Function (F)	Value – (= F/C)
1	Same	Same	Same
2	Fall	Increase	Increase value
3	Same	Increase	
4	Increase	Steeper Increase	
5	Fall	Same	
6	Steeper fall	Fall	

21.7. LIFE CYCLE OF A PRODUCT AND VALUE ENGINEERING

Various phases from idea covering to birth of product, growth phase and maturity phase are involved in product life cycle. Value engineering is not very relevant as compared to maturity phase upto growth phase. In the last phase, attempts are needed to reduce cost while retaining the functions. In such cases, product life cycle becomes flater rather than declining in later phases. Value analysis is useful for products which are:

(*i*) imported
(*ii*) have long design to market item
(*iii*) high court item
(*iv*) critical parts, etc.

Pareto-analysis or ABC analysis may be done to identify high value items. For these items, cost reduction efforts may be focused. Shows life cycle of a product and effect of value oriented product is shown in figure 21.2.

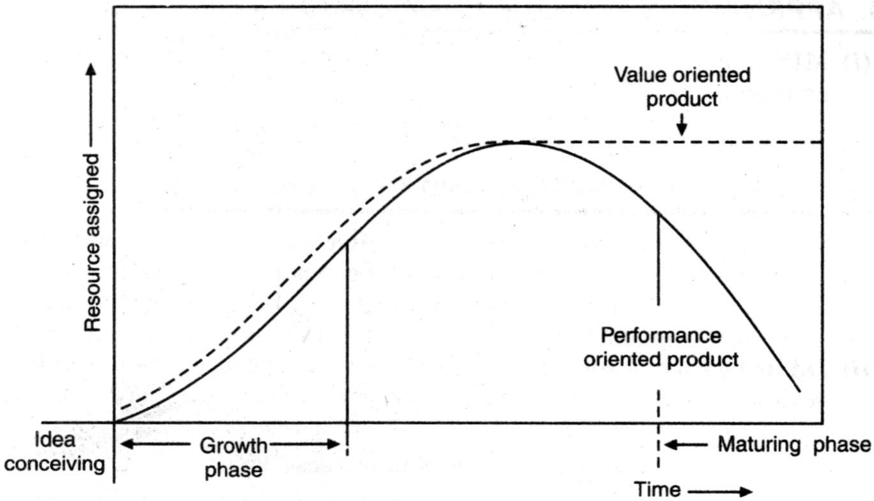

Fig. 21.2. Life Cycle of a Product and Effect of Value Oriented Product

21.8. STEPS IN VALUE ENGINEERING

VE focuses to reduce cost and/or increase the functional property of an item. This may be achieved by using alternative, cheaper material and/or processes. Simple rule of thumb to judge the potential area of improvements are used. If answer is yes to these questions, they go for them:

(*i*) How to reduce cost of material?
(*ii*) How to reduce cost of operation?
(*iii*) How to reduce manpower needed in the production?
(*iv*) How to reduce time of operation/set-up of tools, jigs, etc.?
(*v*) If automation or advanced manufacturing system cost effective, if yes, then should it be adopted now?

There are also many questions related to product performance starting with What, Why, When, Where, What else, How, Does, Do, Is, If, etc. These questions give answer to the VE strategies. Some of the questions are:

(*i*) What is the item?
(*ii*) What does it do?
(*iii*) How much is the cost?

(*iv*) What is the cost of an alternative?
(*v*) What are its other uses?
(*vi*) What machines does the company have?
(*vii*) What special skill is needed in the factory?
(*viii*) What special skill company does not have?
(*ix*) Does the product need all features?
(*x*) Is low cost alternative product available for consumer?
(*xi*) Is the profit margin reasonable?
(*xii*) What is the strategy of competitors?

21.9. APPROACHES IN VALUE ENGINEERING

(*i*) **MISS:** Modify, substitute or subdivided or exchange/eliminate to help change.

M	od
I	fy
S	ubstitute
S	ub-divide

Exchange/eliminate

(*ii*) **DARSIRI:** Data collection, analysis, record ideas, speculate, innovate review and implement.

D	ata collection
A	nalysis
R	ecord ideas
S	peculate
I	nnovate
R	eview
I	mplement

(*iii*) **Profit:** Product return, opportunities by function investigation techniques.

21.10. METHODOLOGY IN VALUE ENGINEERING

Starts with the identification and classification of the items is the starting point value engineering. Then function of the items are identified. Each function is evaluated and compared. Alternative strategy for cost reduction or function improvement is adopted. Fig. 21.3 shows the methodology.

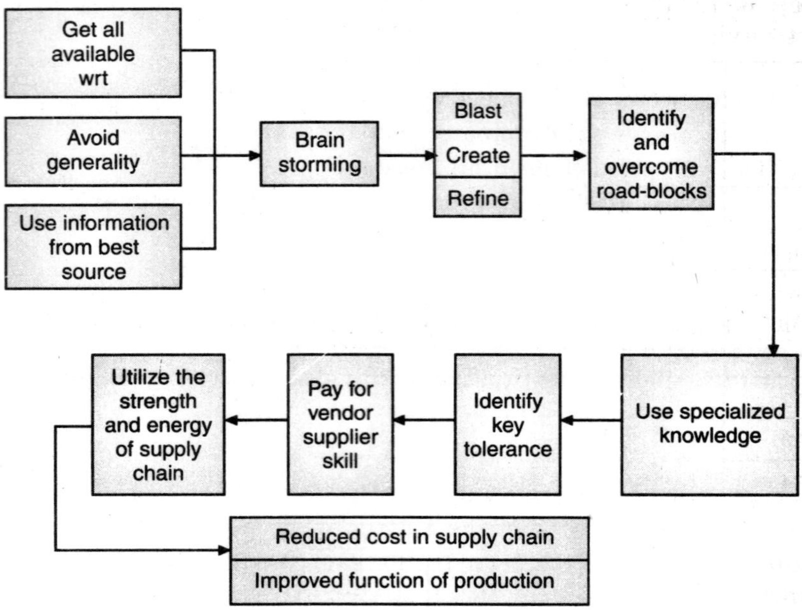

Fig. 21.3. Methodology of VE

21.11. MATRIX METHOD IN VALUE ENGINEERING

The most suitable alternative is related by using a numerical evaluation technique in matrix method.

Step 1: List all functions required to obtain the desired objectives.

Step 2: Generate a relative importance matrix by comparing each factors among them. A 1 is filled if function of a row is more or equally important than the function in column. For example, F1 is more important than F3 but less important then F4 as shown in Table 21.3.

Table 21.3.

Function	F1	F2	F3	F4	F5	Sum of Importance
F1	1	0	1	0	0	2
F2	1	1	1	1	0	4
F3	0	0	1	1	0	2
F4	1	0	0	1	1	3
F5	1	1	1	0	1	4

Step 3: Sum all rows.

Step 4: Arrange functions in descending order of the sum in last column. This is the relative rank of functions. For example, the relative rank is F2, F5, F4, F1 and F3.

Step 5: Create a weight factor (W) for each factor (F_i). W_i is determined on the basis of the impact of F_i on the overall product.

Step 6: Evolve alternatives (A_j) to satisfy the overall objective of the product/service. List cost (C_j) for each A_j.

	Ranked Function (Fi)					Cost (C_j)	Sum of product $G_j = \Sigma(w_i + d_{ij})$	Ratio (G_j/C_j)
	F2	F5	F4	F1	F3			
Weight	0.3	0.25	0.20	0.15	0.1			
Alternative (Aj)	Value of Evaluation Factor (j_{ij})							
A1	3	7	8	4	1	4.98	4.95	0.99
A2	2	6	5	1	4	5.1	3.65	0.71
A3	6	8	4	2	5	5.2	5.4	1.04
A4	5	9	2	3	9	8	5.5	0.68

The value for j_{ij} and C_j in the matrix are hypothetical.

Step 7: Fill evaluation factor (j_{ij}) for each combination of alternative (A_j) and function (F_j) in the evaluation matrix. h_{ij} is between 0 to 10 and depends upon the degree of attainment of the function. For j_{ij} = 0; the alternative is rejected.

For the illustration purpose, we have taken few hypothetical values of j_{ij} and C_j for four alternatives.

Step 8: Calculate the sum of product W_i and j_{ij} for each alternative (A_{ij}). Write it in last column as G_j; for example, alternative A1 has G_j value as $3 \times 0.3 + 7 \times 0.25 + 8 \times 0.2 + 4 \times 0.15 + 1 \times 0.1 = 4.95$

Step 9: Select alternative as one of the following criteria:

S. No.	Criteria	Measure
1.	Lowest cost alternative	Lowest (C_i)
2.	Best product	Highest (G_i)
3.	Best product per unit cost	Highest (G_i/C_i)

21.12. FAST DIAGRAM

FAST (Function Analysis System Technique) diagram is an effective graphical cum-analytical tool in value engineering. Objective is to validate the purpose of a part using questioning method to established the function that is served. Advantages of FAST diagram are:

(*i*) It creates a reference to validate the function of a component.

(*ii*) Graphically represents critical components and their function.

(*iii*) Alternative redrawing of FAST diagram would provide an analyst a list of minimum parts to serve the function at lowest cost.

21.12.1. Steps in FAST Diagram

The steps in FAST diagram are:

Step 1 : Define scope of the analysis. It includes the need (or basic objective) of the part and the interface, which is needed for the part. Interface may not serve the basic objective but may be necessary to make the part functional.

Step 2 : Explode part into smaller components.

Step 3 : Create a well defined list of function for each part, assembly, etc.

Step 4 : Organic functions as per the sequence of questions which starts with how query, then why query, and then when query.

Step 5 : Create FAST diagram using the listed functions in Step 4. A high level function, which is why query type, is placed in the box near the left boundary. Next function is placed in the box which is right of this box. A why query, if placed in second box, should be satisfied by the function in the previous box. This process continues.

Step 6 : Determine critical path of functions. Critical path is the connection of all the cross referenced functions in the FAST diagram.

Step 7 : The functions in Step 3 which are not listed in the fast diagram are the secondary functions. Place them on FAST diagram by answering to a when query. Show these by dotted lines.

Step 8 : All how, why and when query functions are now shown in FAST diagram. Remaining functions are unnecessary or redundant functions. List these functions.

Step 9 : Place component near to the respective function in FAST diagram. One fast diagram is shown in Fig. 21.4.

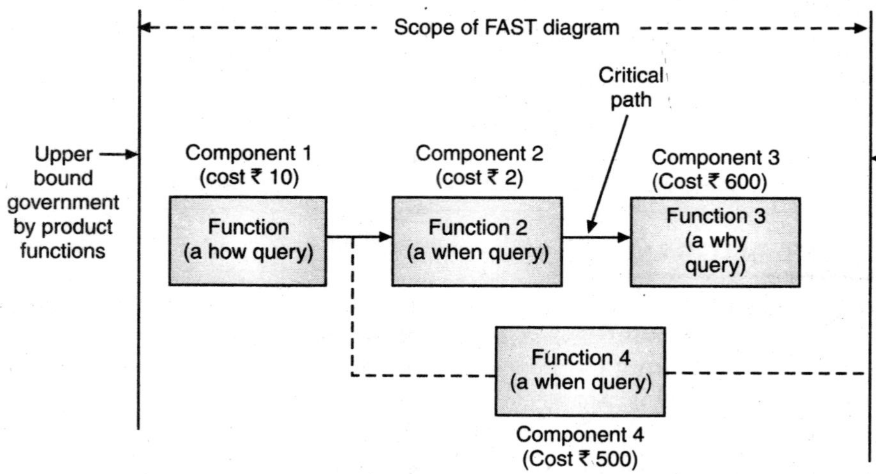

Fig. 21.4. FAST Diagram Showing Component with their Functions.

SUMMARY

1. Value engineering or value analysis is a systematic approach of continuously identifying unnecessary costs in products, processes or systems.
2. Value in general is defined as the ratio of function and cost. This is given by:

$$\text{Value} = \frac{\text{Function}}{\text{Cost}}$$

3. Value engineering is a technique for analyzing the functions of a product, service or a system with the purpose of achieving the required functions at the lowest overall cost consistent. All the requirements, which comprise its value, such as performance, reliability, maintainability appearance etc.
4. FAST (Function Analysis System Technique) diagram is an effective graphical cum-analytical tool in value engineering. Objective is to validate the purpose of a part using questioning method to established the function that is served.

EXERCISES

1. What is value engineering?
2. Explain objective of value engineering.
3. What is cost value?
4. Explain life cycle of a product.
5. Describe steps in value engineering.
6. What is matrix method in value engineering.
7. Explain FAST diagram.

APPENDICES

A APPENDIX

QUALITY MANAGEMENT TERMS

Acceptance	An action by an authorized representative of the acquirer by which the acquirer assumes ownership of products as a partial or complete performance of contract.
Acceptance Criteria	The criteria a product must meet to successfully complete a test phase or meet delivery requirements.
Acceptance Test	Formal testing conducted to determine whether or not a system satisfies its acceptance criteria and to enable the acquirer to determine whether or not to accept the system.
Access	A users ability to communicate with a system.
Acquirer	An organization that procures products for itself or another organization.
Approval	Written notification by an authorized representative of the acquirer that the developers plans, design, or other aspects of the project appear to be sound and can be used as the basis for further work. Such approval does not shift responsibility from the developer to meet contractual requirements.
Architecture	The organizational structure of a system, identifying its components, their interfaces, and a concept of execution among them.
Assembly	A number of parts or subassemblies or any combination thereof joined together to perform a specific function and capable of disassembly.
Associativity	A link between two different functions in a CAD system that assures that a change made in one area is reflected in all other areas. For example, a change to a solid model will be reflected in its

drawing and related CAM program. Bidirectional associativity indicates that updates happen in both directions between functions. For example, a change to a drawing will be reflected in its solids model.

Audit An independent examination of a work product/process or set of work products/processes to assess compliance with specifications, standards, contractual agreements, or other criteria.

Authentication The procedure (essentially approval) used by the approval authority in verifying that specification content is acceptable. Authentication does not imply acceptance or responsibility for the specified item to perform successfully.

Availability The product metric that defines the percentage of time that a product is available and operational for customer use.

Bath-Tub Curve Bath-Tub Curve represents the failure rate of components over the life of the product. Its upward slope at the beginning and end suggests that most components foil either right away (at the beginning of the product life) or towards the end of the expected product life.

Benchmarking An improvement process in which a company measures the performance of its products or processes against that of best-in-class products or companies, determines how the product or company achieved their performance level, and uses the information to improve its own performance.

Best Practice Best Practice is a superior method or innovative practice that contributes to the improved performance of an organization, usually recognized as "best" by other peer organizations.

Blok Diagram Block Diagram is a diagram that shows the operation, interrelationships and interdependencies of components in a system. Boxes or blocks represent the components; connecting lines between the blocks represent interfaces.

Break Even Time Break Even Time — a metric that measures the time from the start of development through production and sales until the product has achieved financial breakeven considering the investment in development and other non-recurring expenses.

Business Process Reengineering Business Process Reengineering (BPR) is the analysis and redesign of workflow within and between enterprises. Authors Michael Hammer and James Champy promoted the idea of BPR as the radical redesign and reorganization of an enterprise to lower costs and increase quality of service.

Certification	A process, which may be incremental, by which a contractor provides evidence to the acquirer that a product meets contractual or otherwise specified requirements.
Components	Components are the named "pieces" of design and/or actual entities of the system/subsystem. In system/subsystem architectures, components consist of subsystems (or other variations), and manual operations.
Components/ Equipment	Repairable assemblies which currently require repairparts support or will require it when introduced into an inventory.
Computer Hardware	Devices capable of accepting and storing computer data, executing a systematic sequence of operations on computer data, or producing control outputs. Such devices can perform substantial interpretation, computation, communication, control, or other logical functions.
Computer Program	A combination of computer instructions and data definitions that enable computer hardware to perform computational or control functions.
Computer Software Configuration Item (CSC)	A functionally or logically distinct part of a computer software components may be top-level or low-level.
Contract	A mutually binding legal relationship obligating the seller to furnish the supplies or services (including construction) and the buyer to pay for them. It includes all types of commitments that obligate the acquirer to an expenditure of appropriate funds and that, except otherwise authorized, are in writing. In addition to bilateral agreements, contracts include, but are not limited to, awards and notices of awards; job orders or task letter issued under purchase orders, under which the contract becomes effective by written acceptance or performance; and bilateral modifications.
	An individual, partnership, company, corporation, association or other service, having a contract with an acquirer for the design, development, manufacture, maintenance, modification, or supply of items under the terms of a contract.

CPM CPM stands for critical path method. It is a mathematically based algorithm for scheduling a set of project activities. It is an important tool for effective management.

Customers Users and suppliers of systems and end-items.

Data Recorded information, regardless of medium or characteristics, of any nature, including administrative, managerial, financial, and technical.

Database Management System An integrated set of computer programs that provide the capabilities needed to establish, modify, make available, and maintain the integrity of a database.

Database A collection of related data stored in one or more computerized files in a manner that can be accessed by users or computer programs via a database management system.

Decomposition Decomposition is the process of dividing the system into its smallest, coherent, self-contained elements. Decomposition is used in systems engineering, software engineering, process mapping and functional analysis system technique.

Deficiencies Deficiencies consist of two types: conditions or characteristics in any hardware or software which are not in compliance with the specified configuration identification; or inadequate (or erroneous) configuration identification which has resulted, or may result, in Cls that do not fulfill approved operational requirements. Deficiencies are corrected by Class II methods of change control or revisions with "change bars". Design. Those characteristics of a system or CSCI that are selected by the developer in response to the requirements, such as definitions of all error messages, others will be implementation related, such as decisions about what software units and logic to use to satisfy the requirements.

Design for Manufacturability 1. Design for Manufacturability (broad definition) is a methodology for designing product's in a way that facilitates the fabrication of the product's components and their assembly into the overall product. In this respect it is synonymous with Design for Manufacturability/ Assembly.

2. Design for Manufacturability (narrow definition) is a methodology for designing product

components in a way that facilitates their fabrication.

Design for Six Sigma Design for Six Sigma (DFSS) is a systematic methodology or quality framework utilizing tools, processes and measurements to enable the design of products and processes that meet customer expectations and can be produced at Six Sigma quality levels. DFSS is built around five connected phases of Define, Measure, Analyze, Design, Verify or DMADV. The tools and methods to support DFSS include voice of the customer (VOC), quality function deployment (QFD), design for manufacturability and assembly, design of experiments (DOE), failure mode and effects analysis (FMEA), process capability studies, design reviews, control plans, etc.

Design Optimization 1. Design Optimization in the broad sense refers to optimizing a design to meet its functional, environmental and lifecycle requirements at a minimum cost.

2. Design Optimization in a narrower sense refers to the use of computer-aided engineering applications which analyze a design and, given constraints and objectives, seek to improve or optimize the design to meet the stated objectives within the stated constraints. These applications will typically use an iterative, goal-seeking cycle to seek design optimization.

Design Records Resign records refers to all technical documentation necessary to define the design, manufacture, packaging, testing, installation, and maintenance of the system.

Design Reviews Design reviews are formal technical reviews conducted during the development of a product to assure that the requirements, concept, product or process satisfies the requirements of that stage of development, the design is sound, the issues are understood, the risks are being managed, any problems are identified, and needed solutions proposed. Typical design reviews include: requirements review, concept/preliminary design review, final design review, and a production readiness/launch review.

Design Simplicity The concept of design simplicity and standardization shall be evident.

Data Validation Testing to assure that the product conforms to defined user needs and requirements. This normally occurs toward the end of the Design Phase following successful design verification and prior to pilot production, beta/market testing, and product launch. Design validation is normally performed on the final product under defined, operating conditions. Multiple validations may be performed if there are different intended uses.

Design Verification Design verification is the process of ensuring the design conforms to specification (design outputs meet design input requirements). Design verification may include: alternate calculations, design reviews, comparison to similar designs, inspection, and system or product testing.

Design-to-Cost A development methodology that treats cost as an independent design parameter. A realistic cost objective is established based on customer affordability, trade-off's are made between the cost objective and other product functions/ parameters, cost models are used to project the cost early in the development cycle, and a variety of techniques such as function analysis and DFM are used to proactively achieve the cost objective.

Detailed Design The conversion of product specifications into designs and their associated process and/or code-to documentation. Detailed design includes design capture, modelling, analysis, test plan development, coding, and design verification and validation.

Develper An organization that develops products ("develops" may include new development, modification, reuse, reengineering, maintenance, or any other activity that results in products) for itself or another organization.

Documented procedure A written description of a course of action to be taken to perform a given task.

Domain The part of the external world, including users and inmates of the system that effects and is affected by the system.

Engineering The management of the engineering and technical effort to transform a conceptual requirement into an operational system. It includes the system performance parameters and preferred system configuration to satisfy the requirement, the planning and control of technical program tasks, integration of the engineering specialties, and the management of a totally integrated effort of design engineering, computer software engineering, test engineering, safety engineering, security engineering, logistics engineering, production engineering, and

	specialty engineering (EMC, environmental, etc.) to meet cost, technical performance and schedule objectives.
Environmental Strees Screening	Environmental Stress Screening (ESS) is a process which applies specific kinds of environmental stresses to products on an accelerated basis, but within their design parameters and limits to cause latent and intermittent flaws to become detectable failures. Also see Highly Accelerated Stress Screening (HASS).
Evaluation	The process of determining whether an item or activity meets specified criteria.
Failure Analysis	Failure Analysis is a collection of techniques to determine the root cause of a component or process defect or failure.
Failure Modes and Effects Analysis	Failure Modes and Effect Analysis (FMEA) is a procedure in which each potential failure mode in every sub-item of an item is analyzed to determine its effect on other sub-items and on the required function of the item. It is used to identify potential failure modes and their associatedy "causes" mechanisms, consider risks of these failure modes, and identify mitigating actions to reduce the probability or impact of the failure.
Failure Reporting and Correctie Action	Failure Reporting and Corrective Action System (FRACAS) is a closed-loop system to capture reports of failure from customers or service technicians in the field or from the factory, analyze these reports, detect trends or problems, and use this analysis to take corrective action in the design, component selection, supplier selection, manufacturing process, or operating manual of the product. Features of a FRACAS system include a database manager, tracking system for document controls, user definable reports which allow selection of data elements and sort options, and search functions.
Failure	A deficiency, defect, non-performance or non-conformance with specified requirements.
Feasibility	Capable of being completed to meet goals. The feasibility of a proposed new product has two dimensions: (*a*) Technical Feasibility (*i.e.,* will the product work?) and (*b*) Financial Feasibility (*i.e.,* will the product make an adequate return on investment for the enterprise?)
Feature	Features are elements of the product that provide

a distinctive benefit to the customer and are often highlighted in describing the benefits of the product to the customer. Features are the differentiating functionality of a product. This functionality may not be available in other products, or it may not be available with the same quality characteristics.

Firmware — The combination of hardware device and computer instructions and/or computer data that resides as read-only software on the hardware device.

Fit — The ability of an item to physically interface or interconnect with or become an integral part of another item.

Formal Testing — The process of conducting testing activities and reporting results in accordance with an approved test plan making provision for customer involvement.

Function — The action or actions which an item is designed to perform.

Functional Area — A distinct group of system performance requirements which, together with all other such groupings forms the next lower-level breakdown of the system on the basis of function.

Functional Requirement — Functional Requirements capture the intended behaviour of the system or product — what the system will do? This behaviour may be expressed as functions, tasks, or services the system or product is required to perform. Therefore, functional requirements do not include performance characteristics, operating conditions or specifications.

Gantt Chart — Gantt Chart is a diagram used in project management, where the x-axis is time and the y-axis shows tasks to be performed to complete the project. Each task is displayed as a horizontal bar spanning the time period during which it is expected to take place. Arrows may be drawn from one task to another to indicate dependencies (when one task can't be begun until another is completed). The Gantt Chart was developed by Charles Gantt in 1917.

Gauge Repeatability and reproducibility — Gauge Repeatability and Reproducibility (GR and R) is the evaluation of a gauging instrument's accuracy by determining whether the measurements taken with it are repeatable and reproducible. Repeatability is the variation in measurement obtained with one measurement instrument when used several times by an appraiser while measuring the identical characteristic on the same part. Reproducibility is the variation in the averages of the measurements made by the different appraisers using the same measuring instrument when measuring the identical characteristic on the same part.

General Availability The point in the product life cycle when production has been ramped-up to sufficient volumes and when product issues have been resolved so that the product is made available to all interested customers.

Group Technology Group Technology is a coding and classification system to identify similarities in part geometry, features, characteristics and processes. It is used to aid in design retrieval, part standardization, manufacturing cell design, and production scheduling.

Hardware Articles made of material, such as aircraft, ships, tools, computers, vehicles, fittings, and their components (mechanical, electrical, electronic, hydraulic, pneumatic). Computer software and technical documentation are excluded.

Hierarchical Design Hierarchical Design is a design methodology where portions of large designs are divided into manageable sections or sub-blocks that may be created, represented symbolically, designed, and then connected together when completed. This methodology allows different parts of the design to be worked on in parallel.

Human Factors Human Factors refers to the characteristics of human beings that are applicable to the design of systems and devices of all kinds. It furthers serious consideration of knowledge about the assignment of appropriate functions for humans and machines, whether people serve as operators, maintainers, or users in the system. And, it advocates systematic use of such knowledge to achieve compatibility in the design of interactive systems of people, machines, and environments to ensure their effectiveness, safety, and ease of performance.

Indirect Costs Costs that are incurred for common or joint objectives that can not be readily identified to a single cost objective and readily treated as a direct cost. A cost that is allocated as opposed to being traced.

Industrial Design	Industrial Design is the design that is done in companies and consultancies by people trained in industrial design, or in art and design schools in general. Industrial design focuses on the physical form and interactive properties as opposed to the functioning of the product or system.
Industrial Engineering	The branch of engineering that is concerned with the efficient production of industrial goods as affected by elements such as plant and procedural design, the management of materials and energy, and the integration of workers within the overall system.
Integrated Logistics Support (ILS)	A disciplined, unified and iterative approach to management and technical activities necessary to: integrate support considerations into system and equipment design, develop support requirements that are related consistently to design, readiness objectives and to each other; acquire required support; and provide required support during the operational phase at minimum cost.
Interface Control	Interface control comprises the delineation of the procedures and documentation, both administrative and technical, contractually necessary for identification of functional and physical characteristics between two or more configuration items which are provided by different contractors/acquiring agencies, and the resolution of the problems thereto.
Interface Design Compatibility	Intra-system and intersystem design compatibility of engineering interfaces shall be delineated as interface requirements in appropriate specifications. Interface control requirements and drawings related to: the major system elements of a prime contractor's contractual responsibility; other equipment, computer software, facilities, and procedural data furnished by the acquirer.
Interface	The functional and physical characteristics required to exist at a common boundary. In development, a relationship among two or more entities in which the entities share, provide, or exchange data. An interface is not a software unit, or other system component; it is a relationship among them.
ISO 9000	ISO 9000 is a set of international standards on quality management and quality assurance

	developed to help companies effectively document the quality system elements to be implemented to maintain an efficient quality system. The standards, initially published in 1987, are not specific to any particular industry, product or service. The standards underwent major revision in 2000 and now include ISO 9000:2000 (definitions), ISO 9001:2000 (requirements) and ISO 9004:2000 (continuous improvement).
ISO/TS 16949:2002	An ISO technical standard titled "Quality management system — particular standards for the application of ISO 9001:2000 for automotive production and relevant service part organizations". This standard replaces QS-9000 and harmonizes requirements for automotive manufacturers internationally.
Item	A non-specific term used to denote any product, including systems, subsystems, assemblies, subassemblies, units, sets, accessories, computer programs, computer software or parts.
Knowldge Management	The overall management process to capture, organize, manage and disseminate knowledge in an organization to improve enterprise effectiveness by avoiding mistakes and avoiding the time to relearn needed knowledge. Since product development is very knowledge intensive, knowledge management offers tremendous leverage and opportunity for improvement.
Knowledge-Based Engineering	Knowledge-Based Engineering is a set of design automation tools that capture design knowledge and rules to automate the design process.
LCL	Lower Control Limit is the lower limit used within statistical process control that define the constraints of common cause variations. When a parameter value falls below the lower control limit, it flags the occurrence of special causes contributing to variation. There is a similar Upper Control Limit and the mean to show the central tendency.
Lean Manufacturing	Lean Manufacturing is an operations philosophy that aims to synchronize production with demand, thereby minimizing inventory and cycle time. Lean Manufacturing is supported during product development with approaches such as robust design, mistake-proofing and standardization.
Life Cycle Cost	Life Cycle Cost is the total cost to the customer of acquiring, operating, and disposing of a

product/system over its full life. These costs include development, acquisition, installation, training, operation, support, and disposal.

Life Cycle A generic term covering all phases of acquisition, operation, and logistics support of an item, beginning with concept definition and continuing through disposal of the item.

Maintainability A characteristic of design and installation which inherently provides for an item to be retained in, or restored to a specified condition within a given period of time, when the maintenance is performed in accordance with prescribed procedures and resources. Synonymous with Serviceability and Supportability.

Manufacturability The characteristic of a produce's design that facilitates tire fabrication of the produce's components and their assembly into the overall product.

Material A generic term covering systems, equipment, stores, supplies and spares, including related documentation, manuals, computer hardware and software.

Memorandum of Agreement The document which specifies agreements between involved parties relative to a specific interface. The MOA delineates responsibilities of each organization and identifies formats and data elements required for a successful interface.

Metrics Measures used to indicate progress or achievement.

Milestone A scheduled event for which some individual is accountable and that is used to measure progress. An important event representing the completion of a major work task or group of work tasks. Milestones are usually scheduled and can be used to measure progress. Reviews are often conducted upon the completion of a milestone.

Module A self-contained part of hardware item designed as a single replaceable unit, with a specific integral electronic function. It should require no installation other than mechanical mounting and completion of electrical connectors.

MTBF Mean Time Between Failure — the average time between failures for a repairable product for a defined unit of measure (*e.g.*, operating hours, cycles, miles, etc.).

Net Present Value Net Present Value is a financial analysis technique that discounts a series of cash inflows (revenue) and outflows (investments and

	expenses) to determine the suitability of an investment. NPV is an evaluation technique used to screen new product development projects as part of portfolio management.
Non-conformance	The failure of a unit or product to conform to specified requirements.
	Non-developmental item is a broad generic term that covers material available from a wide variety of sources with little or no developmental effort by the purchaser.
Opportunity Cost	The economic value of the benefit that is sacrificed when an alternative course of action is taken.
Organization	A unit within a company or other entity (*e.g.*, Government agency or branch of service) within which many projects are managed as a whole. All projects within an organization share a common top-level manager and common policies.
Pareto Analysis/ Diagram	An analysis/diagramming technique using frequency of occurrence to identify and display results generated by each identified cause. This analysis is commonly used to decide where to apply initial effort for maximum effect. See Pareto Principal.
Pareto Principle	The Pareto principle suggests that 20% of a set of independent variables is responsible for 80% of the result. In quantitative terms, for example, 80% of the problems come from 20% of the causes (machines, raw materials, operators etc.). Therefore, effort aimed at the right 20% can solve 80% of the problems.
Part or Identifying Number (PIN)	Part or identifying Number is an alpha-numeric designator which identifies parts, items, or bulk materials that are covered by a specification.
Part	One piece, or two or more pieces joined together which are not normally subjected to disassembly without destruction or impairment of designed use (examples: gear, screws, transistors, capacitors, integrated circuits.
Performance	A quantitative measure characterizing a physical or functional attribute relating to the execution of a mission/operation or function.
PERT	PERT stands for programme evaluation and Review techniques. It a pictorial "Bar Chart" to show various works or jobs required for complex project. It is a very efficient tool and is helpful for management in taking decisions.

Policy A guiding principle, typically established by senior management, which is adopted by an organization or project to influence and determine decisions.

Procedure A series of activities carried out to accomplish a task or operation.

Process An organized set of activities.

Product Lifecycle Management Product Lifecyle Management (PLM) generally refers to the process of managing all data related to the product over its life cycle. It broadens the concept of product data management (PDM) to address managing the product configuration and the availability of product data into its Product Program manager later lifecycle stages of production, operation, support and disposal and to address managing the produce's related process data. ARC Advisory Group defines six components of PLM: innovation and portfolio management, project and program management, collaborative design, product data management, manufacturing process planning, and service and support management. Michigan's PLM Development Consortium defines it as "an integrated, information driven approach to all aspects of a produce's life from its design inception, through its manufacture, deployment and maintenance and culminating in its removal from service and final disposal. PLM is the integration of business systems to manage a produce's lifecycle." CIMdata Inc., says that PLM represents'a business approach to solving the problem of managing the complete set of product and plant definition information and the processes through which it passes. The PLM process includes creating and changing that information, managing it through its life and disseminating and using it throughout the lifecycle of the product."

Product A product is a given set of items. The set could comprise of system, subsystem, hardware, or software items and their documentation.

Program Manager The role at a high enough level in an organization that the primary focus is the long-term vitality of the organization, rather than the short-term project and contractual concerns and pressures. In general, a senior manager for engineering would have responsibility for multiple projects.

Program Evaluation And Review Technique	It is an event-oriented network analysis technique used to estimate project duration when there is a high degree of uncertainty with individual activity duration estimates. PERT applies the Critical Path Method (CPM) to a weighted average duration estimate.
QS-9000	QS-9000 is a quality management standard developed by the Big Three Automakers for the automotive sector. Currently largely replaced by Technical Specification 16949 (ISO/TS 16949).
Quality Assurance	A planned and systematic pattern of all actions necessary to provide adequate confidence that management and technical planning and controls are adequate to: Establish correct technical requirements for design and manufacturing; manage and design activity standards, drawings, specifications, or other documents referenced on drawings, lists, or technical documents.
R-Chart	A statistical process control (SPC) chart that monitors the range (variability) of the process. A sample of parts is collected from the process periodically. The range (maximum minus minimum) of the sample is plotted on the control chart and a determination is made if the process is "under control" or not.
Rapid Manufacturing	Rapid Manufacturing refers to the use of rapid prototyping technologies to directly manufacture low volumes of parts.
Reengineering	The process of examining and altering an existing system to reconstitute it in a new form. May include reverse engineering (analyzing a system and producing a representation at a higher level of abstraction, such as design from code), restructuring (transforming a system from one representation to another at the same level of abstracton), recommendation (analyzing a system and producing user and support documentation), forward engineering (using software products derived from an existing system, together with new requirements, to produce a new system), and translation (transforming source code from one language to another or from one version of a language to another).
Realiability Analysis	A predictive tool used to estimate the "life" of a product. This is usually expressed in terms of hours as "mean time between failure" (MTBF).
Realibility Prediction	Reliability Prediction is the analysis of parts and components in an effort to predict the rate at

which an item will fail. A reliability prediction is one of the most common forms of reliability analyses.

Reliability The probability that an item will continue to function at customer expectation levels at a measurement point, under specified environmental and duty cycle conditions. A function, feature or capability that a product must provide or meet to satisfy the...........

Requirement Customer's needs and enterprise's objectives for a new product. See Product Requirement.

Reverse Engineering 1. Reverse Engineering is the process of capturing the geometry of existing physical objects and then using the data obtained as a foundation for designing a duplicate of the original or an entirely new adaptation. 2. Reverse engineering refers to the procedure of carefully dismantling and inspecting a competitor's product to look for design features that can be incorporated into one's own product.

Risk Management An organized process to identify what can go wrong, to quantify and access associated risks, and to implement/control the appropriate approach for preventing or handling each risk identified. A management process consisting of identification, assessment, mitigation, and management of all project, technical and market risks using formal tools and methods.

Robust Design Design of the product in a manner to desensitize the product to variation including misuse and increase the probability that it will perform as intended.

Robustness The condition of a product or process where its operating parameters remain relatively stable with a minimum of variation even though factors which influence operation or usage, such as wear or environment, change.

Root Cause Analysis Root Cause analysis–Study of original reason for non-conformance with a process. When the root cause is removed or corrected, the non-conformance will be eliminated.

Root Caue A root cause is an antecedent source of a defect such that if it is removed, the defect is decreased or removed itself.

Safety The expectation that a system does not, under defined conditions lead to a state in which human life is endangered.

Simulation Modelling or representation of hardware, software or systems to determine or verify their behaviour, operation or fit. Simulation is used to provide condience that the hardware, software or systems will operate as intended without investing the time or expense of physically constructing the object to verify its operation. In electronic design, simulation is a technique in which the properties of a circuit are represented indirectly by test vectors.

Six Sigma 1. Six Sigma is a statistical measurement of processes that produce less than 3.4 defects or mistakes per million opportunities (or 99.99966% good). 2. Six Sigma is a bottom-line oriented, data-driven quality program. It is based on achieving a level of quality which equates with only 3.4 defects per million opportunities for each product. The six sigma process includes five steps (*a*) define (*b*) measure (*c*) analyze (*d*) improve and (*e*) control.

Software Architecture Software Architecture refers to the high-level structure of software systems. The architecture of a software system identifies a set of components that collaborate to achieve the system goals. The architecture specifies the "externally visible" properties of the component, *i.e.,* those assumptions other components can make of a component, such as its provided services, performance characteristics, fault handling, shared resource usage, etc. It also specifies the relationships among the components and how they interact.

Software Development A set of activities that results in software products. Softwar development may include new development, modification, reuse, re-engineering, maintenance, or any other activities that result in software products.

Specification A document intended primarily for use in procurement, which describes the essential technical requirements for items, materials or services including the procedures for determining whether or not the requirements have been met.

Standard Mandatory requirements employed and enforced to prescribe a disciplined uniform approach to software development, that is, mandatory conventions and practices are in fact standards.

Subcontractor An individual, partnership, corporation, or an association that contracts with an organization (*i.e.,* the prime contractor) to design, develop, and/or manufacture one or more products.

Subsystem An element of a system that, in itself may constitute a system.

Suppliers The term "suppliers" includes contractors, sub-contractors, vendors, developers, sellers or any other term used to identify the source from which products or services are obtained.

Supply Chain Management The procurement, stocking and distribution of components, subassemblies and products throughout the design, manufacturing, and distribution stages, ensuring that the correct components, subassemblies and products are delivered to their appropriate destination at the proper time, the lowest overall cost, and acceptable quality levels.

Surface Mount Technology Surface Mount Technology (SMT) is a method of attaching electronic/electrical components directly to a board substrate rather than through a plated hole.

SWOT Analysis Strengths, Weaknesses, Opportunities and Threats Analysis–a process whereby a group of people determine: (*a*) what strengths do we have? (how can we take advantage of them?); (*b*) what weaknesses do we have? (how can we minimize them?); (*c*) what opportunities are there? (how can we capitalize on them?); (*d*) what threats might prevent us from getting there? (consider technical obstacles, competitive responses, values of people within the organization, etc.). For every obstacle identified, what can we do to overcome or get around it? (This helps to develop contingency plans).

Synthesis The translation of input requirements (including performance, function, and interface) into possible solutions (resources and techniques) satisfying those inputs. Defines a physical architecture of people, product and process solutions for logical groupings of requirements (performance, functions, and interface) and their designs for those solutions.

System A combination of the hardware, software, and firmware.

B APPENDIX

QUALITY ASSURANCE TERMS

A...

Absolute Method: A body of procedures and techniques for which measurement is based entirely on physically defined, fundamental quantities.

Acceptable Quality Level: A limit above which quality is considered satisfactory and below which it is not. In sampling inspection, the maximum percentage of defects or failure that can be considered satisfactory as an average.

Acceptable Quality Range: the interval, between specified upper and lower limits of a sequence of values, within which the values are considered to be satisfactory.

Acceptable Value: An observed or corrected value that falls within the acceptable range. See Corrected value and Observed value.

Acceptance Criteria: Specified limits placed on characteristics of an item, process, or service which are defined in requirements documents. (ASQC Definitions)

Acceptance Sampling: The procedure of drawing samples from a lot or population to determine whether to accept or reject a sampled lot or population.

Accepted Reference Value: A numerical quantity that serves as an agreed-upon basis for comparison, and which is derived as; (1) a theoretical or established quantity based on scientific principles, (2) an assigned value, based on experimental work of some recognized organization, or (3) a consencus quantity based on a collaborative experimental work under the auspices of a scientific or engineering group.

Accreidition Criterion: A requirement that a laboratory must meet to receive authorization and approval to perform a specified task.

Accredited Laboratory: A laboratory which has been evaluated and given approval to perform a specified measurement or task, usually for a specific property or analyte for a specified period of time.

Accrediting Authority: The agency having responsibility and accountability for environmental laboratory accreditation and who grants accredition. For the purposes of NELAC, this is EPA, other federal agencies, or the state.

Accuracy: They degree of agreement between an observed value and an accepted reference value. Accuracy includes a combination of random error (precision) and systematic error (bias) components which are due to sampling and analytical operations; a data quality indicator. EPA recommends that this term not be used and that precision and bias be used to convey the information usually associated with accuracy. See Precision and Bias.

Action Limit: See Control limit.

Adjusted Value: The observed value after adjustment for values of a blank or bias of the measurement system.

Aliquant: A subsample derived by a divisor that divides a sample into a number of equal parts but leaves a remainder; a subsample resulting from such a divisor. See Subsample.

Aliquot: A subsample derived by a divisor that divides a sample into a number of equal parts and leaves no remainder; a subsample resulting from such a division. In analytical chemistry the term aliquot is generally used to define any representative portion of the sample.

Alternate Method: Any body of procedures and techniques of sample collection and/or analysis for a characteristic of interest with is not a reference or approved equivalent method but which has been demonstrated in specific cases to produce results comparable to those obtained from a reference method.

Analysis (chemical): The determination of the qualitative and/or quantitative composition of a substance.

Analysis Duplicates: The subjection of two portions of the same prepared sample, extract or digestate to the determinative step of an analytical method or a measurement system to estimate that step s precision.

Analysis Matrix Spike: The subjection of prepared sample, extract or digestate that has been fortified (spiked) with a known amount of the analyte of interest, to the determinative step of an analytical method to estimate the bias important by the instrumental or determinative procedure.

Analyte: The substance, a property of which is to be measured by chemical analysis.

Analytical Batch: A group of samples, including quality control samples, which are processed together using the sample method. The same lots of reagents, and at the same time or in continuous, sequential time periods. Samples in each batch should be of similar composition and share common internal quality control standards.

Analytical Blank: See Reagent blank.

Analytical Detection Limit (LD): The smallest amount of an analyte that can be distinguished in a sample by a given measurement procedure throughout a given confidence interval (*e.g.*, 0.95) See Method Detection Limit.

Analytical Limit of Discrimination: See Method detection limit.

Analytical Reagent (AR): The American Chemical Society's designation for the highest purity of certain-chemical reagents and solvents. See Reagent grade.

Arithmetic Mean: The sum of all the values of a set of measurements divided by the number of values in the set, usually denoted by x; a measure of central tendency. See Measure of central tendency.

Assessment: The evaluation process used to measure the performance or effectiveness of a system and its elements, used to denote any of the following: audit, performance evaluation, management systems review, peer review, inspection, or surveillance. ANSI/ASQC E4-1994

Assignable Cause: A factor or an experimental variable shown to significantly change the quality of an effect or a result.

Audit: A systematic evaluation to determine the conformance to quantitative specifications of some operational function or activity. See Audit of data quality, Performance evaluation audit, and Technical systems audit, and also Review, and Management systems review.

Audit of Data Quality (ADQ): A qualitative and quantitative evaluation of the documentation and procedures associated with environmental measurements to verify that the resulting data are of acceptable quality.

Audit Sample: See Performance evaluation sample.

Average: See Arithmetic mean.

B...

Background Level (environmental): The concentration of substance in a defined control area during a fixed period of time before, during or after a data gathering operation.

Batch: A quantity of material (*e.g.*, samples) of the same or similar matrix, expected to behave similarly with respect to the procedure(s) being employed and produced or processed in one operation, considered to be a uniform discrete unit.

NELAC defines **batch** as follows: environmental samples which are prepared and/or analyzed together with the same process and personnel, using the same lot(s) of reagents. A **preparation batch** is composed of one to 20 environmental samples of the same NELAC-defined matrix, meeting the above mentioned criteria and with a maximum time between the start of processing of the first and last sample in the batch to be 24 hours. An **analytical batch** is composed of prepared environmental samples (extracts, digestates or concentrates) which are analyzed together as a group. An analytical batch can include prepared samples originating from various environmental matrices and can exceed 20 samples. (Quality Systems)

Batch-lot: The samples collected under sufficiently uniform conditions to be processed as a group. See Batch. Batch size.

Batch-sample: One of the samples drawn from a batch.

Batch-size: The number of samples in a batch-lot.

Bias: The systematic or persistent distortion of a measurement process which deprives the result of representativeness (*i.e.*, the expected sample measurement is different than the sample's true value.) A data quality indicator.

Blank: A sample that has not been exposed to the analyzed sample stream in order to monitor contamination during sampling, transport, storage or analysis. The blank is subjected to the usual analytical and measurement process to establish a zero baseline or background value and is sometimes used to adjust or correct routine analytical results. (AS QC, Definitions of Environmental Quality Assurance Terms, 1996)

Blank Sample: A clean sample or a sample of matrix processed so as to measure artifacts in the measurement (sampling and analysis) process.

C...

Confidence Interval: The numerical interval constructed around a point estimate of a population parameter, combined with a probability statement (the confidence coefficient) linking it to the population's true parameter value. If the same confidence interval construction technique and assumptions are used to calculate future intervals, they will include the unknown population parameter with the same specified probability. See Confidence coefficient.

Confirmation: Verification of the presence of a component through the use of an analytical technique that differs from the original method. These may include:

- Second column confirmation
- Alternate wavelength
- Derivatization
- Mass spectral interpretation
- Alternative detectors or
- Additional cleanup procedures.

Conformity: Fulfilment of specified requirements. (ISO 8402)

Control chart: A graph of some measurement plotted over time or sequence of sampling, together with control limit(s) and, usually, a central line and warning limit(s). See Central line, Control limit and limit.

Control Limit: A specified boundary on a control chart that, if exceeded, indicates a process is out of statistical control, and the process must be stopped, and corrective action taken before proceeding (e.g., for a Shewhart chart the control limits are the mean plus and minus three standard deviations, *i.e.*, the 99.72% confidence level on either side of the central line.)

Control Sample: See quality control sample and Check sample.

Control Standard: See Check standard.

Controlled Variable: A variable that is set at a pre-selected level when a controlled experiment is conducted.

Corrective Action: An action taken to eliminate the causes of an existing nonconformance. deficiency, or other undesirable situation in order to prevent recurrence. (ISO 8402)

Correlation: A measure of association between two variables. See also Correlation coefficient.

Correlation co-efficient: A number between -1 and 1 that indicates the degree of linearity between two variables or sets of numbers. The closer to -1 or +1, the stronger the linear relationship between the two (i.e., the better the correlation.) Values close to zero suggest no correlation between the two variables. The most common correlation coefficient is the product-moment, a measure of the degree of linear relationship between two variables.

Critical-toxicity Range: The interval between the highest concentration at which all test organisms survive and the lowest concentration at which all test organisms die within the test period.

Customer: Any individual or organization for whom items or services are furnished or work performed in response to defined requirements and expectations recipient of a product provided by the supplier. (ISO 8402) Daily standard: synonym for Calibration standard.

D...

Data: Dacts or figures from which conclusions can be inferred.

Data Analysis: The comparison of suitably reduced data with a conceptual model (e.g., a dispersion model) and may include computation of summary statistics, standard errors, confidence intervals, tests of hypotheses, and goodness-of-fit tests.

Data Audit: A qualitative and quantitative evaluation of the documentation and procedures associated with environmental measurements to verify that the resulting data are of acceptable quality (*i.e.*, that they meet specified acceptance criteria.

Data Quality: The totality of features and characteristics of data that bears on their ability to satisfy a given purpose; the sum of the degrees of excellence for factors related to data.

Data Quality Assessment (DQA): The statistical evaluation of a data set to establish the extent to which it meets user-defined application requirements (*i.e.*, DQOs).

Data of Known Quality: Data are of known quality when the qualitative and quantitative components associated with their derivation are documented appropriately for their intended use, and such documentation is verifiable and defensible.

Data Quality Indicators: Quantitative statistics and qualitative descriptors that are used to interpret the degree of acceptability or utility of data to the user. The principal data quality indicators are bias, precision, accuracy (precision and bias are preferred), comparability, completeness, and representativeness.

P...

Proficiency Testing Program: The aggregate of providing rigorously controlled and standardized environmental samples to a laboratory for analysis, reporting of results, statistical evaluation of the results in comparison to peer laboratories and the collective demographics and results summary of all participating laboratories.

Property: A quality or trait belonging and peculiar to a thing; a response variable is a measure of a property. Synonym for Characteristic.

Protocol: A detailed written procedure for a field and/or laboratory operation (*e.g.*, sampling, analysis) which must be strictly adhered to.

Pure Reagent Water: Shall be ASTM Type I or Type II water in which no target analytes or interferences are detected as required by the analytical method.

Q...

Qualified: Status given to an entity when toe capability of fulfilling specified requirements has been demonstrated. (ISO 8402)

Qualitative (Determination or Analysis): The identification of a sample, material, compound or element without any certainty as to its mass, volume or amount. Qualitative results are generally expressed as the presence or absence of a material and are usually not accompanied by confidence statements.

Quality: The sum of features and properties/characteristics of a product or service that bear on its ability to satisfy stated or implied needs. The totality of characteristics of an entity that bear on its ability to satisfy stated and implied needs. (ISO 8402)

The consistent conformance of a product or service to a given set of standards or expectations. (ISO-9000)

Quality (assurance) Assessment: The evaluation of environmental data, comprised of data validation/verification and data quality assessment, to establish whether they meet the quality criteria needed for a specific application.

Quality Assurance (QA): An integrated system of activities involving planning, quality control, quality assessment, reporting and quality improvement to ensure that a product or service meets defined standards of quality with a stated level of confidence.

Quality Assurance Narrative Statement: A description of the quality assurance and quality control activities to be followed for a research project.

Quality Assurance Objectives: The limits on bias, precision, comparability, completeness and representativeness defining the minimal acceptable levels of performance as determined by the data user's acceptable error bounds.

Quality Assurance Project Plan (QAPP): A formal document describing the detailed quality control procedures by which the quality requirements defined for the data and decisions pertaining to a specific project are to be achieved.

Quality Audit: Systematic and independent examination to determine whether quality activities and related results comply with planned arrangements and whether these arrangements are implemented effectively and are suitable to achieve objectives. (ISO 8402)

Quality Circle: A small group of individuals from an organization or unit who have related interests and meet regularly to consider problems or other matters related to the quality of the product or process.

Quality Control (QC): The overall system of technical activities whose purpose is to measure and control the quality of a product or service so that it meets the needs of users. The aim is to provide quality that is satisfactory, adequate, dependable, and economical. Operational techniques and activities that are used to fulfil requirements for quality. (ISO 8402)

Quality Control Chart: See Control chart.

Quality Control Check Sample: See Calibration standard.

Quality Control Sample: An uncontaminated sample matrix spiked with known amounts of analytes from a source independent from the calibration standards. It is generally used to establish Intralaboratory or analyst specific precision and bias or to assess the performance of all or a portion of the measurement system. See also Check sample.

Quality Improvement: Actions taken throughout the organization to increase the effectiveness and efficiency of activities and processes in order to provide added benefits to both the organization and its customer. (ISO 8402)

Quality Loop: Conceptual model of interacting activities that influence quality at the various stages ranging from the identification of needs to the assessment of whether these needs have been satisfied. (ISO 8402)

Quality Management: All activities of the overall management function that determine the quality policy, objectives and responsibilities, and implement them by means such as quality planning, quality control, quality assurance, and quality improvement within the quality system. (ISO 8402)

Quality Management Plan (QMP): A formal document describing the management policies, objectives, principles, organizational authority, responsibilities, accountability, and implementation plan of an agency, organization or laboratory for ensuring quality in its products and utility to its users.

Quality Planning: Activities that establish the objectives and requirements for quality and for the application of quality system elements. (ISO 8402)

Quality Policy: Overall intentions and direction of an organization with regard to quality as formally expressed by top management. (ISO 8402)

Quality System: A structured and documented management system describing the policies, objectives, principles, organizational authority, responsibilities, accountability, and implementation plan of an organization for ensuring quality in its work processes, products (items), and services. The quality system provides the framework for planning, implementing, and assessing work performed by the organization and for carrying out required QA and QC. Organizational structure, procedures, processes, and resources needed to implement quality management. (ISO 8402)

Quantitation Limits: The maximum or minimum levels or quantities of a target variable that can be quantified with the confidence level required by the data user.

Quantitative (determination or analysis): The relatively accurate measurement of the amounts or percentages of one or more components of a sample or material. Depending on the QC operations performed in support of the analysis, qualitative results may be reported with or without estimates of variability.

R...

Random: Lacking a definite plan, purpose or pattern; due to chance.

Random Error: The deviation of an observed value from a true value, which behaves like a variable in that any particular value occurs as though chosen at random from a probability distribution of such errors. The distribution of random error is generally assumed to be normal.

Random Sample or Subsample: A subset of a population or a subset of a sample, selected according to the laws of chance with a randomization procedure.

T...

Test Data: Dee Data.

Test Determination: See Determination.

Test Method: See Method.

Test Portion: A subsample of the proper amount for analysis and measurement of the property of interest. A test portion may be taken from the bulk sample directly, but often preliminary operations, such as mixing or further reduction in particle size, are necessary. See Subsample.

Test Result: A product obtained from performing a test determination. See Test determination.

Test Sample: See Test portion.

Test Specimen: See Test portion.

Test Unit: See Test portion.

Time-proportioned Sample: A composite sample produced by combining samples of a specific size, collected at preselected, uniform time intervals.

Tolerance Chart: A chart in which the plotted quality control data is assessed via a tolerance level (e.g. +/-10% of a mean) based on the precision level judged acceptable to meet overall quality/data use requirements instead of a statistical acceptance criteria (e.g. +/- 3 sigma). (ANSI N42.23-1995, Measurement and Associated Instrument Quality Assurance for Radioassay Laboratories)

Total Quality Management (TQM): The process whereby an entire organization, led by senior management, commits to focusing on quality as a first priority in every activity. TQM implementation creates a culture in which everyone in the organization shares the responsibility for continuously improving the quality of products and services, (*i.e.*, for "doing the right thing, the right way, the first time, on time.") in order to satisfy the customer management approach of an organization, centered on quality, based on the participation of all its members and aiming at long-term success through customer satisfaction, and benefits to all members of the organization and to society. (ISO 8402)

Total Measurement Error: The sum of all the errors that occur from the taking of the sample through the reporting of results; the difference between the reported result.

C APPENDIX

RELIABILITY TERMS

A...

Abs The name for the **absolute value** function in many computer languages; *i.e.*, Abs(x) returns the absolute value of x.

Abscissa The x-coordinate in a two-dimensional Cartesion (rectilinear) coordinate system, normally plotted on the horizontal axis of a graph. The y-coordinate is the **ordinate.**

Absolute value The magnitude of a number, without regard to sign. Symbolised with vertical bars: $|x| = x$ if $x > 0$, $|x| = -x$ if $x < 0$.

Accelerated test Or "accelerated life test". A test or testing process that attempts to uncover latent failures by operating equipment under conditions more severe than would be encountered in the intended environment of use. This testing is "accelerated" in the sense that lesstest time is needed to detect a given number of failures or failure modes. For example, equipment might be operated under temperature extremes, vibration levels, or **duty cycles** beyond its design point. Note that accelerated testing is intended to detect normal failure modes faster, not to uncover failure modes that would not occur in normal operation.

Acceptable quality level The maximum percentage or proportion of defective items in a batch, tested by **sampling,** which can be accepted as a process average.

Acceptance number The maximum number of defective items in a batch, tested by **acceptance sampling,** that will allow the batch to be accepted without further testing. Compare **rejection number.** Acceptance number is always less than rejection number; if, for a particular sample, acceptance number < number of defects < rejection number, then further inspection and/or testing will be required to make an accept/reject decision.

Acceptance sampling Sampling for incoming batches to determine whether to accept the, according to a **sampling plan.** See also **acceptance number.**

Accessibility The ease with which operators and service personnel can access controls, components, etc., required to keep a unit operational or repair it after a failure. Normally applied to physical accessibility, but could also relate to non-physical issues such as the ease-of-use of a computerized interface to test equipment.

Active Device A discrete component whose operation involves the activity of a semiconductor junction. Some examples include integrated circuits and transistors.

Alpha (α) Greek letter alpha. May symbolize the **shape parameter** for a **Weibull distribution** (though β is more common); the **significance level** in **hypothesis testing;** or the probability of a **type I error.**

ASCII American Standard Code for Information Interchange. A universal file format for text files.

Availability The probability that a system is operational when called upon to perform its function. The numerical value of availability is expressed as a probability from 0 to 1. Availability calculation take into account both the failures and the repairs of the system.

B...

Bill of Materials (BOM) Fixed-format ASCII files where each piece of information appears between certain columns.

Burn-in The process of operating devices or equipment often under accelerated voltage, temperature, or load in order to screen out infant mortality failures.

C...

CAGE Commercial and Government Entity Code. Normally a 5 digit code.

CNET 93 A reliability standard developed by France Telecom© that provides reliability models for a wide range of components. CNET 93 is a comprehensive model similar to MIL-HDBK-217, which provides a detailed stress analysis.

Common Cause Failure Group An event or mechanism that can cause two or more failures simultaneously is called a common cause. The failures are referred to as common cause failures. Relax Fault Tree will recognize a common cause failure group of up to 4 events for any single common cause. It also enables you to create a common cause failure group file, to organize our common cause failure information.

Correlation File A file which stores the table correlating your in-house part numbers to part numbers in the Relex-supplied parts libraries.

Cost Analysis Analyzing the cost of a product or system over its entire lifetime, including design, development, and replacement cost. Relex Life Cycle Cost (LCC) enables users to perform cost analyses, using a Cost Breakdown Structure (CBS) tree, factoring in alternatives time intervals, and other variables.

Criticality A relative measure of the consequence of a particular failure mode and its frequency of occurrence. The criticality value is used to rank failure modes when performing of FMECA.

Criticality Matrix A graphical representation of the failure mode and effects, usually graphed as probability of occurrence vs. Severity level. The Relex FMEA Wizard enables users to configure the *X* and *Y* axes of the criticality matrix when setting up Project files to receive FMEA information.

D...

Design FMEA A failure modes effects analysis that studies how a system design reacts to various failure modes. The Relex FMEA software module supports design FMEAs.

Device Any electrical part (*e.g.*, integrated circuit, diode, resistor) with distinct electrical characteristics.

Distribution Parameters Each failure distribution requires the definition of distribution parameters in order to properly analyse reliability data. These parameters vary with the specific failure distributions in use, and can include the mean, standard deviation, characteristic life, shape factor, lower and upper bound, among others.

Duty Cycle Used when computing failure rates based on active and dormant settings. The duty cycle is equal to the percentage of total time or system is in the active environment.

E...

Exponential Distribution Probably the most widely known and used distribution in reliability evaluation of systems. The most important factor fir it to be applicable is that the hazard rate should be constant. Exponential distribution is frequently used for the analysis of time-dependent data when the rate at which events occur does not vary.

F...

Failure Data Analysis A type of reliability analysis used to study a variety of fields, practices, and disciplines, including mechanical, chemical, electrical, electronic, materials, and human failures. It is useful for maintenance planning, developing cost-effective replacement strategies, spare parts forecasting, and warranty analysis.

Failure Mode and Effect Analysis (FMEA) An inductive, bottoms-up method of analyzing system design and performance. At the core of a FMEA is the determination of the effects of various types of failures on a system. Various standards such as SAE J1739, AIAG Potential Failure Mode and Effects Analysis, as well as documents from Ford, Chrysler, and GM, provide instructions and formats on performing FMEAs.

Failure Mode, Effects and Criticality Analysis (FMECA) An inductive, bottoms-up method of analysing system design and performance. At the core of a FMECA is the determination of the effects of various types of failures on a system. A criticality factor, assigned to each failure mode, allows ranking the most severe failure modes and the ones most likely to occur. The analyst may then target these failure modes in order to reduce their probability of occurrence. MIL-STD-1629 provides instructions and formats for performing FMECAs.

Failure Rate The number of failures experienced or expected for a device divided by the total equipment operating time. For constant failure rate items, *i.e.* exponentially distributed failures, the failure rate is the numerical inverse of the Mean Time Between Failures (MTBF).

Fault Tree Analysis A deductive, top-down method of analyzing system design and performance. It involves specifying a top event to analyze, followed by identifying all of the associated elements in the system that could cause that top event to occur. Fault tree analyses are generally performed graphically using a logical structure of AND and OR gates. Sometimes certain elements, or basic events, may need to occur together in order for that top event to occur. In this case, these events would be arranged under an AND gate, meaning that allo the basic events would need to occur to trigger the top event. If the basic events along would trigger the top event, then they would be grouped under an OR gate. The entire system as well as human interactions would be analyzed when performing a fault tree analysis.

First-Year Multiplier Ratio of the first-year failure rate to the steady-state failure rate.

FITS (Also known as failures in time). Describes the failure rate when the units are in failures per billion hours.

FRACAS (Also known as Failure Reporting, Analysis, and Corrective Action System.) A closed-loop system that consists of identifying failures, analyzing the causes of those failures by performing a failure analysis, deciding corrective actions, and implementing those corrective actions back into the design and/or manufacturing process. A good FRACAS system is essential to healthy reliability improvement.

G...

Gumbel-(Lower) An extreme minimum value failure distribution, similar to the Weibull distribution. The Gumbel distribution is related to Weibull since Gumbel is a Type I lower extreme value and Weibull is a Type III (lower).

Gumbel+ (Upper) An extreme maximum value failure distribution. Gumbel + (Upper) i a Type I upper extreme value.

H...

HRD5 A reliability standard developed by British Telecommunications plc© that also provides models for a wide range of components. In general, HRD5 is similar to CNET 93, but provides simpler models and requires fewer data parameters for analysis.

Hybrid A component in a relatively large package that contains other parts mounted on a substrate.

J...

Junction An object used in reliability block diagrams to join multiple connectors together such as in parallel or redundant configurations. Junctions can be set in series, parallel operating, or standby operation.

L...

Lambda-Tau Calculation A calculation method used to determine availability or short-term reliability when the age of a component or system is unknown or indeterminate.

Life Cycle Cost (LCC) A method of analyzing the cost of a piece of equipment over its entire life, including development costs, production costs, warranty costs, repair costs, and disposal costs. Relex Life Cycle Cost (LCC) enables users to perform cost analysis, using a Cost Breakdown Structure (CBS) tree, factoring in alternatives, time intervals, and other variables.

Lognormal Distribution A failure distribution similar to the Normal distribution except that the logarithm of the values of random variables. Rather than the values themselves, are assumed to the normally distributed. Thus all values are positive and the distribution is skewed to the left.

M...

Macros Programs used to automate tedious or time-consuming tasks. Relex Reliability Studio provides complete support for macros, including a macro editor, and ready-made macros to use as reference when creating custom macros.

Maintainability Prediction The measure of the ability of an item to be retained in or restored to a specified condition when skilled personnel perform the maintenance. The Relex Maintainability Prediction module enables you to generate numerous maintainability parameters, including mean time to repair (MTTR), mean maintenance man-hours per repair (MMH/Repair), and many others.

Mean Time Between Failures (MTBF) The mean time expected between failures, typically measured in hours. MTBF is a statistical value and is meant to be the mean over a long period of time and large number of units. For constant failure rate systems, MTBF is the inverse of the failure rate. If failure rate is in failures/million hours, MTBF = 1,000,000/ Failure Rate for components with exponential distributions Technically MTBF should be used only in reference to repairable items, while MTTF (Mean Time to Failure) should be used for non-repairable items, but MTBF is commonly used for both repairable and non-repairable items.

Mean Time to Failure (MTTF) The mean time expected to the first failure of a piece of equipment. It is a statistical value and is meant to be the mean over a long period of time and large number of units. For constant failure rate systems, MTTF is the inverse of the failure rate. If failure rate is in failures/million hours, MTTF = 1,000,000/Failure Rate for components with exponential distributions. Technically MTBF should be used only in reference to repairable items, while MTTF should be used for non-repairable items, but MTBF is commonly used for both repairable and non-repairable items.

Mean Time to Repair (MTTR) The total amount of time spent performing all corrective maintenance repairs divided by the total number of those repairs.

Measures of Reliability When performing a reliability prediction analysis, there are several calculated values that provide measures of reliability. These values include failure rate, mean time between failure (MTBF), reliability and availability.

MIL-HDBK-217 A reliability prediction standard originally developed for defense and aerospace related organizations, but later adopted by many commercial and industrial companies. Many times referred to simply as 217, MIL-HDBK-217 includes mathematical reliability models for nearly all types of electrical and electronic components. These reliability models are based on parameters of the components such as number of pins, number of transistors, power dissipation, and environmental factors. Results from MIL-HDBK-217 are provided as both a failure rate and as an MTBF (Mean Time Between Failures) where the MTBF is the mathematical inverse of the failure rate.

MIL-HDBK-472 The accepted industry standard for maintainability prediction analysis.

MIL-STD-1629 A long recognised standard that describes a method of performing Failure Mode, Effects, and Criticality Analyses. It is used by government, military, and commercial organizations worldwide. MIL-1629 provides calculations for criticality that allows ranking of failure modes dependent on severity classification. The Relex FMAE/FMECA software module supports the calculations included in MIL-STD-1629.

Modes File Contains the associated potential modes of failure and the percentage of time that each mode of failure occurs for a specified part type. It provides contralized storage for all failure modes of all part types. Modes files are typically used for piece-part FMEAs.

Monte Carlo Simulation A simulation method used by Relex RBD when calculating complex reliability block diagrams. A Monte Carlo Simulation will perform random tests on the specified system in order to provide an analysis of the overall reliability and availability of the system. The result of these simulations will determine the reliability and availability of the RBD model.

N...

Net Present Value (Also known as NPV.) The cost of a product or system calculated in the preset-day currency. Relex Life Cycle Cost RLCC) enables NPV calculations, to determine if a particular cost would be better incurred in the present year, or at a future date when costs may be lower.

Normal Distribution An important and widely-used distribution in the field of statistics and probability. All Normal distributions are symmetric, and the mean and standard deviation values are used as its two distribution parameters.

O...

Operating Environment Environment of the selected item in the Relex hierarchical display (tree). When the operating environment is selected for the top level of the tree, it serves as the environment for all levels below it. If an environment is selected in the Properties of a lower level item, that selection overrides the environment at the higher level.

OrCAD Provides digital and analog design, circuit simulation, board layout and other engineering software tools.

P...

Parallel Operating Parallel Operating describes a redundancy of several items performing the same task, but all redundant systems or components are fully operational at all times. In a parallel operating environment, not all units are required for dsuccessful operation.

Passive Device A discrete component that is not active, where an active component is one whose operation involves the activity of a semiconductor junction. Resistors and capacitors are examples of passive devices.

Pi Factors Variables used in reliability prediction equations.

Probabilistic Risk Assessment (Also known as PRA.) Provides a systematic approach for transforming failures, termed initiating events, into risk profiles. A PSA, Probability Safety Assessment, generally deals with issues that are safety related, while a PRA may deal with other non-safety issues. The results of the PRA or PDA provide the resulting probabilities of each risk.

Process FMEA A Failure Modes and Effects Analysis that studies how failure in the manufacturing or service process affects the system operation.

R...

Rayleigh Distribution A special case of the Weibull failure distribution where the Beta (slope) value is known to equal 2.

Redundancy The existence of more than one piece of equipment, any of which could perform a given function. These multiple pieces of equipment are used to help improve the reliability and availability of the system. Redundancy is generally expressed as the number of pieces of equipment required and the total number available. For example, 2 out of 3 means that there are a total of 3 pieces of equipment and 2 pieces are required for proper operation of the system.

Reliability The ability of an item to perform a required function under stated conditions for a stated period of time. The numerical value of reliability is expressed as a probability from 0 to 1 and is also sometimes known as the probability of mission success. Reliability is the probability, assuming the system was operating at time zero, that it continues to operate until time t.

Reliability Allocations performed early in a design to allocate a given failure rate or MTBF goal across the equipment in the system. Various weights are assigned to each piece of equipment during the reliability allocations process.

Reliability Analysis A method of analyzing a system design evaluating reliability-related issues. The goal is to improve the field reliability of the system.

Reliability Block Diagram (RBD) A visual representation of a system that contains redundancy in its elements. Redundancy is used for mission critical functions where a single-point failure is not acceptable and reliability needs to be improved. Generally represented in a graphical manner, a simple RBD model may be calculated using analytical solutions, while a more complex RBD may be calculated using a Monte Carlo simulation. Results may include reliability, availability, failure rate, and MTBF.

Reliability Calculations Performed in order to quantify the reliability of a peace of hardware or software prior to and during development and production.

Reliability Centered Maintenance (RCM)

Reliability Prediction A primary component of reliability analysis is referred to as the failure rate, or the number of failures expected during a certain period of time. Calculation of equipment failure rate, and the related MTBF (Mean Time Between Failure), is the basis of performing a reliability prediction analysis.

Reliability Software A general term that may be used to describe software that performs reliability predictions per MIL-HDBK-217 and Bellcore, FMEA and FMECA analysis, reliability block diagram Reliability software packages may also be used in FRACAS systems, life cycle cost analysis, and maintainability predictions.

Risk Priority Number (RPN) Used in FMEA analysis to rank the importance of different types of failure. RPN = Severity × Occurrence × Detection.

S...

SAE Society of Automotive Engineers. Organization that developed the SAE ARP5580 that outlines a general approach for FMEA analysis. The Relex FMEA/FMECA software module supports the SAE ARP 5580 FMEA/ FMECA standards.

Sensitivity Analysis Used in life cycle cost (LCC) analysis to determine how robust a particular alternative is, and how sensitive the overall cost is to certain variables.

Series Series describes a redundancy of several items performing the same task, but all redundant systems or components are full operational at all times and are required for successful operation.

Single Point Failure Used to describe an item that, if failed, would cause a failure of the system.

Software Reliability A method of analyzing and predicting the reliability of a piece of software. There many methods available to quantify the reliability of software. Software reliability is a relatively new field of study, compared to hardware reliability.

SPICE (Also known as Simulation Program with Integrated Circuit Emphasis.) A circuit simulation technology from Berkeley, California. SPICE has been developed into software products for sale from software manufacturers such as Intusoft and OrCAD among others.

Standby Nonoperating A type of redundancy recognized by Relex in the Relex RBD software module. Standby Nonoperating describes a redundancy of several items performing the same task, but all redundant systems, or components are not fully operational at all times. Only those that are required are in an "on" state. In a Standby Nonoperating environment, not all units are required for successful operation.

Steady-State Failure Rate The constant failure rate after one year of operation.

T...

Tasks, Basic A simple, one-step procedure. When tasks are created and stored for use in a maintainability prediction, they generally include a description and an associated time along with other fields that may be available for further definition of the task.

Tasks, Super (Also known as a maintenance action). A collection of basic tasks which have been combined together to represent a larger task. For example, replacing a video card could be classified as a super task. It would include several of the following basic tasks: removal of screw, removal of card, replacement of card and replacement of screw. The Relex Maintainability Prediction software module includes a Tasks file of standard maintenance tasks based on the tasks and repair times from MIL-HDBK-471. It also allows the creation of user-defined Tasks files.

Telcordia The latest version, "Reliability Prediction Procedure for Electronic Equipment-A Telcordia Technologies Special Report," SR-332, Issue 1, May 2001, updates the calculation models, and serves as the update to Bellcore Issue 6.

Types of Redundancies There are generally three different types of configurations that would be used in system design. These configuration types are: Series, Parallel Operating, and Standby Non-Operating.

U...

Uniform Distribution The simplest of all failure distributions. It can be interpreted as meaning that the random variable can only exist in the range *A* to *B* (lower to upper) and that the probability of occurrence is directly proportional to the interval length.

Unit An assembly of devices (e.g., circuit pack, module, plug-in, racks, and power supplies).

User-Defined Values Files Allows user to create custom environments, quality levels, categories and subcategories. Once a User-Defined Values file is enabledin a Project file, the environments, quality levels, categories and subcategories defined in the User-Defined Values file can be used within the Project file.

W...

WeiBayes A special case of the Weibull failure distribution where the Beta (slope) value can be defined by the user.

Weibull Distribution A failure distribution that is very useful in reliability activities because it can be used to model many other life distributions. By adjusting the beta factor, or shape parameter, of the Weibull distribution, it can be made to model a decreasing, constant, or increasing hazard rate. The Weibull distribution provides reasonably accurate failure analysis and failure forecasts with extremely small samples.

D APPENDIX

MAINTENANCE TERMS

Actuarial analysis – Statistical analysis of failure data to determine the age-reliability characteristics of an item.

Applications Parts List – A list of all parts required to perform a specific maintenance activity. Typically set up as a standard list attached to a Standard Job for Routine Tasks. Not to be confused with a Bill of Materials.

Apprentice – a tradesperson (or craftsperson) in training

Asset – unlike in the accounting definition, in maintenance this is commonly taken to be any item of physical plant or equipment.

Asset Management – the systematic planning and control of a physical resource throughout its life. This may include the specification, design, and construction of the asset, its operation, maintenance and modification while in use, and its disposal when no longer required.

Asset Register – a list of all the Assets in a particular workplace, together with information about those assets, such as manufacturer, vendor, make, model, specifications etc.

Availability – the proportion of total time that an item of equipment is capable of performing its specified functions, normally expressed as a percentage. It can be calculated by dividing the equipment available hours by the total number of hours in any given period. One of the major sources of disagreement over the definition of availability is whether downtime should be divided by total hours, or by Scheduled Operating Time. For example, if your plant is only scheduled to operate 5 days a week, should downtime incurred over the weekend be included in the calculation of availability? The view I take is that one of the prime goals of any organization should be to maximize its Return on Assets. This can only be achieved by reducing the total downtime, regardless of whether this downtime was scheduled or not. For this reason, I prefer to use a definition of downtime that considers **all** downtime, as a proportion of total time, not scheduled operating time.

Available Hours – the total number of hours that an item of equipment is capable of performing its specified functions. It is equal to the total hours in any given period, less the downtime hours.

Average Life – how long, on average, a component will last before it suffers a failure. Commonly measured by Mean Time Between Failures.

Backlog – Work which has not been completed by the nominated 'required by date'. The period for which each Work Order is overdue is defined as the difference between the current date and the 'required by date'. All work for which no 'required by' date has been specified is generally included on the backlog. Backlog is generally measured in "crew-weeks", that is, the total number of labor hours represented by the work on the backlog, divided by the number of labor hours available to be worked in an average week by the work crew responsible for completing this work. As such, it is one of the common Key Performance Indicators used in maintenance.

Benchmarking – the process of comparing performance with other organizations, identifying comparatively high performance organizations, and learning what it is they do that allows them to achieve that high level of performance.

Bill of Materials – a list of all the parts and components that make up a particular asset. Not to be confused with an Applications Parts List.

Breakdown – a specific type of failure, where an item of plant or equipment is completely unable to function.

Call-out – To summon a tradesperson to the workplace during his normal non-working time so that he can perform a maintenance activity (normally an emergency maintenance task)

Component – a subassembly of an Asset, usually removable in one piece and interchangeable with other, standard components (eg. Truck engine).

Computerized Maintenance Management System – a computerized system to assist with the effective and efficient management of maintenance activities through the application of computer technology. It generally includes elements such as a computerized Work Order system, as well as facilities for scheduling Routine Maintenance Tasks, and recording and storing Standard Jobs, Bills of Materials and Applications Parts Lists, as well as numerous other features.

Condition Based Maintenance – an equipment maintenance strategy based on measuring the condition of equipment in order to assess whether it will fail during some future period, and then taking appropriate action to avoid the consequences of that failure. The condition of equipment could be monitored using Condition Monitoring, Statistical Process Control techniques, by monitoring equipment performance, or through the use of the Human Senses. The terms Condition Based Maintenance, On-Condition Maintenance and Predictive Maintenance can be used interchangeably.

Condition Monitoring – the use of specialist equipment to measure the condition of equipment. Vibration Analysis, Tribology and Thermography are all examples of Condition Monitoring techniques.

Conditional Probability of Failure – The probability that an item will fail during a particular age interval, given that it survives to enter that age.

Contract Acceptance Sheet – A document that is completed by the appropriate Contract Supervisor and Contractor to indicate job completion and acceptance. It also forms part of the appraisal of the contractors performance.

Corrective Maintenance – Any maintenance activity which is required to correct a failure that has occurred or is in the process of occurring. This activity may consist of repair, restoration or replacement of components.

Craftsperson – Alternative to Tradesperson. A skilled maintenance worker who has typically been formally trained through an apprenticeship program.

Criticality – The priority rank of a failure mode based on some assessment criteria.

Defect – A term typically used in the maintenance of mobile equipment. A defect is typically a potential failure or other condition that will require maintenance attention at some time in the future, but which is not currently preventing the equipment from fulfilling its functions.

Discard task – The removal and disposal of items or parts.

Downtime – the time that an item of equipment is out of service, as a result of equipment failure. The time that an item of equipment is available, but not utilized is generally not included in the calculation of downtime.

Economic Life – the total length of time that an asset is expected to remain actively in service before it is expected that it would be cheaper to replace the equipment rather than continuing to maintain it. In practice, equipment is more often replaced for other reasons, including: because it no longer meets operational requirements for efficiency, product quality, comfort etc., or because newer equipment can provide the same quality and quantity of output more efficiently.

Emergency Maintenance Task – a maintenance task carried out in order to avert an immediate safety or environmental hazard, or to correct a failure with significant economic impact.

Engineering Work Order – the prime document used to initiate an engineering investigation, engineering design activity or engineering modifications to an item of equipment.

Environmental Consequences – a failure has environmental consequences if it could cause a breach of any known environmental standard or regulation.

Equipment Life – this term often isn't very useful, in a practical sense. For example, if I was to tell you that my Aunt has an axe that she uses for chopping firewood, and in the last 40 years it has had 2 new axeheads and 5 new handles, how would you define the life of the axe? Perhaps it makes more sense to talk about Component Life. Also see Economic Life, Useful Life and Average Life for some more practical definitions.

Equipment Maintenance Strategies – the choice of routine maintenance tasks and the timing of those tasks, designed to ensure that an item of equipment continues to fulfil its intended functions.

Estimated Plant Replacement Value – the estimated cost of capital works required to replace all the existing assets with new assets capable of producing the same quantity and quality of output. This is a key value often used in benchmarking activities.

Estimating Index – the ratio of Estimated Labor Hours required to complete the work specified on Work Orders to the Actual Labor Hours required to complete the work specified on those Work Orders, commonly expressed as a percentage. This is a commonly used measure of Labor productivity, particularly when there are well-defined Estimating standards. A figure of greater than 100% for the Estimating Index indicates a higher than standard level of productivity, while a figure of less than 100% indicates a lower than standard level of productivity.

Expert System – a software based system which makes or evaluates decisions based on rules established within the software. Typically used for fault diagnosis.

Fail-safe – an item is fail-safe if, when the item itself incurs a failure, that failure becomes apparent to the operating workforce in the normal course of events.

Failure – an item of equipment has suffered a failure when it is no longer capable of fulfilling one or more of its intended functions. Note that an item does not need to be completely unable to function to have suffered a failure. For example, a pump that is still operating, but is not capable of pumping the required flow rate, has failed. In Reliability Centered Maintenance terminology, a failure is often called a Functional Failure. Would you classify a planned equipment shutdown as a failure? Would you classify a routine equipment shutdown at shift change as a failure? Under this definition, the answer in the first case would be yes, but in the second case would be no. The justification for the inclusion of planned shutdowns as failures is that a failure, as defined, causes a disruption to the desired steady-state nature of the production process, and therefore should, ideally, be avoided.

Failure Code – a code typically entered against a Work Order in a CMMS which indicates the cause of failure (eg. lack of lubrication, metal fatigue etc.)

Failure Consequences – a term used in Reliability Centered Maintenance. The consequences of all failures can be classified as being either Hidden, Safety, Environmental, Operational, or Non-Operational.

Failure Effect – a description of the events that occur after a failure has occurred as a result of a specific Failure Mode. Used in Reliability Centered Maintenance, FMEA and FMECA analyses.

Failure Finding Interval – the frequency with which a Failure Finding Task is performed. Is determined by the frequency of failure of the Protective Device, and the desired availability required of that Protective Device.

Failure Finding Task – Used in Reliability Centered Maintenance terminology. A routine maintenance task, normally an inspection or a testing task, designed to determine, for Hidden Failures, whether an item or component has failed. A failure finding task should not be confused with an On-Condition Task, which is intended to determine whether an item is **about** to fail. Failure Finding tasks are sometimes referred to as Functional Tests.

Failure Mode – any event which causes a failure.

Failure Modes, Effects and Criticality Analysis – a structured method of assessing the causes of failures and their effect on production, safety, cost, quality etc.

Failure Modes and Effects Analysis – a structured method of determining equipment functions, functional failures, assessing the causes of failures and their failure effects. The first part of a Reliability Centered Maintenance analysis is a Failure Modes and Effects Analysis.

Failure Pattern – the relationship between the Conditional Probability of Failure of an item, and its age. Failure patterns are generally applied to Failure Modes. Research in the airline industry established that there are six distinct failure patterns. The type of failure pattern that applies to any given failure mode is of vital importance in determining the most appropriate equipment maintenance strategy. This fact is one of the key principles underlying Reliability Centered Maintenance.

Forward Workload – All known backlog work and work which is due or predicted to become backlog work within a pre-specified future time period.

Function – The definition of what we want an item of equipment to do, and the level of performance which the users of the equipment require when it does it. Note that an item of equipment can have many functions, commonly split into Primary and Secondary Functions. Note also that the level of performance specified is that required by the users of the equipment, which may be quite different to the original design, or maximum, performance capability for the equipment.

Functional Failure – Used in Reliability Centered Maintenance terminology. The inability of an item of equipment to fulfil one or more of its functions. Interchangeably used with Failure.

Gantt Chart – A bar chart format of scheduled activities showing the duration and sequencing of activities.

Go-line – Used in relation to mobile equipment. Equipment which is available, but not being utilized is typically parked on the Go-line. This term is used interchangeably with Ready Line.

Hazop – a structured process, originally developed by ICI following the Flixborough disaster, intended to proactively identify equipment modifications and/or safety devices required in order to avoid any significant safety or environmental incident as a result of equipment failure. Similar, in some respects to Reliability Centered Maintenance, but not as rigorous as Reliability Centered Maintenance in identifying underlying causes of failure, and does not consider, in any depth, the possibility of avoiding such incidents through applying appropriate Proactive Maintenance tasks.

Hidden Failure – a failure which, on its own, does not become evident to the operating crew under normal circumstances. Typically, protective devices which are not fail-safe (examples could include standby plant and equipment, emergency systems etc.)

Infant Mortality – The relatively high conditional probability of failure during the period immediately after an item returns to service.

Inherent Reliability – A measure of the reliability of an item, in its present operating context, assuming adherence to ideal equipment maintenance strategies.

Inspection – Any task undertaken to determine the condition of equipment, and/or to determine the tools, labour, materials, and equipment required to repair the item.

Key Performance Indicators – A select number of key measures that enable performance against targets to be monitored.

Life – that strange experience you have all day, every day. In a maintenance context, you may want to look at Equipment Life.

Life Cycle Costing – a process of estimating and assessing the total costs of ownership, operation and maintenance of an item of equipment during its projected equipment life. Typically used in comparing alternative equipment design or purchase options in order to select the most appropriate option.

Logistic support analysis (LSA) – A methodology for determining the type and quantity of logistic support required for a system over its entire lifecycle. Used to determine the cost effectiveness of asset based solutions.

Maintainability – the ease and speed with which any maintenance activity can be carried out on an item of equipment. May be measured by Mean Time to Repair. Is a function of equipment design, and maintenance task design (including use of appropriate tools, jigs, work platforms etc.).

Maintainability Engineering – The set of technical processes that apply maintainability theory to establish system maintainability requirements, allocate these requirements down to system elements and predict and verify system maintainability performance.

Maintenance – any activity carried out on an asset in order to ensure that the asset continues to perform its intended functions, or to repair the equipment. Note that modifications are not maintenance, even though they may be carried out by maintenance personnel.

Maintenance Engineering – a staff function whose prime responsibility is to ensure that maintenance techniques are effective, that equipment is designed and modified to improve maintainability, that ongoing maintenance technical problems are investigated, and appropriate corrective and improvement actions are taken. Used interchangeably with Plant Engineering and Reliability Engineering.

Maintenance Policy – a statement of principle used to guide Maintenance Management decision making.

Maintenance Schedule – a list of planned maintenance tasks to be performed during a given time period, together with the expected start times and durations of each of these tasks. Schedules can apply to different time periods (i.e. Daily Schedule, Weekly Schedule etc.)

Maintenance Strategy – a long-term plan, covering all aspects of maintenance management which sets the direction for maintenance management, and contains firm action plans for achieving a desired future state for the maintenance function.

Mean Time Between Failures – a measure of equipment reliability. Equal to the number of failures in a given time period, divided by the total equipment uptime in that period.

Mean Time To Repair – a measure of maintainability. Equal to the total equipment downtime in a given time period, divided by the number of failures in that period.

Model Work Order – A Work Order stored in the CMMS which contains all the necessary information required to perform a maintenance task. (see also Standard Job)

Modification – any activity carried out on an asset which increases the capability of that asset to perform its required functions.

No Scheduled Maintenance – an Equipment Maintenance Strategy, where no routine maintenance tasks are performed on the equipment. The only maintenance performed on the equipment is Corrective Maintenance, and then only after the equipment has suffered a failure. Also described as a Run-to-Failure strategy.

Non-Destructive Testing – testing of equipment, which does not destroy the equipment, to detect abnormalities in physical, chemical or electrical characteristics. For some reason which escapes me, vibration analysis and tribology are not generally considered to be NDT techniques, even though they meet the above criteria. Techniques which **are** considered to be NDT techniques are ultrasonic thickness testing, dye penetrant testing, x-raying, and electrical resistance testing.

Non-Operational Consequences – a failure has non-operational consequences if the only impact of the failure is the direct cost of the repair (plus any secondary damage caused to other equipment as a result of the failure.

Non-routine Maintenance – Any maintenance task which is not performed at a regular, pre-determined frequency.

Operating Context – the operational situation within which an asset operates. For example, is it a stand-alone piece of plant, or is it one of a duty-standby pair? Is it part of a batch manufacturing process or a continuous production process? What is the impact of failure of this item of equipment on the remainder of the production process? The operating context has enormous influence over the choice of appropriate equipment maintenance strategies for any asset.

Operating Hours – the length of time that an item of equipment is actually operating.

Operational Consequences – a failure has operational consequences if it has a direct adverse impact on operational capability (lost production, increased production costs, loss of product quality, or reduced customer service)

Operational Efficiency – used in the calculation of Overall Equipment Effectiveness. The actual output produced from an asset in a given time period divided by the output that would have been produced from that asset in that period, had it produced at its rated capacity. Normally expressed as a percentage.

Outage – a term used in some industries (notably power generation) which is equivalent to a shutdown.

Overall Equipment Effectiveness – a term initially coined in connection with Total Productive Maintenance. It provides a measure of overall asset productivity. Is generally expressed as a percentage, and can be calculated by multiplying Availability by Utilization by Operational Efficiency by Quality Rate.

Overhaul – a comprehensive examination and restoration of an asset to an acceptable condition.

P-F Interval – a term used in Reliability Centered Maintenance. The time from when a Potential Failure can first be detected on an asset or component using a selected Predictive Maintenance task, until the asset or component has failed. Reliability Centered Maintenance principles state that the frequency with which a Predictive Maintenance task should be performed is determined solely by the P-F Interval.

Percent Planned Work – the percentage of total work (in labor hours) performed in a given time period which has been planned in advance.

Planned Maintenance – any maintenance activity for which a pre-determined job procedure has been documented, for which all labour, materials, tools, and equipment required to carry out the task have been estimated, and their availability assured before commencement of the task.

Plant Engineering – a staff function whose prime responsibility is to ensure that maintenance techniques are effective, that equipment is designed and modified to improve maintainability, that ongoing maintenance technical problems are investigated, and appropriate corrective and improvement actions are taken. Used interchangeably with Maintenance Engineering and Reliability Engineering.

Potential Failure – a term used in Reliability Centered Maintenance. An identifiable condition which indicates that a functional failure is either about to occur, or in the process of occurring.

Predictive Maintenance – an equipment maintenance strategy based on measuring the condition of equipment in order to assess whether it will fail during some future period, and then taking appropriate action to avoid the consequences of that failure. The condition of equipment could be monitored using Condition Monitoring, Statistical Process Control techniques, by monitoring equipment performance, or through the use of the Human Senses. The terms Condition Based Maintenance, On-Condition Maintenance and Predictive Maintenance can be used interchangeably.

Preventive Maintenance – an equipment maintenance strategy based on replacing, overhauling or remanufacturing an item at a fixed interval, regardless of its condition at the time. Scheduled Restoration tasks and Scheduled Discard tasks are both examples of Preventive Maintenance tasks.

Primary Function – a term used in Reliability Centered Maintenance. The primary functionality required of an asset - the reason the asset was acquired. For example it is likely that the primary function of a pump is to pump a specified liquid at a specified rate against a specified head of pressure.

Priority – the relative importance of a task in relation to other tasks. Used in scheduling work orders.

Proactive Maintenance – Any tasks used to predict or prevent equipment failures.

Probabalistic Risk Assessment – A "top-down" approach used to apportion risk to individual areas of plant and equipment, and possibly to individual assets so as to achieve an overall target level of risk for a plant, site or organization. These levels of risk are then used in risk-based techniques, such as Reliability Centered Maintenance and Hazop, to assist in the development of appropriate equipment maintenance strategies, and to identify required equipment modifications.

Probabalistic Safety Assessment – Similar to Probabalistic Risk Assessment, except focused solely on Safety related risks.

Project Evaluation & Review Technique (PERT) Chart – Scheduling tool which shows in flow chart format the interdependencies between project activities.

Protective Device – Devices and assets intended to eliminate or reduce the consequences of equipment failure. Some examples include standby plant and equipment, emergency systems, safety valves, alarms, trip devices, and guards.

Purchase Requisition – The prime document raised by user departments authorizing the purchase of specific materials, parts, supplies, equipment or services from external suppliers.

Purchase Order – The prime document raised by an organization, and issued to an external supplier, ordering specific materials, parts, supplies, equipment or services.

Quality Rate – used in the calculation of Overall Equipment Effectiveness. The proportion of the output from a machine or process which meets required product quality standards. Normally specified as a percentage.

Ready Line – Used in relation to mobile equipment. Equipment which is available, but not being utilized is typically parked on the Ready Line. This term is used interchangeably with Go-Line.

Redesign – a term which, in Reliability Centered Maintenance, means any one-off intervention to enhance the capability of a piece of equipment, a job procedure, a management system or people's skills.

Reliability – the capability of an asset to continue to perform its intended functions. Normally measured by Mean Time Between Failures.

Reliability Centered Maintenance – A structured process, originally developed in the airline industry, but now commonly used in all industries to determine the equipment maintenance strategies required for any physical asset to ensure that it continues to fulfill its intended functions in its present operating context. A number of books have been written on the subject, but none better than Moubray's book, RCM II.

Reliability Engineering – a staff function whose prime responsibility is to ensure that maintenance techniques are effective, that equipment is designed and modified to improve maintainability, that ongoing maintenance technical problems are investigated, and appropriate corrective and improvement actions are taken. Used interchangeably with Plant Engineering and Maintenance Engineering.

Repair – any activity which returns the capability of an asset that has failed to a level of performance equal to, or greater than, that specified by its Functions, but not greater than its original maximum capability. An activity which increases the maximum capability of an asset is a modification.

Restoration – any activity which returns the capability of an asset that has **not** failed to a level of performance equal to, or greater than, that specified by its Functions, but not greater than its original maximum capability. Not to be confused with a modification or a repair.

Return on Assets – an accounting term. Let's not get into a lengthy discussion of the relative merits of various accounting standards, how assets should be valued (book value, replacement value, depreciation rates and methods etc.), and differences between tangible and intangible assets. This is the stuff that accountants have wet dreams over, but not maintenance engineers. In practical terms, as it impacts on maintenance, Return on Assets is the profit attributable to a particular plant or factory, divided by the amount of money invested in plant and equipment at that plant or factory. It is normally expressed as a percentage. As such, it is roughly equivalent (in principle - please excuse the pun!) to the interest rate that you get on money invested in the bank, except that in this case the money is invested in plant and equipment.

Risk – The potential for the realization of the unwanted, negative consequences of an event. The product of conditional probability of an event, and the event outcomes.

Rotable – a term often used in the maintenance of heavy mobile equipment. A rotable component is one which, when it has failed, or is about to fail, is removed from the asset and a replacement component is installed in its place. The component that has been removed is then repaired or restored, and placed back in the maintenance store or warehouse, ready for re-issue.

Routine Maintenance Task – any maintenance task that is performed at a regular, predefined interval.

Run-to-Failure – No Scheduled Maintenance - an Equipment Maintenance Strategy, where no routine maintenance tasks are performed on the equipment. The only maintenance performed on the equipment is Corrective Maintenance, and then only after the equipment has suffered a failure. Also described as a No Scheduled Maintenance strategy.

Safety Consequences – a failure has safety consequences if it causes a loss of function or other damage that could hurt or kill someone.

Schedule Compliance – one of the Key Performance Indicators often used to monitor and control maintenance. It is defined as the number of Scheduled Work Orders completed in a given time period (normally one week), divided by the total number of Scheduled Work Orders that should have been completed during that period, according to the approved Maintenance Schedule for that period. It is normally expressed as a percentage, and will always be less than or equal to 100%. The closer to 100%, the better the performance for that time period.

Scheduled Maintenance – any maintenance work that has been planned and included on an approved Maintenance Schedule.

Scheduled Discard Task – a maintenance task to replace a component with a new component at a specified, pre-determined frequency, regardless of the condition of the component at the time of its replacement. An example would be the routine replacement of the oil filter on a motor vehicle every 6,000 miles. The frequency with which a Scheduled Discard task should be performed is determined by the Useful Life of the component.

Scheduled Operating Time – the time during which an asset is scheduled to be operating, according to a long-term production schedule.

Scheduled Restoration Task – a maintenance task to restore a component at a specified, pre-determined frequency, regardless of the condition of the component at the time of its replacement. An example would be the routine overhaul of a slurry pump every 1,000 operating hours. The frequency with which a Scheduled Restoration task should be performed is determined by the Useful Life of the component.

Scheduled Work Order – a Work Order that has been planned and included on an approved Maintenance Schedule.

Secondary Damage – Any additional damage to equipment, above and beyond the initial failure mode, that occurs as a direct consequence of the initial failure mode.

Secondary Function – a term used in Reliability Centered Maintenance. The secondary functionality required of an asset - generally not associated with the reason for acquiring the asset, but now that the asset has been acquired, the asset is now required to provide this functionality. For example a secondary function of a pump may be to ensure that all of the liquid that is pumped is contained within the pump (ie. the pump doesn't leak). An asset may have tens or hundreds of secondary functions associated with it.

Shutdown – that period of time when equipment is out of service.

Shutdown Maintenance – Maintenance that can only be performed while equipment is shutdown.

Standard Job – A Work Order stored in the CMMS which contains all the necessary information required to perform a maintenance task. (see also Model Work Order)

Standing Work Order – a work order that is left open either indefinitely or for a pre-determined period of time for the purpose of collecting labor hours. costs and/or history for tasks for which it has been decided that individual work orders should not be raised. Examples would include Standing Work Orders raised to collect time spent at Safety Meetings, or in general housekeeping activities.

Stores Issue – the issue and/or delivery of parts and materials from the store or warehouse.

Stores Requisition – The prime document raised by user departments authorizing the issue of specific materials, parts, supplies or equipment from the store or warehouse.

Terotechnology – the application of managerial, financial, engineering and other skills to extend the operational life of, and increase the efficiency of, equipment and machinery.

Thermography – the process of monitoring the condition of equipment through the measurement and analysis of heat. Typically conducted through the use of infra-red cameras and associated software. Commonly used for monitoring the condition of high voltage insulators and electrical connections, as well as for monitoring the condition of refractory in furnaces and boilers, amongst other applications.

Total Asset Management – an integrated approach (yet to be developed!) to Asset Management which incorporates elements such as Reliability Centered Maintenance, Total Productive Maintenance, Design for Maintainability, Design for Reliability, Value Engineering, Life Cycle Costing, Probabalistic Risk Assessment and others, to arrive at the optimum Cost-Benefit-Risk asset solution to meet any given production requirements.

Tradesperson – Alternative to Craftsperson. A skilled maintenance worker who has typically been formally trained through an apprenticeship program.

Tribology – the process of monitoring the condition of equipment through the analysis of properties of its lubricating and other oils. Typically conducted through the measurement of particulates in the oil, or the measurement of the chemical composition of the oil (Spectographic Oil Analysis). Commonly used for monitoring the condition of large gearboxes, engines and transformers, amongst other applications.

Total Productive Maintenance – a company-wide equipment management program, with its origins in Japan, emphasizing production operator involvement in equipment maintenance, and continuous improvement approaches. Numerous books have been written on the subject, including Nakajima's authoritative introduction, and a more recent Western hemisphere update by Willmott.

Total System Support (ToSS) – The composite of all considerations needed to assure the effective and economical support of a system throughout its programmed life-cycle.

Unplanned Maintenance – any maintenance activity for which a pre-determined job procedure has **not** been documented, or for which all labor, materials, tools, and equipment required to carry out the task have been **not** been estimated, and their availability assured before commencement of the task.

Unscheduled Maintenance – any maintenance work that has **not** been included on an approved Maintenance Schedule prior to its commencement.

Uptime – strangely enough, the opposite of downtime. It is defined as being the time that an item of equipment is in service and operating.

Useful Life – the maximum length of time that a component can be left in service, before it will start to experience a rapidly increasing probability of failure. The Useful Life determines the frequency with which a Scheduled Restoration or a Scheduled Discard task should be performed. Note that for the concept of the Useful Life of a component to hold true, components must, at some consistent point in time, experience a rapidly increasing probability of failure. Research in the airline industry showed that, in this industry at least, this was only true for 11% of the components in modern aircraft.

Utilization – the proportion of available time that an item of equipment is operating. Calculated by dividing equipment operating hours by equipment available hours. Generally expressed as a percentage.

Value Engineering – a systematic approach to assessing and analyzing the user's requirements of a new asset, and ensuring that those requirements are met, but not exceeded. Consists primarily of eliminating perceived "non-value-adding" features of new equipment.

Vibration Analysis – the process of monitoring the condition of equipment, and the diagnosis of faults in equipment through the measurement and analysis of vibration within that equipment. Typically conducted through hand-held or permanently positioned accelerometers placed on key measurement points on the equipment. Commonly used on most large items of rotating equipment, such as turbines, centrifugal pumps, motors, gearboxes etc.

Work Order – The prime document used by the maintenance function to manage maintenance tasks. It may include such information as a description of the work required, the task priority, the job procedure to be followed, the parts, materials, tools and equipment required to complete the job, the labor hours, costs and materials consumed in completing the task, as well as key information on failure causes, what work was performed etc.

Work Request – The prime document raised by user departments requesting the initiation of a maintenance task. This is usually converted to a work order after the work request has been authorized for completion.

Workload – the amount of labor hours required to carry out specified maintenance tasks.

Appendix-E

List of Indian Standards on Quality

I. Management

ISO 9000-1: 1993	Quality Management and Quality Assurance Standards Part 1: Guidelines for selection and use.
ISO 9000-2: 1993	Quality Management and Quality Assurance Standards Part 2: Generic guidelines for application of ISO 9001. ISO 9002, and ISO 9003.
ISO 9000-3:1991	Quality Management and Quality Assurance Standards. Part 3: Guidelines for the application of ISO 9001 to the development, supply and maintenance of software.
ISO 9000-4:1993	Quality Management and Quality Assurance Standards Part 4: Application for Dependability Management.
ISO 9001:1993	Quality systems—Model for quality assurance in design/development, production, installation, installation and servicing.
ISO 9002:1993	Quality systems—Model for quality assurance in production, installation and servicing.
ISO 9003:1993	Quality systems—Model for quality assurance in the final inspection and test.
ISO 9004-1:1993	Quality management and quality system elements Part 1: Guidelines.
ISO 9000-2:1993	Quality System—Model for quality assurance in production, installation and serireing.
ISO 9004-3:1991	Quality management and quality system elements. Part 2: Guidelines for processed materials.
ISO 9000-4:1993	Quality management and quality system elements. Part 4: Guidelines for quality improvement.
ISO 10011-1:1990	Guidelines for auditing quality systems. Part 1: Auditing.
ISO 10011-2:1991	Guidelines for auditing quality systems. Part 2: Qualification criteria for quality system auditors.
ISO 10011-3:1991	Guidelines for auditing quality systems. Part 3: Management of audit programmes.
ISO 10012-1:1992	Quality assurance requirements for measuring equipment. Part I: Metrological qualification system for measuring equipment.

II. Quality System Certification Models in Different Countries

DVNQA	United Kingdom		The Netherlands		United States
AOQC	Associated Official Quality Certification Ltd.	BSI	Quality Assurance	ABS	Quality Evaluations
ASTA	Certification Services Prudential Chambers	BVQI	Bureau Veritas Quality	AGA	Americal Gas Association
BSI BASE	Quality Assurance British Approvals Service for Electrical Cables	DNVI	International Ltd. Det norske Veritas	BVQI	Bureau Varitas Quality International
BVQI CICS	Bureau Vertias Quality Int. Ceramic Industry Certification Scheme Ltd.	DS	Industry Danish Standards Assoc.	CGSB	Canadian General Standards Board
CQAAC	Construction Quality Assurance Arcade Chambers	KEMA	Keuring Van Eledtrotechnische Materialen	DNV LRQA	Industrial Services Lioyd's Register Quality Assurance
DNVQA LRQA LPCB	Det Norske Veritas Quality Assurance Ltd. Lioyd's Register Quality Loss Prevention Certification Board Ltd.	KIWA LRQA SCTC	Lioyd's Register Q.A. Stiching Certificate	QMI	Quality Management Institute
NICQA QSRMC TRADA (UK) YQAF	National Inspection Council QA Ltd. Quality Scheme for Ready Mixed Concrete Certification Service UK Certification Authority for Reinforcing Steels Yarsley Quality Assured Firms Ltd.	SOBA TUV VEG VPGI	Textielverzorign Certex Stiching Onderwoeken Beproeving Appendages Nederland Gasinstituut Stichting VPGI Certificost	QSR TUV ULV Incotte	Quality Systems Registers Rbeinland on North America Underwriter's Lshoratories USA Inc.

III. List of Related URL's

Name of Co.	Subjects	Website
American Foundrymen's Society	Professional services for foundry industries	www.afsinc.org
Flow Science Inc. co.	Foundary software	www.flow3d.com
Magma Foundary Technologies Inc.	Simulation software Co.	www.magmasoft.com
Nova cast AB Co.	Foundary software	www.novacast.se
UES Software Inc.	Foundary software	www.use-software.com
CT-Castech, Inc. Co.	Foundary software	www.castech.fi
Autodesk Inc.	3D CAD Systems suppliers	www.autodesk.com
Baystate Technologies	3D CAD Systems suppliers	www.cadkey.com
Nihon Unisys Ltd.	3D CAD Systems suppliers	www.unisys.co.jp
Cimatron Ltd.	3D CAD Systems suppliers	www.cimatron.com
Hoshin Planning Software	Consultants for TQM	www.rootcause.com
American Society for Quality	TQM	www.asq.org
Quality American Inc.	TQM Consultants	www.qualityamerica.com
Inst. of British Foundrymen	Professional Society Foundry industries in UK	www.ibf.org.uk
Quality Management Corp. Consultants for SPC.	TQM Consultants	www.qmscorporation.com
North West Analytical Inc.	SQC Consultants	www.nwquality.com
Iron Casting Research Institute	Research Institute for iron casting in USA	www.icri.com

Other URL/ES of general nature on quality management and control

www.myunolarc.nasa.gov/dfc/tqc./html

www.omni.bus.ed.ac.uk

www.dfca.larc.nasagov/dfc/qtec.html

www.skyenet.com

www.qadas.com

www.work911.com

www.tqmarticles.htm

www.burstnet.com

www.nursingnet.org

www.ibpinc.com/career/qualityassurance.stm

❒❒❒

Appendix-F

ASTM International

ASTM **International** (ASTM), originally known as **the American Society for Testing and Materials,** is an international standards organization that develops and publishes voluntary consensus technical standards for a wide range of materials, products, systems, and services. The organization's headquarters is in West Conshohocken, Pennsylvania, about 5 miles northwest of Philadelphia.

ASTM predates other standards organizations such as BS1 (1901), DIN (1917) and AFNOR (1926), but differs from these in that it is not a national standards body, that role being taken in the USA by ANSI. However, ASTM has a dominant role among standards developers in the USA, and claims to be the world's largest developer of standards. Using a consensus process, ASTM supports thousands of volunteer technical committees, which draw their members from around the world and collectively develop and maintain more than 12,000 standards.

ASTM International publishes the *Annual Book of ASTM Standards* each year in print, CD and online versions. The online version was available by subscription and cost was based upon usage. For 2008, the complete set of books or CDs cost almost USD $9000 and included 81 volumes. For 2010, the complete set of books or CDs cost almost USD $9700 and included 82 volumes.

History

A group of scientists and engineers, led by Charles Benjamin Dudley, formed the American society for Testing and Materials in 1898 to address the frequent rail breaks plaguing the fast-growing railroad industry. The group developed a standard for the steel used to fabricate rails.

The organization celebrated its Centennial in 1998 with an examination of the group's mission and discussion about their relevance in today's global environment.

In 2001, ASTM changed its name to ASTM International to reflect global participation in ASTM and worldwide use of its standards.

In 2005, ASTM in partnership with Citation technologies, created the first ever dynamic, online web library for Environmental Due Diligence standards.

In 2008, ASTM published new Biodiesel Blend Specifications.

In 2009, ASTM in conjunction with Citation Technologies resolved the challenge of Federal agencies not updating ASTM standards within 29, 40. and/or 49 CFR·

In 2009, a joint effort by standards development organizations AAMI, ANSI, ASTM, and DIN created a single, centralized database for medical device standards.

Standards

The standards produced by ASTM International fall into six categories:

- The **Standard Specification,** that defines the requirements to be satisfied by subject of the standard.
- The **Standard Test Method,** that defines the way a test is performed and the precision of the result. The result of the test may be used to assess compliance with a Standard Specification.
- The **Standard Practice,** that defines a sequence of operations that, unlike a Standard Test Method, does not produce a result.
- The **Standard Guide,** that provides an organized collection of information or series of options that does not recommend a specific course of action.
- The **Standard Classification,** that provides an arrangement or division of materials, products, systems, or services into groups based on similar characteristics such as origin, composition, properties, or use. The **Terminology Standard,** that provides agreed definitions of terms used in the other standards.

The quality of the standards is such that they are frequently used worldwide.

The *Annual Book of ASTM Standards* covers 15 sections of interest plus a master index:

1. Iron and Steel Products.
2. Non-ferrous Metal Products.
3. Metals Test Methods and Analytical Procedures.
4. Construction.
5. Petroleum Products, Lubricants, and Fossil Fuels.
6. Paints, Related Coatings, and Aromatics.
7. Textiles.
8. Plastics.
9. Rubber.
10. Electrical Insulation and Electronics.
11. Water and Environmental Technology.
12. Nuclear, Solar, and Geothermal Energy.
13. Medical Devices and Services.
14. General Methods and Instrumentation.
15. General Products, Chemical Specialties, and End Use Products.
16. Index to all sections and volumes.

ASTM Standards can be purchased as a digital library subscription or individually from ASTM and other qualified standards providers. When maintaining a large standards library, often digital subscriptions are used to simplify staying current on standards and to remain in compliance with all copyright laws. A technical library at a university may also have copies of standards to review.

Membership and Organization

Membership in the organization is open to anyone with an interest in its activities. Standards are developed within committees, and new committees are formed as needed, upon request of interested members. Membership in most committees is voluntary and is initiated by the member's own request, not by appointment nor by invitation. Members are classified as users, producers, consumers, and "general interest". The latter include academics and consultants. Users include industry users, who may be producers in the context of other technical committees, and end-users such as consumers. In order to meet the requirements of antitrust laws, producers must constitute less than 50% of every committee or subcommittee, and votes are limited to one per producer company. Because of these restrictions, there can be a substantial waiting-list of producers seeking organizational memberships on the more popular committees. Members can, however, participate without a formal vote and their input will be fully considered.

As of 2007, more than 30,000 members, including over 1100 organizational members from more than 120 countries, including 120 members in China as of 2005.

ASTM International presents several awards for contributions to standards authorship, including the ASTM International Award of Merit (the organization's highest award).

Standards Compliance

ASTM International has no role in requiring or enforcing compliance with its standards. The standards, however, may become mandatory when referenced by an external contract, corporation, or government.

- In the United States, ASTM standards have been adopted, by incorporation or by reference, in many federal, state, and municipal government regulations. The National Technology Transfer and Advancement Act, passed in 1995, requires the federal government to use privately developed consensus standards whenever possible. The Act reflects what had long been recommended as best practice within the federal government.
- Other governments (local and worldwide) also have referenced ASTM standards.
- Corporations doing international business may choose to reference an ASTM standard.

Appendix-G

NORMAL DISTRIBUTION

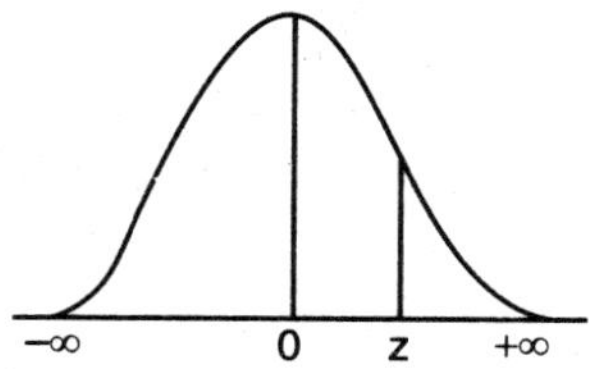

	.00	.01	.02	.03	.04	.05	.06	.07	.08	.09
.0	.5000	.5040	.5080	.5120	.5160	.5199	.5239	.5279	.5319	.5359
.1	.5398	.5438	.5478	.5517	.5557	.5996	.5636	.5675	.5714	.5753
.2	.5793	.5832	.5871	.5910	.5948	.5987	.6026	.6064	.6103	.6141
.3	.6179	.6217	.6255	.6293	.6331	.6368	.6406	.6443	.6480	.6517
.4	.6554	.6591	.6628	.6664	.6700	.6737	.6772	.6808	.6844	.6879
.5	.6915	.6950	.6985	.7019	.7054	.7088	.7123	.7157	.7090	.7224
.6	.7257	.7291	.7324	.7357	.7389	.7422	.7454	.7486	.7517	.7549
.7	.7580	.7611	.7642	.7673	.7704	.7734	.7764	.7794	.7823	.7852
.8	.7881	.7910	.7939	.7967	.7995	.8023	.8015	.8078	.8106	.8133
.9	.8159	.8186	.8212	.8238	.8264	.8289	.8315	.8340	.8365	.8389
1.0	.8413	.8438	.8461	.8485	.8508	.8531	.8554	.8577	.8599	.8621
1.1	.8643	.8665	.8686	.8708	.8729	.8749	.8770	.8790	.8810	.8830
1.2	.8849	.8869	.8888	.8907	.8925	.8944	.8962	.8980	.8997	.9015
1.3	.9032	.9049	.9066	.9082	.9099	.9115	.9131	.9147	.9162	.9177
1.4	.9192	.9207	.9222	.9236	.9215	.9265	.9279	.9292	.9306	.9319
1.5	.9332	.9345	.9357	.9370	.9382	.9394	.9406	.9418	.9429	.9441
1.6	.9452	.9463	.9474	.9484	.9495	.9505	.9515	.9525	.9535	.9545
1.7	.9554	.9564	.9573	.9582	.9591	.9599	.9608	.9616	.9625	.9633
1.8	.9641	.9649	.9656	.9664	.9671	.9678	.9686	.9693	.9699	.9706
1.9	.9713	.9719	.9726	.9732	.9738	.9744	.9750	.9756	.9761	.9767
2.0	.9772	.9778	.9783	.9788	.9793	.9795	.9803	.9808	.9812	.9817
2.1	.9821	.9826	.9830	.9834	.9834	.9838	.9842	.9846	.9850	.9857
2.2	.9861	.9864	.9868	.9871	.9875	.9878	.9881	.9884	.9887	.9890
2.3	.9893	.9896	.9898	.9901	.9904	.9906	.9909	.9911	.9913	.9916
2.4	.9918	.9920	.9922	.9925	.9927	.9929	.9931	.9932	.9934	.9936

2.5	.9938	.9940	.9941	.9943	.9945	.9946	.9948	.9949	.9951	.9952
2.6	.9953	.9955	.9956	.9957	.9959	.9960	.9961	.9962	.9963	.9964
2.7	.9965	.9966	.9967	.9968	.9969	.9970	.9971	.9972	.9973	.9974
2.8	.9974	.9975	.9976	.9977	.9977	.9978	.9979	.9979	.9980	.9981
2.9	.9981	.9982	.9982	.9983	.9984	.9984	.9985	.9985	.9986	.9986
3.0	.9987	.9987	.9987	.9988	.9988	.9989	.9989	.9989	.9990	.9990
3.1	.9990	.9991	.9991	.9991	.9992	.9992	.9992	.9992	.9993	.9993
3.2	.9993	.9993	.9994	.9994	.9994	.9994	.9994	.9995	.9995	.9995
3.3	.9995	.9995	.9995	.9996	.9996	.9996	.9996	.9996	.9996	.9997
3.4	.9997	.9997	.9997	.9997	.9997	.9997	.9997	.9997	.9997	.9998

Appendix-H

Cumulative Poisson Probabilities

	x													
np	**0**	**1**	**2**	**3**	**4**	**5**	**6**	**7**	**8**	**9**	**10**	**11**	**12**	**13**
.05	.951	.999	.1000											
.10	.905	.995	1.000											
.15	.861	.990	.999	.1000										
.20	.819	.982	.999	.1000										
.25	.779	.974	.998	.1000										
.30	.741	.963	.996	.1000										
.35	.705	.951	.994	.1000										
.40	.670	.938	.992	.999	.1000									
.45	.638	.925	.989	.999	.1000									
.50	.607	.910	.986	.998	.1000									
.55	.577	.894	.982	.998	.1000									
.60	.549	.878	.977	.997	.1000									
.65	.522	.861	.972	.996	.999	.1000								
.70	.497	.844	.966	.994	.999	.1000								
.75	.472	.827	.959	.993	.999	.1000								
.80	.449	.809	.953	.991	.999	.1000								
.85	.427	.791	.945	.989	.998	.1000								
.90	.407	.772	.937	.987	.998	.1000								
.95	.387	.754	.929	.984	.997	.1000								
1.0	.368	.736	.920	.981	.996	.999	.1000							
1.1	.333	.699	.900	.974	.995	.999	.1000							
1.2	.301	.663	.879	.966	.992	.998	.1000							
1.3	.273	.627	.857	.957	.989	.998	.1000							
1.4	.247	.592	.833	.946	.986	.997	.999	.1000						
1.5	.223	.558	.809	.934	.981	.996	.999	.1000						
1.6	.202	.525	.783	.921	.976	.994	.999	.1000						
1.7	.183	.493	.757	.907	.970	.992	.998	.1000						
1.8	.165	.463	.731	.891	.964	.990	.997	.999	.1000					
1.9	.150	.434	.704	.875	.956	.986	.997	.999	.1000					
2.0	.135	.406	.677	.857	.947	.983	.995	.999	.1000					
2.2	.111	.355	.623	.819	.928	.975	.993	.998	.1000					
2.4	.091	.308	.570	.779	.904	.964	.988	.997	.999	.1000				
2.6	.074	.267	.518	.736	.877	.951	.983	.995	.999	.1000				
2.8	.061	.231	.469	.692	.848	.935	.986	.992	.998	.999	.1000			

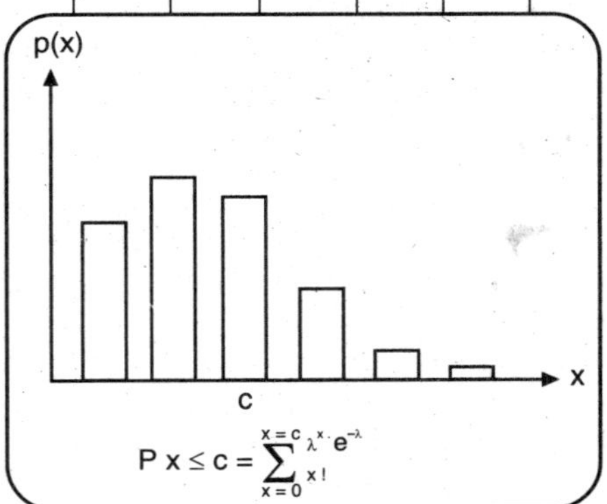

							x							
np	**0**	**1**	**2**	**3**	**4**	**5**	**6**	**7**	**8**	**9**	**10**	**11**	**12**	**13**
3.0	.050	.199	.423	.647	.815	.916	.966	.988	.996	.999	.1000			
3.2	.041	.171	.380	.603	.781	.875	.955	.983	.994	.998	.1000			
3.4	.033	.147	.340	.558	.744	.871	.972	.977	.992	.997	.999	.1000		
3.6	.027	.126	.303	.515	.706	.844	.927	.969	.988	.996	.999	.1000		
3.8	.022	.107	.269	.473	.81	.909	.960	.984	.994	.998	.999	.1000		
4.0	.018	.092	.238	.433	.629	.785	.889	.949	.979	.992	.997	.999	.1000	
4.2	.015	.078	.210	.395	.590	.753	.867	.936	.972	.989	.996	.999	1.000	
4.4	.012	.066	.185	.359	.551	.720	.844	.921	.964	.985	.994	.998	.999	1.000
4.6	.010	.056	.163	.326	.513	.683	.818	.905	.995	.980	.992	.997	.999	1.000
4.8	.008	.048	.143	.294	.476	.651	.791	.887	.994	.975	.990	.996	.999	1.000
5.0	.007	.040	.125	.265	.440	.616	.762	.867	.932	.968	.986	.995	.998	.999
5.2	.006	.034	.109	.238	.406	.581	.732	.845	.918	.960	.982	.993	.997	.999
5.4	.005	.029	.095	.213	.373	.546	.702	.822	.903	.951	.977	.990	.996	.999
5.6	.004	.024	.082	.191	.342	.512	.670	.797	.886	.941	.972	.988	.995	.998
5.8	.003	.021	.072	.170	.313	.478	.638	.771	.867	.929	.965	.984	.993	.997
6.0	.002	.017	.062	.151	.285	.446	.606	.744	.847	.916	.957	.980	.991	.996
6.2	.002	.015	.054	.134	.259	.41	.574	.716	.826	.902	.949	.975	.989	.995
6.4	.002	.012	.046	.119	.235	.384	.542	.687	.803	.886	.939	.969	.986	.994
6.6	.001	.010	.040	.105	.213	.355	.511	.658	.780	.869	.927	.963	.982	.992
6.8	.001	.009	.034	.093	.192	.327	.480	.628	.755	.850	.915	.955	.978	.990
7.0	.001	.007	.030	.082	.173	.301	.450	.599	.729	.830	.901	.947	.973	.987
7.2	.001	.006	.025	.072	.156	.276	.420	.569	.703	.810	.887	.937	.967	.984
7.4	.001	.005	.022	.063	.140	.253	.392	.539	.676	.788	.871	.926	.961	.980
7.6	.001	.004	.019	.055	.0125	.231	.365	.510	.648	.765	.854	.915	.954	.976
7.8	.000	.004	.016	.048	.112	.210	.338	.481	.620	.741	.835	.902	.945	.971
8.0	.000	.003	.014	.042	.100	.191	.313	.453	.593	.707	.816	.888	.936	.966
8.2	.000	.003	.012	.037	.089	.174	.290	.425	.565	.692	.796	.873	.926	.960
8.4	.000	.002	.010	.032	.079	.157	.267	.399	.537	.666	.774	.857	.915	.952.
8.6	.000	.002	.009	.028	.070	.142	.246	.373	.509	.640	.752	.840	.903	.945
8.8	.000	.001	.007	.024	.062	.128	.226	.348	.482	.614	.729	.822	.890	.936
9.0	.000	.001	.006	.021	.055	.116	.207	.324	.456	.587	.706	.803	.876	.926
9.2	.000	.001	.005	.018	.049	.104	.189	.301	.430	.561	.682	.783	.861	.916
9.4	.000	.001	.005	.016	.043	.093	.173	.279	.404	.535	.658	.763	.845	.904
9.6	.000	.001	.004	.014	.038	.084	.157	.258	.380	.509	.633	.741	.828	.892
9.8	.000	.001	.003	.012	.033	.075	.143	.239	.356	.483	.608	.719	.810	.879
10.0	0	.000	.003	.010	.029	.067	.130	.220	.333	.458	.583	.697	.792	.864
10.2	0	.000	.002	.009	.026	.060	.118	.203	.311	.433	.558	.674	.772	.849
10.4	0	.000	.002	.008	.023	.053	.107	.186	.29	.409	.533	.650	.752	.834
10.6	0	.000	.002	.007	.020	.048	.097	.171	.269	.385	.508	.627	.732	.817
10.8	0	.000	.001	.006	.017	.042	.087	.157	.250	.363	.484	.603	.710	.799
11.0	0	.000	.001	.005	.015	.038	.079	.143	.232	.341	.460	.579	.689	.781
11.2	0	.000	.001	.004	.013	.033	.071	.131	.215	.319	.436	.555	.667	.762
11.4	0	.000	.001	.004	.012	.029	.064	.119	.198	.299	.413	.532	.644	.743
11.6	0	.000	.001	.003	.010	.026	.057	.106	.183	.279	.391	.508	.622	.723
11.8	0	.000	.001	.003	.009	.023	.051	.099	.169	.260	.369	.485	.599	.702
12.0	0	.000	.001	.002	.088	.020	.046	.090	.155	.242	.347	.462	.576	.682
12.2	0	0	0.000	0.002	0.007	0.018	0.041	0.081	0.142	0.225	0.327	0.439	0.553	0.660
12.4	0	0	0.000	0.002	0.006	0.016	0.037	0.073	0.131	0.209	0.307	0.417	0.530	0.639
12.6	0	0	0.000	0.001	0.005	0.014	0.033	0.066	0.120	0.194	0.288	0.395	0.508	0.617
12.8	0	0	0.000	0.001	0.004	0.012	0.029	0.060	0.109	0.179	0.269	0.374	0.485	0.595

							x							
np	**0**	**1**	**2**	**3**	**4**	**5**	**6**	**7**	**8**	**9**	**10**	**11**	**12**	**13**
13.0	0	0	0.000	0.001	0.004	0.011	0.026	0.054	0.100	0.16	0.252	0.353	0.463	0.573
13.2	0	0	.000	.001	.003	.009	.023	.049	.091	.153	.235	.333	.441	.551
13.4	0	0	.000	.001	.003	.008	.020	.044	.083	.141	.219	.314	.420	.529
13.6	0	0	.000	.001	.002	.007	.018	.039	.075	.130	.204	.295	.399	.507
13.8	0	0	.000	.001	.002	.006	.016	.035	.068	.119	.189	.277	.378	.486
14.0	0	0	0	.000	.002	.006	.014	.032	.062	.109	.176	.260	.358	.464
14.2	0	0	0	.000	.002	.005	.013	.028	.056	.100	.163	.244	.339	.443
14.4	0	0	0	.000	.001	.004	.011	.025	.051	.092	.151	.228	.320	.423
14.6	0	0	0	.000	.001	.004	.014	.032	.062	.084	.139	.213	.302	.402
14.8	0	0	0	.000	.001	.003	.009	.020	.042	.077	.129	.198	.285	.383
15.0	0	0	0	.000	.001	.003	.008	.018	.037	.070	.118	.185	.268	.363
15.2	0	0	0	.000	.001	.002	.007	.016	.034	.064	.109	.172	.251	.344
15.4	0	0	0	.000	.001	.002	.006	.014	.030	.058	.100	.160	.236	.326
15.6	0	0	0	.000	.001	.002	.005	.013	.027	.053	.092	.148	.221	.308
15.8	0	0	0	0	.000	.002	.005	.011	.025	.048	.084	.137	.207	.291
16.0	0	0	0	0	.000	.001	.004	.010	.022	.043	.077	.127	.193	.275
16.2	0	0	0	0	.000	.001	.004	.009	.020	.039	.071	.117	.180	.259
16.4	0	0	0	0	.000	.001	.003	.008	.018	.035	.065	.108	.168	.243
16.6	0	0	0	0	.000	.001	.003	.007	.016	.032	.059	.100	.156	.228
16.8	0	0	0	0	.000	.001	.002	.006	.014	.029	.054	.092	.145	.214
17.0	0	0	0	0	.000	.001	.002	.005	.013	.026	.049	.085	.135	.201
17.2	0	0	0	0	.000	.001	.002	.005	.011	.024	.045	.078	.125	.188
17.4	0	0	0	0	.000	.001	.002	.004	.010	.021	.041	.071	.116	.176
17.6	0	0	0	0	0	.000	.001	.004	.009	.019	.037	.065	.107	.164
17.8	0	0	0	0	0	.000	.001	.003	.008	.017	.033	.060	.09	.153
18.0	0	0	0	0	0	.000	.001	.003	.007	.015	.030	.055	.092	.143
18.2	0	0	0	0	0	.000	.001	.003	.006	.014	.027	.050	.085	.133
18.4	0	0	0	0	0	.000	.001	.002	.006	.012	.025	.046	.078	.123
18.6	0	0	0	0	0	.000	.001	.002	.005	.011	.022	.042	.082	.115
18.8	0	0	0	0	0	.000	.001	.002	.004	.010	.020	.038	.066	.106
19.0	0	0	0	0	0	.000	.001	.002	.004	.009	.018	.035	.061	.098
19.2	0	0	0	0	0	0	.000	.001	.003	.008	.017	.032	.056	.091
19.4	0	0	0	0	0	0	.000	.001	.003	.007	.015	.029	.051	.084
19.6	0	0	0	0	0	0	.000	.001	.003	.006	.013	.026	.047	.078
19.8	0	0	0	0	0	0	.000	.001	.002	.006	.012	.024	.043	.072
20.0	0	0	0	0	0	0	.000	.001	.002	.005	.011	.021	.039	.066

Appendix-I

Single Sampling Table for AOQL = 2%

Lot sie	Process average			Process average			Process average			Process average			Process average			Process average		
	0 to 0.04%			0.05 to 0.40%			0.41 to 80%			0.81 to 1.20%			1.21 to 1.60%			1.61 to 2.00%		
	n	c	p_t%	n	c	p_t%	n	c	p_t%	n	c	p_t%	n	c	p_t%	n	c	p_t%
1–15	All	0	...	All	0	...	All	0	...	All	0	...	All	0	...	All	0	...
16 – 50	14	0	13.6	14	0	13.6	14	0	13.6	14	0	13.6	14	0	13.6	14	0	13.6
51 – 100	16	0	12.4	16	0	12.4	16	0	12.4	16	0	12.4	16	0	12.4	16	0	12.4
101 – 200	17	0	12.2	17	0	12.2	17	0	12.2	17	0	12.2	35	1	10.5	35	1	10.5
201 – 300	17	0	12.3	17	0	12.3	17	0	12.3	37	1	10.2	37	1	10.2	37	1	10.2
301 – 400	18	0	11.8	18	0	11.8	38	1	10.0	38	1	10.0	38	1	10.0	60	2	805
401 – 500	18	0	11.9	18	0	11.9	39	0	9.8	39	1	9.8	60	2	8.6	60	2	8.6
501 – 600	18	0	11.9	18	0	11.9	39	1	9.8	39	1	9.8	60	2	8.6	60	2	8.6
601 – 800	18	0	11.9	40	1	9.6	40	1	9.6	65	2	8.0	65	2	8.0	85	3	7.5
801 – 1000	18	0	12.0	40	1	9.6	40	1	9.6	65	2	8.1	65	2	8.2	90	3	7.4
1,001 – 2,000	18	0	12.0	41	1	9.4	65	2	8.2	65	2	9.2	95	3	7.0	120	4	6.5
2,001 – 3,000	18	0	12.0	41	1	9.4	65	2	8.2	95	3	7.0	120	4	6.5	180	6	5.8
3,001 – 4,000	18	0	12.0	42	1	9.3	65	2	8.2	95	3	7.0	155	5	6.0	210	7	5.5
4,001 – 5,000	18	0	12.0	42	1	9.3	70	2	7.5	125	4	6.4	155	5	6.0	245	8	5.3
5,000 – 7,000	18	0	12.0	42	1	9.3	95	3	7.0	125	4	6.4	185	6	5.6	280	9	5.1
7,001 – 10,000	42	1	9.3	70	1	7.5	95	3	7.0	155	5	6.0	220	7	5.4	350	11	4.8
10,001 – 20,000	42	1	9.3	70	2	7.6	95	3	7.0	190	6	5.6	290	9	4.9	460	14	4.4
20,001 – 50,000	42	1	9.3	70	2	7.6	125	4	6.4	220	7	5.4	395	12	4.5	720	21	3.9
50,001 – 100,000	42	1	9.3	95	3	7.0	160	5	5.9	290	9	4.9	505	15	4.2	955	27	3.7

Appendix-J

SAMPLING TABLE

Minimum acceptable reliability (at 90% confidence level)	Actual reliability Product (with10% Producers risk)	Sample size	Number of failures allowed
0.80	0.99	11	0
	0.98	18	1
	0.96	25	2
	0.95	32	3
	0.94	38	4
	9.93	45	5
0.85	0.99	15	0
	0.98	25	1
	0.97	34	2
	0.96	43	3
	0.95	52	4
	0.95	60	5
0.90	0.99	23	0
	0.99	38	1
	0.98	52	2
	0.97	65	3
	0.97	78	4
	9.97	91	5
0.95	0.998	45	0
	0.993	77	1
	0.990	105	2
	0.987	132	3
	0.985	158	4
	9.983	184	5
0.97	0.999	60	0
	0.996	130	1
	0.994	180	2
	0.992	220	3
	0.991	270	4
	9.991	310	5

0.99	0.999	230	0
	0.999	390	1
	0.998	530	2
	0.997	670	3
	0.997	800	4
	9.997	930	5

❐❐❐

Appendix-K

Reliability and Survial

Failures per 100,000 hr.	Mean Time to Failure	Hours of Operation: 5	10	20	50	100	200	300	500	1000	2000	3000	5000	7500	10,000	15,000
		Probability, Percent, of Survial, P														
1	100,000	99.995	99.99	99.98	99.95	99.9	99.8	99.7	99.5	99.0	98.0	97.0	95.1	92.8	90.5	86.1
2	50,000	99.990	99.98	99.96	99.90	99.8	99.6	99.4	99.0	98.0	96.1	94.2	90.5	86.1	81.9	74.1
3	33,000	99.985	99.97	99.94	99.85	99.7	99.4	99.1	98.5	97.0	94.2	91.4	86.1	79.9	74.1	63.8
4	25,000	99.980	99.96	99.92	99.80	99.6	99.2	98.8	98.0	96.1	92.3	88.7	81.7	74.1	67.0	54.9
5	20,000	99.975	99.95	99.90	99.75	99.5	99.0	98.5	97.5	95.1	90.5	86.1	77.9	68.7	60.7	47.2
6	16,700	99.970	99.94	99.88	99.70	99.4	98.8	98.2	97.0	94.2	88.7	83.5	74.1	63.8	54.9	40.7
7	14,300	99.965	99.93	99.86	99.65	99.3	98.6	97.9	96.6	93.2	86.9	81.1	70.5	59.2	49.7	35.0
8	12,500	99.960	99.92	99.84	99.60	99.2	98.4	97.6	96.1	92.3	85.2	78.7	67.0	54.9	44.9	30.1
9	11,100	99.955	99.91	99.82	99.55	99.1	98.2	97.3	95.6	91.4	83.5	76.3	63.8	50.9	40.7	25.9
10	10,000	99.95	99.9	99.8	99.5	99.0	98.0	97.0	95.1	90.5	81.9	74.1	60.7	47.2	36.8	22.3
20	5,000	99.90	99.8	99.6	99.0	98.0	96.1	94.2	90.5	81.9	67.0	54.9	36.8	22.3	13.5	5.0
30	3,330	99.85	99.7	99.4	98.5	97.0	98.0	97.0	95.1	90.5	81.9	74.1	60.7	47.2	36.8	22.3
40	2,500	99.80	99.6	99.2	98.0	96.1	92.3	88.7	81.9	67.0	44.9	30.1	13.5	5.0	1.8	0.2
50	2,000	99.75	99.5	99.0	97.5	95.1	90.5	86.1	77.9	60.7	36.8	22.3	8.2	2.4	0.7	0.1

60	1,670	99.70	99.4`	98.8	97.0	94.2	88.7	83.5	74.1	54.9	30.1	16.5	5.0	1.1	0.3
70	1,430	99.65	99.3	98.6	96.6	93.2	86.9	81.1	70.5	49.7	24.7	12.3	3.0	0.5	0.1
80	1,250	99.60	99.2	98.4	96.1	92.3	85.2	78.7	67.0	44.9	20.2	9.1	1.8	0.3	
90	1,110	99.55	99.1	98.2	95.6	91.4	83.5	76.3	63.8	40.7	16.5	6.7	1.1	0.1	
100	1,000	99.5	99.0	98.0	95.1	90.5	81.9	74.1	60.7	36.8	13.5	5.0	.07		
200	500	99.0	98.0	96.1	90.5	81.9	67.0	54.9	36.8	13.5	1.8	0.3			
300	333	98.5	97.0	94.2	86.1	74.1	54.9	40.7	22.3	5.0	.03				
400	250	98.0	96.1	92.3	81.9	67.0	44.9	30.1	13.5	1.8					
500	200	97.5	95.1	90.5	77.9	60.7	36.8	22.3	8.2	0.7					
600	167	97.0	94.2	88.7	47.1	54.9	30.2	16.6	5.0	0.3					
700	143	96.6	93.2	86.9	70.5	49.7	24.7	12.3	3.0	0.1					
800	125	96.1	92.3	85.2	67.0	44.9	20.2	9.1	1.8						
900	111	95.6	91.4	83.5	63.7	40.7	16.5	6.7	1.1						
1000	100	95.1	90.5	81.9	60.7	36.8	13.5	5.0	0.7						

Appendix-L

SAMPLE QUESTION PAPERS

SAMPLE QUESTION PAPER-I

Time: 3 Hours ***Total Marks: 100***

Note: Attempt all the questions

1. **Attempt any four parts of the following:** **(5×4=20)**
   (*a*) Describe evolution of quality control.
   (*b*) What is TQM? Explain in detail.
   (*c*) How do you assess customer needs?
   (*d*) Why is it necessary to have control on purchased product?
   (*e*) Explain quality aspects in sales.
   (*f*) Explain the basic steps to design an organisation structure for quality management.
2. **Attempt any two parts of the following:** **(10×2=20)**
   (*a*) Explain Kaizen concept.
   (*b*) How is formulation of design specification done?
   (*c*) How is the evaluation of suppliers done? Explain.
3. **Attempt any four parts of the following:** **(5×4=20)**
   (*a*) What do you understand by quality control.
   (*b*) Explain benchmarking.
   (*c*) Explain the importance of reliability in everyday life.
   (*d*) What is JIT? Explain.
   (*e*) Explain Quality system ISO: 9000.
   (*f*) Explain operating characteristics of quality curves.
4. **Attempt any two parts of the following:** **(10×2=20)**
   (*a*) How is claims analysis is done?
   (*b*) Discuss the factors to be considered for organisational structure for quality management.
   (*c*) Explain the difference between MTTR and MTTF.
5. **Attempt any four parts of the following:** **(5×4=20)**
   (*a*) What is descriptive statistics? Explain.
   (*b*) Describe Taguchi Quality.
   (*c*) Explain various causes of failures.
   (*d*) What is the need to develop sources of production? Discuss.
   (*e*) Explain CL, UCL and LCL.
   (*f*) Explain objectives of TQM.

❑❑❑

SAMPLE QUESTION PAPER-II

Time: 3 Hours ***Total Marks: 100***

Note: Attempt all the questions

1. Attempt any four parts of the following: (5×4=20)

(*a*) Explain quality concept change.
(*b*) Explain Kaizen concept.
(*c*) Explain product development cycle.
(*d*) What is strategic sourcing? Explain.
(*e*) What is after-sales service? Explain.
(*f*) Explain cost of quality.

2. Attempt any two parts of the following: (10×2=20)

(*a*) Describe implementation techniques of Kaizen.
(*b*) Explain general considerations for a good design.
(*c*) Explain procurement process.

3. Attempt any four parts of the following: (5×4=20)

(*a*) Explain objectives of TQM.
(*b*) What is PDSA cycle?
(*c*) Explain quality circles in detail.
(*d*) Explain SQC.
(*e*) Discuss application of various charts.
(*f*) Discuss identification and analysis of defects.

4. Attempt any two parts of the following: (10×2=20)

(*a*) What do you understand by guarantee of a product.
(*b*) What is a quality value and how does it corelate with quality cost?
(*c*) What is zero defect? Explain.

5. Attempt any four parts of the following: (5×4=20)

(*a*) Explain CL, UCL and LCL.
(*b*) What is mean and range chart? Explain.
(*c*) Explain detail of the capacity verification of a supplier.
(*d*) Explain reliability control and maintainaility.
(*e*) Describe Taguchi Quality.
(*f*) Explain the difference between MTTR and MTTF.

❑❑❑

UNIVERSITY QUESTION PAPERS

B.Tech.

Sixth Semester Examination, 2006-07

TOTAL QUALITY MANAGEMENT (TME-602)

Time: 3 Hours **Total Marks: 100**

Remark: ***This question paper is as per old syllabus***

Note: Attempt *all* questions. Marks are shown against each question.

1. Attempt any four parts : **(5×4=20)**

(*a*) Enlist the five definitions of quality.

(*b*) Explain evaluation of Total Quality Management (TQM) concept from quality control.

(*c*) Explain concept of external customer and internal customer giving example from business and industry.

(*d*) Compare centralized purchasing and decentralized purchasing process.

(*e*) Explain vendor rating procedure.

(*f*) Distinguish between quality control and inspection.

2. Attempt any two parts : **(10×2=20)**

(*a*) Briefly describe various quality costs. How do quality costs differ between manufacturing and service organization.

(*b*) Write short note on Pareto analysis.

(*c*) How attitude of top management does affect the quality of output. Differentiate between vision statement and mission statement of an organization.

3. Attempt any two parts : **(10×2=20)**

(*a*) Explain any two of following :

(*i*) Normal distribution

(*ii*) Binomial distribution

(*iii*) Poisson distribution.

(*b*) The following table shows the number of defective items discovered in 10 items taken on 10 consecutive days of a month.

Data :	1	2	3	4	5	6	7	8	9	10
Defects per unit	1	2	4	16	4	3	5	20	16	4

Construct control chart for number of defects per unit from the above data.

(*c*) M/s Gida Electrical Ltd. produces tube lights. The following data is available on the number of lumens for 40 watt tubelights.

Sample Number	1st Tube Light	2nd Tube Light	3rd Tube Light	4th Tube Light
1	604	612	566	600
2	597	601	607	603
3	581	570	585	592
4	620	605	595	588
5	590	614	608	604

Draw $\bar{X}$ and R charts and state process control position.
(It is given that for sample size of 4, $D_4 = 2.282$, $D_3 = 0$, $A_2 = 0.729$)

4. Attempt any two parts : (10×2=20)

(*a*) What do you understand by acceptance sampling/ Describe briefly double sampling plan.

(*b*) Describe OC curve and explain following terms :

(*i*) Manufacturer's Risk
(*ii*) Purchaser's Risk
(*iii*) AQL
(*iv*) LTPD

(*c*) Find the probability of a lot being accepted if it has a incoming quality 5% defective a sample size of 40 and acceptance number is 1.

5. Attempt any two parts : (10×2=20)

(*a*) Explain simplification and standardization. Illustrate each with suitable examples.

(*b*) Describe major elements of ISO : 9000.

(*c*) Write short notes on :

(*i*) Quality loss function
(*ii*) MRP.

❐❐❐

B.Tech.

Seventh Semester Examination, 2011-12

TOTAL QUALITY MANAGEMENT (EME-041)

Time: 3 Hours ***Total Marks: 100***

Note: Attempt *all* questions. Marks allotted are indicated against every part of each question.

1. Attempt any two parts of the following:

(*a*) (*i*) Discuss the journey of quality program from "inspection and testing" era to today's most mordern era. (7)

(*ii*) Explain term "TQM". (3)

(*b*) (*i*) Enlist different methods to evaluate supplies. Explain any one in detail. (6)

(*ii*) Explain how to assess the quality in sales and services. (4)

(*c*) (*i*) What do you mean by capacity of a supplier? What are the different measure to assess it? (3+4)

(*ii*) What do you mean by warranty and guarantee? How are they being analysed and set? (4)

2. Attempt any two parts of the following:

(*a*) (*i*) What do you mean by quality loss function? How does it explain the effect of quality and associated decisions on the life of common people as well as industrial personnel? (6)

(*ii*) Working as an individual, determine different obstacles to implement TQM in a manufacturing organisation. (4)

(*b*) What are different types of organizational structures? Compare them on the basis of inventories, quality, production planning and control, set-up and throughput issues with suitable examples. (10)

(*c*) Discuss the different causes of operator errors and managerial approaches to overcome the same. (10)

3. Attempt any two parts of the following:

(*a*) (*i*) What do you mean by process capability index? (2)

The specification from the manufacturing of a particular type of metal coating call for the temperature of the drying oven to be 380±15°F. The company that is considering using coating run tests by taking a large number of reading about mean temperature setting was found to be 2.06°F. What is the process capability index? (4)

(*ii*) Explain different situation that may appear in a control chart while ploting data of samples on control limits. (4)

(*b*) (*i*) Show relationship between sample and population. Also explain how this relation help us to develop control charts. (4)

(*ii*) What are α and β in relation to type I and type II error? (3)

(*iii*) Standard deviation for three samples are 9.6, 10.2, 9.8. If all these three samples are drawn as single sample, what will be the standard deviation? (3)

(*c*) The following table gives the average and range in kg° for tensile tests on an improved plastic cord. The subgroup size is 4. Determine the trial control limits. If points are out of control, assume assignable causes and calculate revised limits and central line. (10)

Subgroup Number	$\bar{X}$	R	Subgroup Number	$\bar{X}$	R
1	476	32	14	482	22
2	466	23	15	506	23
3	484	32	16	496	23
4	466	26	17	478	25
5	470	24	18	484	24
6	494	24	19	506	23
7	486	28	20	476	25
8	496	23	21	485	29
9	488	24	22	490	25
10	482	26	23	463	22
11	498	25	24	469	27
12	464	24	25	474	22
13	484	24			

4. Attempt any two parts of the following :

(*a*) Explain :

(*i*) Fault isolation and self diagnostics;

(*ii*) Parts standardization and interchangeability

(*iii*) Modularization and accessibility, and

(*iv*) Repair and/or replacement. (2.5 × 4)

□□□

B.Tech.
Seventh Semester Examination, 2013-14
TOTAL QUALITY MANAGEMENT (EME-041)

Time: 3 Hours ***Total Marks: 100***

Note: Attempt *all* questions. Marks allotted are indicated against every part of each question.

1. **Attempt any four of the following:** **(5×4=20)**
   (*a*) What do you mean by evolution of quality?
   (*b*) Write the principles of TQM.
   (*c*) Explain the methods of manufacturing in the light of flexible and pull manufacturing. What are the three main considerations of modern manufacturing?
   (*d*) What is the role of suppliers in modern manufacturing? Explain the criteria for selecting the suppliers.
   (*e*) What is strategic sourcing? Explain.
   (*f*) Explain quality aspects in sales. What is the role of after-sales efforts in ensuring maximum customer satisfaction?

2. **Attempt any four of the following:** **(5×4=20)**
   (*a*) Discuss the factors to be considered for organizational structure for quality management.
   (*b*) What do you understand by quality functions? Explain.
   (*c*) What is quality value and how does it correlate with quality cost?
   (*d*) Briefly describe the various quality costs, which cost should a company concentrate most on? Give reasons.
   (*e*) Explain the dimensions of quality.
   (*f*) Human factor is most important element in quality of a product. Justify.

3. **Attempt any two of the following:** **(10×2=20)**
   (*a*) What do you mean by variables and attributes? Explain $\bar{X}$ and *R*-charts.
   (*b*) What do you mean by fraction defective? Why *p*-chart even though much inferior as compared to the $\bar{X}$ and *R*-charts is effectively used in diagnosis of causes of trouble? Explain p-chart.
   (*c*) Control charts for $\bar{X}$ and R are maintained on a certain dimension of a manufactured part, measured in cm. The subgroup size is 4. The values of $\bar{X}$ and R are computed for each subgroup. After 20 subgroups $\sum\bar{X} = 41.283$, and $\sum R = 0.280$. Compute 3-sigma limits for the $\bar{X}$ and R-charts, and estimate the value of σ, on the assumption that the process is in statistical control. ($A_2 = 0.729$, $D_2 = 4.698$, $D_3 = 0.00$, $D_4 = 2.282$)

4. **Attempt any two of the following:** **(10×2=20)**
   (*a*) Explain the process of identification and analysis of defects.
   (*b*) Explain a cause-and-effect (Ishikawa) diagram to identify a process defect.
   (*c*) Write short notes on:
      (*i*) Reliability of components assembled in series and parallel systems.
      (*ii*) Quality circle.

5. **Attempt any two of the following:** **(10×2=20)**
   (*a*) What are the seven wastes identified by Shigeo Shingo, as being the targets of continuous improvement in production processes?
   (*b*) What is ISO 9000? Explain its salient features.
   (*c*) Write short notes on:
      (*i*) JIT Technique, (*ii*) Taguchi Quality Loss Function.

□□□

B.Tech.
Seventh Semester Examination, 2016-17
QUALITY MANAGEMENT (NOE-072)

Time: 3 Hours ***Total Marks: 100***

SECTION-A

1. **Attempt all parts. All parts carry equal marks. Write answer of each in short.** **(10×2 = 20)**
   (*a*) What do you mean by Quality?
   (*b*) Define MTTE.
   (*c*) What do you understand guarantee and warrentee?
   (*d*) Define Sampling.
   (*e*) Define control chart?
   (*f*) What do you mean by reliability?
   (*g*) What is analysis of claims?
   (*h*) Define the meaning of prototype.
   (*i*) What do you mean by Maintainability?
   (*j*) What do you understand by quality cost?

SECTION-B

2. **Attempt any five questions from this section.** **(5×10=50)**
   (*a*) Define Quality Management. What are the dimensions of quality?
   (*b*) What is process capability study? How it is measured?
   (*c*) What do you mean by manufacturing? Explain the methods and techniques of manufacturing.
   (*d*) Discuss Taguchi approach to ensure quality from traditional quality assurance procedures.
   (*e*) Does operator attitude affect quality? Explain.
   (*f*) What do you mean by quality circles? Explain the objectivies and structure of quality circle.
   (*g*) Explain the construction and analysis of X-chart and R-chart.
   (*h*) What is Deming 14 Point Programme?

SECTION-C

Note: Attempt any two questions from this section. **(2×15=30)**

3. Explain evaluation of Total Quality Management (TQM) concept from quality control.

4. What do you mean by defect study? Discuss identification and analysis of defects.

5. What is the role of training as envisaged in ISO-9000 standards?

REFERENCES

1. Pandey O.N, Aneja B., Reliability Engineering and Quality Management, Kataria and Sons, 3rd Edn, 2011
2. Summers D.C.S., Quality Management, PHI, Second Edition,
3. Bounds G., Beyond Total Quality Management, McGraw Hill, 1994.
4. Menon H.G., TQM in New Product Manufacturing, McGraw Hill, 1992.
5. Pandey O.N., Aneja B., Industrial Management, Kataria and Sons, 2011.
6. Charles E. Ebeling, *Reliability and Maintainability Engineering*, Tata McGraw-Hill Edition, 2000.
7. Lal H., Total Quality Management, New Age Intl. Pub., 1990.
8. P.L. Jain, *Quality Control and Total Quality Management,* 2007.
9. Bhadhuri B. Basu S.K., Terotechnology2: Reliability Engineering and Maintanence Management, Asia, books, 2003.
10. G.H. Sandler, *System Reliability Engineering*, Prentice Hall.
11. Pandey O.N. Aneja B., Engineering and Managerial Economics, Kataria and Sons, 2010
12. *E. Balagurusamy*, *Reliability Engineering,* Tata McGraw-Hill.
13. S.K. Sinha & B.K. Kale, *Life Testing and Reliability Estimation*, Wiely Eastern Ltd.
14. Dhillon B.S., Quality Control, Reliability, and Engineering Design, Marcel Dekker, Inc., New York, 1985.
15. Dhillon B.S. and Singh C., Engineering Reliability: New Techniques and Applications, Wiley-Interscience, John Wiley & Sons, Inc., New York, 1981.
16. Dummer G.W.A. and Griffin N., *Electronic Equipment Reliability*, John Wiley & Sons, Inc., New York, 1960.
17. Govil A.K., *Reliability Engineering*, Tata McGraw Hill Publishing Company Limited, New Delhi, 1983.
18. Klaassen K.B. and Jack C.L.van Peppen, *System Reliability,* Chapman and Hall, Inc., New York, 1989.
19. Tillman FA., Hwang C.L. and Kuo W., *Optimization of Systems Reliability*, Marcel Dekker, Inc., New York, 1980.
20. Trivedi K.S., *Probability and Statistics with Reliability, Queuing and Computer Science Application*, Prentice-Hall, Inc., Englewood Cliffs, New Jersey, 1982.
21. Ammer Deans, *Materials Management*, Irwin Homewood, Illinois, (1980).
22. Khanna.O.P., *Industrial Engineering*, Dhanpat Rai Publications, New Delhi.
23 Maynard, H.B., *Industrial Engineering Handbook*, Mc.Graw Hill Book Company, New York.
24. Sharma.B.R., *Environmental and Pollution Awareness.*
25. T.R.Banga, *Industrial Engineering and Management.* Khanna Publisher, Delhi.
26 Telsang Martand, *Industrial Engineering and Production Management*, S.Chand and Company., 2005

INDEX

D...

E...

F...

G...

H...

I...

J...

K...

L...

M...

N...

O...

P...

Win Prizes

ATTENTION READERS

Win Prizes

We request for your frank assessment regarding various aspects of the book as given below :

QUALITY MANAGEMENT

By: O.N. Pandey and Bhupesh Aneja

Please Fill up the Sheet in Neat and Clean Writing

(i) What topics of your syllabus, important for your examination point of view, are not covered in this book ?

....

....

....

....

(ii) What are the chapters wherein the treatment of the subject matter is not systematic or organised or updated ?

....

....

....

....

(iii) Have you come across misprints/mistakes/factual inaccuracies in the book? Please specify the chapters and the page numbers.

....

....

....

....

(iv) What other books on the same subject have you found/heard better than the present book? Please specify in terms of price and quality.

....

....

....

....

(v) Further suggestions and comments for the improvement of this book.

....

....

....

....

(PS: If need be please attach a separate sheet to write your views)

OTHER DETAILS

(i) Who recommended you this book ? (Please tick (✓) in the box near the option relevant to you.

☐ Teacher ☐ Friends ☐ Bookseller

....

....

....

....

(ii) Name of the recommending teacher, his designation and address :

....

....

....

....

(iii) Name and address of the bookseller you purchased the book from :

....

....

....

....

(iv) Name and address of your institution (Please mention the University or Board as the case may be)

....

....

....

....

(v) Your name and complete postal address :

....

....

....

....

(vi) Course you are enrolled in :

....

....

....

....

The best assessment will be awarded each month: The award will be in the form of our publications as decided by the Editorial Board.

Please mail the filled up coupon at your earliest to :

S.K. KATARIA & SONS®
4885/109, Prakash Mahal, Dr. Subhash Bhargav Lane,
Opposite Delhi Medical Association, Daryaganj, New Delhi–110002
☎: +91-11-23243489, 23269384
Email : katariabooks@yahoo.com
katariabook@gmail.com